Spatio-Temporal Methods in Environmental Epidemiology

CHAPMAN & HALL/CRC
Texts in Statistical Science Series

Series Editors
Francesca Dominici, *Harvard School of Public Health, USA*
Julian J. Faraway, *University of Bath, UK*
Martin Tanner, *Northwestern University, USA*
Jim Zidek, *University of British Columbia, Canada*

Texts in Statistical Science

Spatio-Temporal Methods in Environmental Epidemiology

Gavin Shaddick

University of Bath, UK

James V. Zidek

University of British Columbia, Canada

CRC Press
Taylor & Francis Group
Boca Raton London New York

CRC Press is an imprint of the
Taylor & Francis Group, an **informa** business

A CHAPMAN & HALL BOOK

CRC Press
Taylor & Francis Group
6000 Broken Sound Parkway NW, Suite 300
Boca Raton, FL 33487-2742

First issued in paperback 2020

© 2016 by Taylor & Francis Group, LLC
CRC Press is an imprint of Taylor & Francis Group, an Informa business

No claim to original U.S. Government works

ISBN-13: 978-1-4822-3703-0 (hbk)
ISBN-13: 978-0-367-78346-4 (pbk)

Visit the Taylor & Francis Web site at
http://www.taylorandfrancis.com

and the CRC Press Web site at
http://www.crcpress.com

To Jo, Lynne and Cynthia

Contents

---

†Single dagger denotes sections that contain particularly technical material.

List of Figures

List of Tables

Preface

Motivated by increased societal concerns about environmental hazards this book explores the interface between environmental epidemiology (EE) and spatio–temporal (ST) modelling. Its aim is to promote the interface between statisticians and practitioners in order to allow the rapid advances in the field of spatio–temporal statistics to be fully exploited in assessing risks to health. The aim of EE is to understand the adverse health effects of environmental hazards and to estimate the risks associated with those hazards. Such risks have traditionally been assessed either over time at a fixed point in space or over space at a fixed point in time. ST modelling characterises the distribution of those hazards and associated risks over both geographical locations and time. Understanding variation and exploiting dependencies over both space and time greatly increases the power to assess those relationships.

Motivated by real life problems, this book aims to provide both a link between recent advances in ST methodology and epidemiological applications and to provide a means to implement these methods in practice. The book recognises the increasing number of statistical researchers who are collaborating with environmental epidemiologists. Many excellent books on spatial statistics and spatial epidemiology were available when this book was written, including Banerjee, Carlin, and Gelfand (2015), Cressie (1993), Cressie and Wikle (2011), Diggle and Ribeiro (2007), Le and Zidek (2006), Schabenberger and Gotway (2000), Stein (1999), Waller and Gotway (2004), Lawson (2013) and Elliott, Wakefield, Best, and Briggs (2000). This selection provides an excellent resource however none specifically addresses the interface between environmental epidemiology and spatio–temporal modelling. This is the central theme of this book and we believe it promotes a major new direction in environmental epidemiology where there is an increasing need to consider spatio-temporal patterns in data that can be used to borrow strength over space and time and to reduce bias and inefficiency.

The genesis of the book was a thematic year at the Statistical and Applied Mathematical Sciences Institute (SAMSI) on 'Space–time Analysis for Environmental Mapping, Epidemiology and Climate Change'. It was the basis of a course in spatio–temporal epidemiology and was successfully used in a 13-week graduate statistics course at the University of British Columbia. A key feature of the book is the coverage of a wide range of topics from an introduction to epidemiological principles and the foundations of ST modelling, with specific focus on their application, to new directions for research. This includes both traditional and Bayesian approaches, the latter providing an important unifying framework for the integration of ST modelling into EE and which is key to many of the approaches presented within the book. Coverage of current research topics includes visualisation and mapping; the analysis of high-dimensional data; dealing with stationary fields; the combination of deterministic and statistical models; the design of monitoring networks and the effects of preferential sampling.

Throughout the book, the theory of spatial, temporal and ST modelling is presented in the context of its application to EE and examples are given throughout. These examples are provided together with embedded R code and details of the use of specific R packages and other software, including WinBUGS/OpenBUGS and modern computational methods such as integrated nested Laplace approximations (INLA). Additional code, data and examples are provided in the online resources associated with the book. A link can be found at http://www.crcpress.com/product/isbn/9781482237030.

As a text, this book is intended for students in both epidemiology and statistics. The original course at the University of British Columbia (UBC) was intended for graduate level students in statistics and epidemiology and lasted for 13 weeks, with two 90-minute lectures and a two-hour lab session each week.

The main topics covered included:

- Types of epidemiological studies; cohort, case–control, ecological.
- Measures of risk; relative risks, odds ratios, absolute risk, sources of bias, assessing uncertainty.
- Bayesian statistics and computational techniques; Markov Chain Monte Carlo (MCMC) and Integrated nested Laplace approximations (INLA).
- Regression models in epidemiology; Logistic and Poisson generalised linear models, generalised additive models, hierarchical models, measurement error models.
- Temporal models; time series models, temporal auto-correlation, smoothing splines.
- Spatial models; point processes, area and point referenced methods, mapping, geostatistical methods, spatial regression, non-stationary models, preferential sampling.
- Spatial–temporal models; separable models, non-separable models, modelling exposures in space and time, correction for ecological bias.

In addition to the material covered in this course, the book contains details of many other topics several of which are covered in greater technical depth. Sections that contain particularly technical material are noted with a dagger (†).

Many universities operate a 15–week schedule and this book was initially designed for such courses. Three examples include: (i) a course for epidemiologists with the emphasis on the application of ST models in EE; (ii) a course for biostatisticians, covering underlying principles of modelling and application; and (iii) an advanced course on more theoretical aspects of spatio–temporal statistics and its application in EE. Further information on possible course structures, together with exercises, lab projects and other material can be found in the online resources.

To conclude, we would like to express our deepest gratitude to all of those who have contributed to this book both in terms of specific assistance in its creation and also in a much wider sense. This includes our many co-authors and past and present students. In regard to the latter, special thanks goes out to the Class of 2013 at UBC who provided invaluable feedback when the material in the book was first presented. A huge thank you goes to Song Cai, whose work as a lab instructor was instrumental in developing the exercises and projects. Thank you also to those who attended the shortcourses on this topic at Telford (RSS), Toronto (SSC) and Cancun (ISBA).

The book was written jointly in the Department of Mathematical Sciences at the University of Bath and the Department of Statistics at the University of British Columbia, both as home institutions but also as hosts of visits by the two authors. We thank the staff and faculty of these two departments for their support and guidance. We would like to thank Yang 'Seagle' Liu (UBC), Yiping Dou (UBC) and Yi Liu (Bath) for allowing us to include their code and other material in the book and in the online resources. Similarly, thanks go to Ali Hauschilt (UBC), Kathryn Morrison and Roni English for proofreading and Millie Jobling (Bath) for comments and help in preparing the text. Many thanks also to Duncan Lee for providing data and code for some of the examples. The Peter Wall Institute for Advanced Studies (PWIAS, UBC) generously supported an opportunity for both authors to work together at PWIAS and to finalise the book.

We appreciate the support of CRC/Chapman and Hall and particularly Rob Calver, senior acquisitions editor at the Taylor and Francis Group, who played a key role in encouraging and supporting this work throughout its production. Many thanks also to Suzanne Lassandro. Special thanks go to the anonymous reviewers of the book who provided extremely helpful and insightful comments on early drafts.

The first author would like to give special thanks to Paul Elliott and Jon Wakefield for stimulating and maintaining his interest in the subject and to Jon for continued statistical enlightenment. Finally, he would like to give special thanks to his mum (Cynthia) for everything, from beginning to end, and to Jo for encouragement, patience, support and continued good nature during a period in which plenty was required. From both authors, very special thanks go to Lynne for continual encouragement, support and providing essential sustenance. Above all, she took on the virtually impossible role of ensuring that this transatlantic co-authorship actually resulted in a book!

Bath, UK   *Gavin Shaddick*
Vancouver, Canada   *James V. Zidek*
February 2015

Abbreviations

ACF	Autocorrelation function
AIC	Akaike information criterion
AR	Autoregressive process
ARIMA	Autoregressive regressive integrated moving average
ARMA	Autoregressive regressive moving average
Be–Ne–Lux	Belgium, Netherlands, Luxembourg region
BIC	Bayesian information criteria
BLUP	Best linear unbiased predictor
BSP	Bayesian spatial predictor
CapMon	Canadian Acid Precipitation Monitoring Network
CAR	Conditional autoregressive approach
CAR	Conditional autoregressive
CDF	Cumulative distribution function
CI	Confidence interval or in Bayesian analysis, credible interval
CMAQ	Community multi–scale air quality model
CMT	Chemical transport models
CO	Carbon monoxide
CRF	Concentration response function
CSD	Census subdivision
CV	Cross validation
DIC	Deviance information criterion
DLM	Dynamic linear model
DT	Delauney triangle
ERF	Exposure response function
EU	European union
GCV	Generalised cross validation
GEV	Generalized extreme value distribution
GF	Gaussian field
GMRF	Gaussian Markov random field
GPD	Generalized Pareto distribution
Hg	Mercury
ICD	International Classification Codes for Diseases
IRLS	Interactively re–weighted least squares
KKF	Kriged Kalman filter
LHS	Latin hybercube sampling
MAR	Multivariate autoregressive process

MCAR	Multivariate CAR
MESA	Multi–ethnic study of atherosclerosis
mg/kg or mg^{-1}	Micrograms per kilogram
μgm^{-3}	Micrograms per meter cubed
MRF	Markov random field
MSE	Mean squared error
NADP	National Atmospheric Deposition Program
NCS	Natural cubic splines
NO$_2$	Nitrogen dioxide
NUSAP	Numerical–Units–Spread–Assessment–Pedigree
O$_3$	Ozone
PACF	Partial autocorrelation function
Pb	Chemical symbol for lead
PDF	Probability density function
PM$_{10}$	Airborne particulates of diameter less than 10 μgm^{-3}
PM$_{2.5}$	Airborne particulates of diameter less than 2.5 μgm^{-3}
POT	Peak over threshold
ppb	Particles per billion
ppm	Particles per million
PYR	Person year at risk
RR	Relative risk
RS	Relative sensitivity
SAR	Simultaneous autoregressive
SIR	Susceptible–infective–recovered model for infectious diseases
SMR	Standard mortality ratio
SO$_2$	Sulphite or sulphur dioxide
SPDE	Stochastic partial differential equations
UTM	Universal Transverse Mercator (coordinate system)
VAR	Vector autoregressive process
VOC	Volatile organic compounds
WHO	World Health Organization

The Authors

Dr Gavin Shaddick is a Reader in Statistics in the Department of Mathematical Sciences at the University of Bath. He has a PhD in Statistics and Epidemiology from Imperial College, London and an MSc in Applied Stochastic Systems from University College, London. His research interests include the theory and application of Bayesian statistics to the areas of spatial epidemiology, environmental health risk and the modelling of spatio-temporal fields of environmental hazards. Of particular interest are computational techniques that allow the implementation of complex statistical models to real life applications where the scope over both space and time may be very large.

Dr Jim Zidek is Professor Emeritus at the University of British Columbia. He received his MSc and PhD from the University of Alberta and Stanford University, both in Statistics. His research interests include the foundations of environmetrics, notably on the design of environmental monitoring networks, and spatio-temporal modelling of environmental processes. His contributions to statistics have been recognised by a number of awards including Fellowships of the ASA and IMI, the Gold Medal of the Statistical Society of Canada and Fellowship in the Royal Society of Canada, one of that country's highest honours for a scientist.

Chapter 1

Why spatio–temporal epidemiology?

1.1 Overview

Spatial epidemiology is the description and analysis of geographical data, specifically health data in the form of counts of mortality or morbidity and factors that may explain variations in those counts over space. These may include demographic and environmental factors together with genetic, and infectious risk factors (Elliott & Wartenberg, 2004). It has a long history dating back to the mid-1800s when John Snow's map of cholera cases in London in 1854 provided an early example of geographical health analyses that aimed to identify possible causes of outbreaks of infectious diseases (Hempel, 2014). Since then, advances in statistical methodology together with the increasing availability of data recorded at very high spatial and temporal resolution has lead to great advances in spatial and, more recently, spatio–temporal epidemiology.

These advances have been driven in part by increased awareness of the potential effects of environmental hazards and potential increases in the hazards themselves. Over the past two decades, population predictions based on conventional demographic methods have forecast that the world's population will rise to about 9 billion in 2050, and then level off or decline. However, recent analyses using Bayesian methods have provided compelling evidence that such projections may vastly underestimate the world's future population and instead of the expected decline, population will continue to rise (Gerland et al., 2014). Such an increase will greatly add to the anthropogenic contributions of environmental contamination and will require political, societal and economic solutions in order to adapt to increased risks to human health and welfare. In order to assess and manage these risks there is a requirement for monitoring and modelling the associated environmental processes that will lead to an increase in a wide variety of adverse health outcomes. Addressing these issues will involve a multi-disciplinary approach and it is imperative that the uncertainties that will be associated with each of the components can be characterised and incorporated into statistical models used for assessing health risks (Zannetti, 1990).

In this chapter we describe the underlying concepts behind investigations into the effects of environmental hazards and particularly the uncertainties that are present at each stage of the process. This leads to a discussion of the reasons why considering

1

diseases and exposures over both space and time are becoming increasingly important in epidemiological analyses. In this book we advocate a Bayesian approach to
modelling and later in this chapter we consider a general framework for modelling
spatio–temporal data and introduce the notation that will be used throughout the
book. Different types of spatial data are introduced together with a brief summary of
the effect that the underlying generating mechanism will have on subsequent modelling, a subject that is further developed in later chapters. Throughout this chapter,
concepts and theory are presented together with examples.

1.2   Health-exposure models

An analysis of the health risks associated with an environmental hazard will require a
model which links exposures to the chosen health outcome. There are several potential sources of uncertainty in linking environmental exposures to health, especially
when the data might be at different levels of aggregation. For example, in studies of
the effects of air pollution, data often consists of health counts for entire cities with
comparisons being made over space (with other cities experiencing different levels
of pollution) or time (within the same city) whereas exposure information is often
obtained from a fixed number of monitoring sites within the region of study.

Actual exposures to an environmental hazard will depend on the temporal trajectories of the population's members that will take individual members of that population through a sequence of micro-environments, such as a car, house or street
(Berhane, Gauderman, Stram, & Thomas, 2004). Information about the current state
of the environment may be obtained from routine monitoring or through measurements taken for a specialised purpose. An individual's actual exposure is a complex
interaction of behaviour and the environment. Exposure to the environmental hazard
affects the individual's risk of certain health outcomes, which may also be affected
by other factors such as age and smoking behaviour.

1.2.1   Estimating risks

If a study is carefully designed, then it should be possible to obtain an assessment
of the magnitude of a risk associated with changes in the level of the environmental
hazard. Often this is represented by a relative risk or odds ratio, which is the natural result of performing log–linear and logistic regression models respectively. They
are often accompanied by measures of uncertainty, such as 95% confidence (or in
the case of Bayesian analyses, credible) intervals. However, there are still several
sources of uncertainty which cannot be easily expressed in summary terms. These
include the uncertainty associated with assumptions that were implicitly made in any
statistical regression models, such as the shape of the dose–response relationship (often assumed to be linear). The inclusion, or otherwise, of potential confounders and
unknown latencies over which health effects manifest themselves will also introduce
uncertainty. In the case of short-term effects of air pollution for example, a lag (the
difference in time between exposure and health outcome) of one or two days is often

chosen (Dominici & Zeger, 2000) but the choice of a single lag doesn't acknowledge the uncertainty associated with making this choice. Using multiple lags in the statistical model may be used but this may be unsatisfactory due to the high correlation amongst lagged exposures, although this problem can be reduced by using distributed lag models (Zannetti, 1990).

1.2.2 A new world of uncertainty

The importance of uncertainty has increased dramatically as the twentieth century ushered in the era of post-normal science as articulated by Funtowicz and Ravetz (Funtowicz & Ravetz, 1993). Gone were the days of the solitary scientist runnning carefully controlled bench-level experiments with assured reproducibility, the hallmark of good classical science. In came a science characterized by great risks and high levels of uncertainty, an example being climate science with its associated environmental health risks. Funtowicz–Ravetz post-normality has two major dimensions (Aven, 2013): (i) decision stakes or the value dimension (cost–benefit) and (ii) system uncertainties. Dimension (i) tends to increase with (ii); just where certainty is needed the most, uncertainty is reaching its peak.

Post-normal science called for a search for new approaches to dealing with uncertainty, ones that recognised the diversity of stakeholders and evaluators needed to deal with these challenges. That search led to the recognition that characterising uncertainty required a dialogue amongst this extended set of peer reviewers through workshops and panels of experts. Such panels are convened by the US Environmental Protection Agency (EPA) who may be required to debate the issues in a public forum with participation of outside experts (consultants) employed by interest groups such as in the case of air pollution the American Lung Association and the American Petroleum Producers Association.

1.3 Dependencies over space and time

Environmental epidemiologists commonly seek associations between an environmental hazard Z and a health outcome Y. A spatial association is suggested if measured values of Z are found to be large (or small) at locations where counts of Y are also large (or small). Similarly, temporal associations arise when large (or small) values of Y are seen at times when Z are large (or small). A classical regression analysis might then be used to assess the magnitude of any associations and to assess whether they are significant. However such an analysis would be flawed if the pairs of measurements (of exposures), Z and the health outcomes, Y, are spatially correlated, which will result in outcomes at locations close together being more similar than those further apart. In this case, or in the case of temporal correlation, the standard assumptions of stochastic independence between experimental units would not be valid.

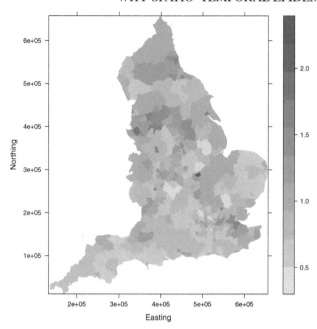

Figure 1.1: Map of the spatial distribution of risks of hospital admission for a respiratory condition, chronic obstructive pulmonary disease (COPD), in the UK for 2001. The shades of blue correspond to standardised admission rates, which are a measure of risk. Darker shades indicate higher rates of hospitalisation allowing for the underlying age–sex profile of the population within the area.

An example of spatial correlation can be seen in Figure 1.1 which shows the spatial distribution of the risk of hospital admission for chronic obstructive pulmonary disease (COPD) in the UK. There seem to be patterns in the data with areas of high and low risks being grouped together suggesting that there may be spatial dependence that would need to be incorporated in any model used to examine associations with potential risk factors.

1.3.1   Contrasts

Any regression-based analysis of risk requires contrasts between low and high levels of exposures in order to assess the differences in health outcomes between those levels. A major breakthrough in environmental epidemiology came from recognising that time series studies could, in some cases, supply the required contrasts between levels of exposures while eliminating the effects of confounders to a large extent.

This is now the standard approach in short-term air pollution studies (Katsouyanni et al., 1995; Peng & Dominici, 2008) where the levels of a pollutant,

Z, varies from day-to-day within a city while the values of confounding variables, X, for example the age–sex structure of the population or smoking habits, do not change over such a short time period. Thus if Z is found to vary in association with short-term changes in the levels of pollution then relationships can be established. However, the health counts are likely to be correlated over time due to underlying risk factors that vary over time. It is noted that this is of a different form than that for communicable diseases where it may be the disease itself that drives any correlation in health outcomes over time. This leads to the need for temporal process models to be incorporated within analyses of risk. In addition, there will often be temporal patterns in the exposures. Levels of air pollution for example are correlated over short periods of time due to changes in the source of the pollution and weather patterns such as wind and temperature.

Periods of missing exposure data can greatly affect the outcomes of a health analysis, both in terms of reducing sample size but also in inducing bias in the estimates of risk. There is a real need for models that can impute missing data accurately and in a form that can be used in health studies. It is important that, in addition to predicting levels of exposure when they are not available, such models should also produce measures of associated uncertainty that can be fed into subsequent analyses of the effect of those exposures on health.

Example 1.1. *Daily measurements of particulate matter*

An example of temporal correlation in exposures can be seen in Figure 1.2, which shows daily measurements of particulate matter over 250 days in London in 1997. Clear auto-correlation can be seen in this series of data with periods of high and low pollution. There are also periods of missing data (shown by triangles along the x-axis) where measurements were not available.

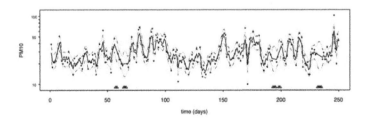

Figure 1.2: Time series of daily measurements of particulate matter (PM_{10}) for 250 days in 1997 in London. Measurements are made at the Bloomsbury monitoring site in central London. Missing values are shown by triangles. The solid black line is a smoothed estimate produced using a Bayesian temporal model and the dotted lines show the 95% credible intervals associated with the estimates.

It is noted that classical time series composition and analysis is primarily interested in modelling the behaviour of a response variable over time rather than its relationship to a set of explanatory variables which is at the heart of environmental epidemiology. However the classical theory can play a key role in learning the nature of any serial dependence in outcomes, both in health and exposures, and for constructing suitable models that incorporate such dependence.

Until recently epidemiological studies have considered measurements over space or time but rarely both. Increased power can be gained by combining space and time when processes evolve over both of these domains. We then have contrasting levels in both the spatial and temporal domains. When the spatial fields are temporally independent, replicates of the spatial field become available. Spatial dependence is then easier to model. However the likely presence of temporal dependence leads to a need to build complex dependence structures. At the cost of increased complexity, such models may utilise the full benefit of the information contained in the spatiotemporal field. This means that dependencies across both space and time can be exploited in terms of 'borrowing strength'. For example, values could be predicted at unmonitored spatial locations or at future times to help protect against predicted overexposures.

Example 1.2. *Spatial prediction of NO_2 concentrations in Europe*

In this example we see the result of using a spatial model to predict levels of nitrogen dioxide (NO_2) across Europe (Shaddick, Yan, et al., 2013). Measurements were available from monitoring sites at approximately 400 sites situated throughout Europe and these data were used to predict concentrations for every 1km $\times$ 1km geographical grid cell within the region. In this case, a Bayesian model was fit within WinBUGS and posterior predictions were imported (via R) to ESRI ArcGIS for mapping. The results can be seen in Figure 1.3.

In addition to the issues associated with correlation over space and time, environmental epidemiological studies will also face a major hurdle in the form of *confounders*. If there is a confounder, X, that is the real cause of adverse health effects there will be problems if it is associated with both Z and Y. In this case, apparent relationships observed between Z and Y may turn out to be spurious. It may therefore be important to model spatio–temporal variation in confounding variables in addition to the variables of primary interest.

1.4   Examples of spatio–temporal epidemiological analyses

Environmental exposures will vary over both space and time and there will potentially be many sources of variation and uncertainty. Statistical methods must be able

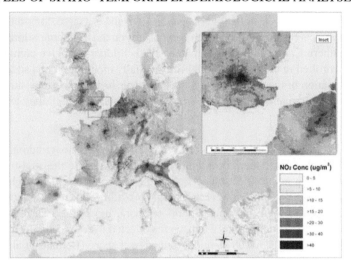

Figure 1.3: Predictions of nitrogen dioxide (NO_2) concentrations throughout Europe. The predictions are from a Bayesian spatial model and are the medians of the posterior distributions of predictions based on measurements from approximately 400 monitoring sites.

to acknowledge this variability and uncertainty and be able to estimate exposures at varying geographical and temporal scales in order to maximise the information available that can be linked to health outcomes in order to estimate the associated risks. In addition to estimates of risks, such methods must be able to produce measures of uncertainty associated with those risks. These measures of uncertainty should reflect the inherent uncertainties that will be present at each of the stages in the modelling process.

This has led to the application of spatial and temporal modelling in environmental epidemiology, in order to incorporate dependencies over space and time in analyses of association. The value of spatio–temporal modelling can be seen in two major studies that were underway at the time this book was being written: (i) the Children's Health Study in Los Angeles and (ii) the MESA Air (Multi-Ethnic Study of Atherosclerosis Air Pollution) study.

Example 1.3. *Children's health study – Los Angeles*

Children may suffer increased adverse effects to air pollution compared to adults as their lungs are still developing. They are also likely to experience higher exposures as they breathe faster and spend more time outdoors engaged in strenuous activity. The effects of air pollution on children's health is therefore a very important health issue.

The Children's Health Study began in 1993 and is a large, long-term study of the effects of chronic air pollution exposures on the health of children living in Southern California. Approximately 4000 children in twelve communities were enrolled in the study although substantially more have been added since the initiation of the study. Data on the children's health, their exposures to air pollution and many other factors were recorded annually until they graduated from high school.

This study is remarkable as the complexity of such longitudinal studies has generally made them prohibitively expensive. While the study was observational in nature, i.e. subjects could not be randomised to high or low exposure groups, children were selected to provide good contrast between areas of low and high exposure. Spatio–temporal modelling issues had to be addressed in the analysis since data were collected over time and from a number of communities which were distributed over space (Berhane et al., 2004).

A major finding from this study was that:

Current levels of air pollution have chronic, adverse effects on lung growth leading to clinically significant deficit in 18-year-old children. Air pollution affects both new onset asthma and exacerbation. Living in close proximity to busy roads is associated with risk for prevalent asthma. Residential traffic exposure is linked to deficit in lung function growth and increased school absences. Differences in genetic makeup affect these outcomes. (http://hydra.usc.edu/scehsc/about-studies-childrens.html)

Example 1.4. *Air pollution and cardiac disease*

The MESA Air (Multi-Ethnic Study of Atherosclerosis and Air Pollution) study involves more than 6000 men and women from six communities in the United States. The study started in 1999 and continues to follow participants' health as this book is being written.

The central hypothesis for this study is that long-term exposure to fine particles is associated with a more rapid progression of coronary atherosclerosis. Atherosclerosis is sometimes called hardening of the arteries and when it affects the arteries of the heart, it is called coronary artery disease. The problems caused by the smallest particles is their capacity to move through the gas exchange membrane into the blood system. Particles may also generate anti-inflammatory mediators in the blood that attack the heart.

Data are recorded both over time and space and so the analysis has been designed to acknowledge this. The study was designed to ensure the necessary contrasts needed for good statistical inference by taking random spatial samples of subjects from six very different regions. The study has yielded a great deal of new knowledge about the effects of air pollution on human health. In particular, exposures to chemicals and other environmental hazards appear to have a very serious impact on cardiovascular health.

> *Results from MESA Air show that people living in areas with higher levels of air pollution have thicker carotid artery walls than people living in areas with cleaner air. The arteries of people in more polluted areas also thickened faster over time, as compared to people living in places with cleaner air. These findings might help to explain how air pollution leads to problems like stroke and heart attacks. (http://depts.washington.edu/mesaair/)*

1.5 Bayesian hierarchical models

Bayesian hierarchical models are an extremely useful and flexible framework in which to model complex relationships and dependencies in data and they are used extensively throughout the book. In the hierarchy we consider, there are three levels;

(i) *The observation, or measurement, level; $Y|Z,X_1,\theta_1$.*
    Data, Y, are assumed to arise from an underlying process, Z, which is unobservable but from which measurements can be taken, possibly with error, at locations in space and time. Measurements may also be available for covariates, X_1. Here θ_1 is the set of parameters for this model and may include, for example, regression coefficients and error variances.

(ii) *The underlying process level; $Z|X_2,\theta_2$.*
    The process Z drives the measurements seen at the observation level and represents the true underlying level of the outcome. It may be, for example, a spatio–temporal process representing an environmental hazard. Measurements may also be available for covariates at this level, X_2. Here θ_2 is the set of parameters for this level of the model.

In this book we advocate a Bayesian approach and so there will be an additional level at which distributions are assigned to all unknown quantities.

(iii) *The parameter level; $\theta = (\theta_1,\theta_2)$.*
    This contains models for all of the parameters in the observation and process level and may control things such as the variability and strength of any spatio–temporal relationships.

Here the notation $Y|X$ means that the distribution of Y is conditional on X.

This book involves models for both health counts and exposures and each of these can be framed in the context of a hierarchical model. To avoid ambiguity between the two, we use $Y^{(1)}$, $X^{(1)}$, $Z^{(1)}$, $\theta^{(1)}$ for health models and $Y^{(2)}$, $X^{(2)}$, $Z^{(2)}$, $\theta^{(2)}$ for exposure models.

When health or exposure models are considered separately the $Y^{()}$ notation is dropped for clarity of exposition. Also, it is noted that we do not generally consider cases where health counts from routinely available data sources may not be an accurate reflection of the underlying health of the population at risk, i.e. it is assumed that $Y^{(1)} = Z^{(1)}$. In practice this might not be an entirely accurate assumption due to misclassification, migration or data anomalies.

1.5.1   A hierarchical approach to modelling spatio–temporal data

We now describe an implementation of this approach for modelling spatial–temporal data. A spatial–temporal random field, $Z_{st}, s \in \mathscr{S}, t \in \mathscr{T}$, is a stochastic process over a region and time period. This underlying process is not directly measurable, but realisations of it can be obtained by taking measurements, possibly with error. Monitoring will only report results at N_T discrete points in time, $T \in \mathscr{T}$ where these points are labelled $T = \{t_0, t_1, \dots, t_{N_T}\}$. The same will be true over space, since where air quality monitors can actually be placed may be restricted to a relatively small number of locations, for example on public land, leading to a discrete set of N_S locations $S \in \mathscr{S}$ with corresponding labelling, $S = \{s_0, s_1, \dots, s_{N_T}\}$.

As described above, there are three levels to the hierarchy that we consider. The observed data, $Z_{st}, s = 1, \dots, N_S, t = 1, \dots, N_T$, at the first level of the model are considered conditionally independent given a realisation of the underlying process, Z_{st}. The second level describes the true underlying process as a combination of two terms: (i) an overall trend, μ_{st} and (ii) a random process, ω_{st}. The trend, or mean term, μ_{st} represents broad scale changes over space and time which may be due to changes in covariates that will vary over space and time. The random process, ω_{st} has spatial–temporal structure in its covariance. In a Bayesian analysis, the third level of the model assigns prior distributions to the hyperparameters from the previous levels.

$$
\begin{aligned}
Y_{st} &= Z_{st} + v_{st} \\
Z_{st} &= \mu_{st} + \omega_{st}
\end{aligned}
$$

$$(1.1)$$

where the v_{st} is an independent random, or measurement, error term, μ_{st} is a space–time mean field (trend) and ω_{st} is a spatial–temporal process.

Throughout the book, where time and space are considered separately the notation is simplified to reflect the single domain, e.g. Y_s and Y_t are used, reserving Y_{st} for occasions where both space and time are under consideration.

1.5.2   Dealing with high-dimensional data

Due to both the size of the spatio–temporal components of the models that may now be considered and the number predictions that may be be required, it may be

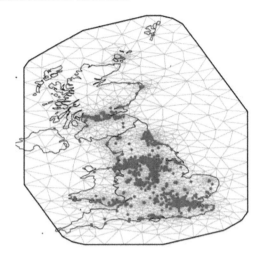

Figure 1.4: Triangulation for the locations of black smoke monitoring sites within the UK for use with the SPDE approach to modelling point-referenced spatial data with INLA. The mesh comprises 3799 edges and was constructed using triangles that have minimum angles of 26 and a maximum edge length of 100 km. The monitoring locations are highlighted in red.

computationally impractical to perform Bayesian analysis using packages such as WinBUGS (Lunn, Thomas, Best, & Spiegelhalter, 2000) or bespoke MCMC in any straightforward fashion. This can be due to both the requirement to manipulate large matrices within each simulation of the MCMC and issues of convergence of parameters in complex models (Finley, Banerjee, & Carlin, 2007).

Throughout the book, we show examples of recently developed techniques that perform 'approximate' Bayesian inference based on integrated nested Laplace approximations (INLA) and thus do not require full MCMC sampling to be performed (Rue, Martino, & Chopin, 2009). INLA has been developed as a computationally attractive alternative to MCMC. In a spatial setting such methods are naturally aligned for use with areal level data rather than the point level. However recent developments allow a Gaussian field (GF) with a Matern covariance function to be represented by a Gaussian Markov Random Field (GMRF) (Lindgren, Rue, & Lindström, 2011). This is available within the R-INLA package and an example of its use can be see in Figure 1.4 which shows a triangulation of the locations of black smoke (a measure of particulate air pollution) monitoring sites in the UK. The triangulation is part of the computational process which allows Bayesian inference to be performed on large sets of point-referenced spatial data.

1.6   Spatial data

Three main types of spatial data are commonly encountered in environmental epidemiology. They are (i) lattice, (ii) point-referenced and (iii) point-process data.

(i) Lattices refer to situations in which the spatial domain consists of a discrete set of 'lattice points'. These points may index the corners of cells in a regular or irregular grid. Alternatively, they may index geographical regions such as administrative units or health districts (see for example Figure 1.1), This is an important topic in spatio–temporal epidemiology and detailed discussions can be found in Gotway and Young (2002); Cressie and Wikle (2011) and Banerjee et al. (2015). We denote the set of all lattice points by $\mathscr{L}$ with data available at a set of N_L points, $l \in L$ where $L = l_1, ..., l_{N_L}$. In many applications, such as disease mapping, L is commonly equal to $\mathscr{L}$. A key feature of this class is its neighbourhood structure; a process that generates the data at a location has a distribution that can be characterised in terms of its neighbours.

(ii) Point-referenced data are measured at a fixed, and often sparse, set of 'spatial points' in a spatial domain or region. That domain may be continuous, $\mathscr{S}$ but in the applications considered in this book the domain will be treated as discrete both to reduce technical complexity and to reflect the practicalities of siting monitors of environmental processes. For example, when monitoring air pollution, the number of monitors may be limited by financial considerations and they may have to be sited on public land. Measurements are available at a selection of N_S sites, $s \in S$ where $S = s_1, ..., s_{N_S}$. Sites would usually be defined in terms of their geographical coordinates such as longitude and latitude, i.e. $s_l = (a_l, b_l)$.

iii) Point-process data consists of a set of points, S, that are randomly chosen by a spatial point process (Diggle, 2013) . These points could mark, for example, the incidence of a disease such as childhood leukaemia (Gatrell, Bailey, Diggle, & Rowlingson, 1996). Despite the importance of spatial point process modelling we do not cover this topic and its range of applications in this book. The reader is directed to Diggle (1993) and Diggle (2013) for further reading on this subject.

Example 1.5. *Visualising spatial data*

In this example, we consider ways in which spatial data can be visualised. This is an important topic which encompasses aspects of model building, including the assessment of the validity of modelling assumptions, and the presentation of results. In this section we provide a brief introduction to the subject and show some examples of how spatial data can be visualised using a variety of R packages and particularly how R can be used to interact with Google maps in order to display spatial data.

We illustrate this by mapping measurement of lead concentrations in the Meuse River flood plain. The Meuse River is one of the largest in Europe and the subject of much study (Ashagrie, De Laat, De Wit, Tu, & Uhlenbrook,

2006). A comprehensive dataset was collected in its flood plain in 1990 and provides valuable information on the concentrations of a variety of elements in the river. The information is measured at 155 sampling sites within the flood plain.

The following questions will be of interest to health researchers and others:

- How much spatial variation is there in lead concentrations?
- Are there distinct spatial trends in the concentration levels?
- Are there enough sampling sites to appropriately characterise this random field?
- Can levels be mapped between the sampling sites for example at a center of human activity located between sites?

The following R code shows how the data on lead concentrations in the Meuse River flood plain can be displayed on a Google map of the area.

```
> library(sp)
> library(rgdal)

> data(meuse)

### We first assign a reference system used in the
    Netherlands
> proj4string(meuse_sp) <- CRS('+init=epsg:28992')
### Then we convert it to another for Google mapping that
### requires the latitude - longitude scale
> meuse_ll <- spTransform(meuse_sp,
                                CRS("+proj=longlat +datum=WGS84"
                                ))
###Finally we write the result that can be read by Google
    maps
> writeOGR(meuse_ll, "meuse.kml", "meuse", driver="KML")
```

Figure 1.5 shows the result of opening the meuse.kml file in Google maps. It shows the sampling sites marked with map tacks. Google's Street View then lets an observer see the map tacks. Clicking on one of the visible map tacks reveals the sample data record for that site within Street View. The ability to interact with Google maps is not restricted to just visualising data but can be used to overlay the results of statistical analyses, such as spatial predictions, onto maps. This subject is covered in detail in Bivand, Pebesma, and Gómez-Rubio (2008).

High concentrations of lead can be observed at that site of 91 milligrams per kilogram (mg/kg) or equivalently parts per million (ppm), with the conversion being one-to-one. Turning to look in the opposite direction, another tack could be seen that is associated with a measurement of about half of the first's lead concentration level. There is great interest in the observed levels as lead in soil can get into the food chain and cause adverse health reactions, including retarded neurological development in children (Reagan & Silbergeld, 1990).

(a) Sampling sites near Meuse River   (b) Map tack opens to show sample

Figure 1.5: Google Earth and Google Street Map provide useful ways of visualising spatial data. Here we see (a) the location at which samples were taken in the Meuse River flood plain and (b) the information that was collected.

1.7   Good spatio–temporal modelling approaches

Often spatio–temporal models are purpose-built for a particular application and then presented as a theoretical model. It is then reasonable to ask what can be done with that model in settings other than those in which it was developed. More generally, can it be extended for use in other applications? There are a number of key elements which are common to good approaches to spatio–temporal modelling. The approaches should do the following:

- Incorporate all sources of uncertainty. This has led to the widespread use of Bayesian hierarchical modelling in theory and practice.

- Have an associated practical theory of data-based inference.

- Allow extensions to handling multivariate data. This is vital as it may be a mix of hazards that cause negative health impacts. Even in the case where a single hazard is of interest, the multivariate approach allows strength to be borrowed from the other hazards which are correlated with the one of concern (Sun, 1998).

- Be computationally feasible to implement. This is of increasing concern as we see increasingly large domains of interest. One might now reasonably expect to see a spatial domain with thousands of sites and thousands of time points.

- Come equipped with a design theory that enables measurements to be made optimally for estimating the process parameters or for predicting unmeasured process values. Good data are fundamental to good spatio–temporal modelling, yet this aspect is commonly ignored and can lead to biased estimates of exposures and thus risk (Shaddick & Zidek, 2014; Zidek, Shaddick, & Taylor, 2014).

- Produce well calibrated error bands. For example, a 95% band should contain predicted values 95% of the time, i.e. they have correct *coverage probabilities*. This is important not only in substantive terms, but also in model checking. There may be questions about the formulation of a model, for example of the precise nature of the spatio–temporal process that is assumed, but that may be of secondary importance if good empirical performance of the model can be demonstrated. This criteria can really challenge the perceived efficacy of a model; "All models are wrong but they can useful." as George Box famously remarked (Box & Draper, 1987).

1.8 Summary

This book provides a comprehensive treatment of methods for spatio–temporal modelling and their use in epidemiological studies. From this book the reader will have gained an understanding of the following topics:

- The basic concepts of epidemiology and the estimation of risks associated with environmental hazards.

- Hierarchical modelling with a Bayesian framework.

- The theory of spatial, temporal and spatio–temporal processes needed for environmental health risk analysis.

- Fundamental questions related to the nature and role of uncertainty in environmental epidemiology and methods that may help answer those questions.

- Important areas of application within environmental epidemiology together with strategies for building the models that are needed and coping with challenges that arise.

- Methods and software for the analysis and visualisation of environmental and health. Examples of R and WinBUGS code are given throughout the book and, together with data for the examples, the code is included in the online resources.

- A variety of exercises, both theoretical and practical, to assist in the development of the skills needed to perform spatio–temporal analyses.

- New frontiers and areas of current and future research.

Chapter 2

Modelling health risks

2.1 Overview

The estimation of health risks is most commonly performed using regression models with the aim of developing a realistic model of the system in question, to identify the variables of interest and to estimate the strength of their effect on a health outcome. Though an epidemiological regression model cannot itself prove causality (that can only really be provided by randomised experiments), it can indicate the change in the response variable that might be associated with changes in exposure which is a very useful tool in understanding and developing insight into possible causal relationships.

This chapter considers different methods for expressing the risks associated with exposures and their appropriateness given the type of study that is being performed. A number of aspects of model fitting are introduced, including the choice of distribution (e.g. Normal, Poisson or Binomial), the choice of the regression (e.g. linear, log–linear or logistic) and how the actual computation is carried out. There are many reasons for performing a regression analysis and the intended use of the regression equation will influence the manner in which the model is constructed and interpreted. In the applications considered here, the interest is in explanation and prediction, but not prediction as is used in the stock market where prediction of future events alone is often important.

2.2 Types of epidemiological study

Epidemiology is the study of why and how often diseases occur in different groups of people. A key feature of epidemiology is the measurement of disease outcomes in relation to a population at risk. Population studies aim to compare diseased and non-diseased individuals according to a common characteristic, such as age, sex, or exposure to a risk factor. Cohort studies observe the progress of individuals over time and are useful for investigating and determining aetiological factors. A study may follow two groups with one group exposed to some risk factor while the other group is not exposed with the aim of seeing if, for example, exposure influences the occurrence of certain diseases. For rare conditions very large samples will be required to observe any cases. One insurmountable problem with cohort studies is that if the cohort has to be set up from scratch there may be a long delay before analysis can take place.

One solution to the problem of the small number of people with the condition of interest is the case–control study. In this type of study we take a group of people with the disease, the cases, and a second group without the disease, the controls. We then find the exposure of each subject to the possible causative factor and see whether this differs between the two groups. The advantages of this method of investigation are that it is relatively quick, as the event of interest has already happened, and it is cheap, as the sample size can be small. However, there can also be difficulties with this method in terms of the selection of cases, the selection of controls and obtaining the data.

In collecting a set of controls we want to look to gather a group of people who do not have the disease in question, but who are otherwise comparable to our cases. We must first decide upon the population from which they are to be drawn. There are many sources of controls; two obvious ones being the general population and patients with other diseases. The latter is usually preferred because of its accessibility. However, these two populations are not the same. While it is easier to use hospital patients as controls, there may be bias introduced because the factor of interest may be associated with other diseases.

Having defined the population from which to choose the sample we must bear in mind that there are many factors, such as age and sex, which affect exposure to risk factors. The most straightforward way to choose the sample is to take a large random sample of the control population, ascertain all the sample's relevant characteristics, and then adjust for differences during the analysis, using for example logistic regression models. The alternative is to try to match a control to each case, so that for each case there is a control of the same age, sex, etc. If we do this, we can then compare our cases and controls knowing that the effects of these intervening variables are automatically adjusted for. If we wish to exclude a case we must exclude its control too, or the groups will no longer be comparable. It is possible to have more than one control per case, but this can lead to the analysis becoming more complicated.

2.3 Measures of risk

Here we consider measures of risk that are commonly estimated in epidemiological studies and that can be estimated using regression models. The choice of where risks are expressed, in terms of either relative risk or odds ratio, will largely be driven by the type of study that is being analysed. Here we briefly describe three common designs: population, cohort and case–control studies. Each will produce a different type of outcome and, due to the intrinsic properties of their designs, different methods will be required for estimating risks.

Let D and $\bar{D}$ denote 'disease' and 'not disease', and E and $\bar{E}$ denote 'exposed' and 'unexposed' respectively. The risk of an event is the probability that an event will occur within a given period of time. The risk of an individual having a disease $(P(D))$ is estimated by the frequency with which that condition has occurred in a similar population in the past.

2.3.1 Relative risk

To compare risks for people with and without a particular risk factor, we look at the ratio. Suppose the risk for the exposed group is $\pi_1 = P(D|E)$ and the risk for the unexposed group is $\pi_0 = P(D|\bar{E})$.

The *relative risk* is

$$RR = \frac{\pi_1}{\pi_0}$$

		Exposure to Risk Factor		
		Unexposed ($\bar{E}$)	Exposed (E)	
	Absent ($\bar{D}$)	n_{00}	n_{01}	n_{0+}
Disease	Present (D)	n_{10}	n_{11}	n_{1+}
		n_{+0}	n_{+1}	n_{++}

Table 2.1: Number of diseased/non-diseased individuals by to a risk factor.

Given the number of diseased/non-diseased individuals who are in the exposed/non-exposed groups, as seen in Table 2.1, we can calculate the relative risk. The overall incidence, irrespective of exposure, will be n_{1+}/n_{++}.

From Table 2.1, we have

$$RR = \frac{n_{11}/n_{+1}}{n_{10}/n_{+0}} = \frac{n_{11}n_{+0}}{n_{+1}n_{10}}$$

In a population study, the total sample size n_{++} is fixed; in a cohort study, the margins n_{+0} and n_{+1} are fixed; and in a case–control study, the margins n_{0+} and n_{1+} are fixed.

Note that we cannot estimate the relative risk directly from a case–control study, as the n_{++} individuals are not selected at random from the population.

To construct confidence intervals we use

$$Var(\log RR) = \frac{1}{n_{11}} - \frac{1}{n_{+1}} + \frac{1}{n_{10}} - \frac{1}{n_{+0}}$$

Assuming that $\log RR$ is normally distributed, a 95% CI for log RR is

$$\log RR \pm 1.96 \sqrt{\frac{1}{n_{11}} - \frac{1}{n_{+1}} + \frac{1}{n_{10}} - \frac{1}{n_{+0}}}$$

Example 2.1. *The risk of lung cancer associated with radon exposure in miners*

Table 2.2 gives the number of cases of lung cancer (diseased) in uranium miners in Beverlodge, Canada (Lubin et al., 1995) who were exposed to radon (exposed) and the number who were not exposed (unexposed).

Table 2.2: Number of cases of lung cancer in uranium miners who were exposed and unexposed to radon in a mine in Beverlodge, Canada. Here the relative risk is 1.58 (95% CI; 0.90 – 2.76).

		Exposure to Risk Factor		
		No radon ($\bar{E}$)	Radon (E)	
	No lung cancer ($\bar{D}$)	22204	30415	52619
Disease	Lung cancer (D)	18	39	57
		22222	30454	52676

The relative risk of lung cancer is therefore $\frac{39/30415}{18/22204} = 1.58$, i.e. there is a 58% increase in the risk of lung cancer for miners (from this mine) when they are exposed to radon. The log of the relative risk is 0.46 and the 95% CI on the log scale is $-0.10 - 1.02$. Therefore on the original scale of the data it is $0.90 - 2.76$, indicating that the observed increase in risk is non-significant.

2.3.2 Population attributable risk

If a relative risk is large but very few people are exposed to the risk factor then the effect on the population will not be large, despite serious consequences to the individual. The effect of a risk factor on community health can be measured by *attributable risk*, which is related to the relative risk and the percentage of the population affected.

The attributable risk is the proportion of the population risk that can be associated with the risk factor:

$$\text{AR} = \frac{\text{population risk} - \text{unexposed risk}}{\text{population risk}} = \frac{P(D) - P(D|\bar{E})}{P(D)} = \frac{\theta(\text{RR}-1)}{1+\theta(\text{RR}-1)}$$

where θ is the proportion of the population who are exposed to the risk factor, $P(E)$.

Example 2.2. *Population attributable risk of lung cancer associated with radon exposure*

Continuing the example of the miners exposed to radon, if the proportion of the population of miners who are exposed to radon is 5% then the attributable risk is 0.03 whereas if the proportion exposed rose to 10%, then the attributable risk would rise to 0.06. As more miners are exposed, the proportion of the risk for the entire population that can be attributed to radon exposure increases.

2.3.3 Odds ratios

An alternative to the risk is the *odds* of an event. If the probability of an event is p then the odds is

$$\frac{p}{(1-p)}$$

This can be useful since it is not constrained to lie between 0 and 1. We often use the *log odds*:

$$\log\left(\frac{p}{(1-p)}\right)$$

Another way to compare people with and without a particular risk factor is to use the *odds ratio*. The odds ratio for disease given exposure is

$$OR = \frac{\text{odds of disease given exposed}}{\text{odds of disease given unexposed}} = \frac{P(D|E)\,(1-P(D|\bar{E}))}{P(D|\bar{E})\,(1-P(D|E)}$$

For a 2×2 table as in Table 2.1 the odds ratio is

$$OR = \frac{n_{00}n_{11}}{n_{10}n_{01}}$$

The odds ratio is a useful measure since it is independent of the prevalence of the condition.

Example 2.3. *Odds ratio of lung cancer associated with radon exposure in miners*

From the example of the miners exposed to radon seen in Table 2.2, the odds ratio will be $(39/30415)/(18/22204) = 1.58$. In this case the odds ratios and the relative risk are very similar, as the risk of disease is small.

2.3.4 Relationship between odds ratios and relative risk

The odds ratio gives a reasonable estimate of the relative risk when the proportion of subjects with the disease is small.

The risk of disease for an exposed person is

$$\pi_1 = \frac{P(D \cap E)}{P(E)}$$
$$\approx \frac{P(E|D)P(D)}{P(E|\bar{D})}$$

Similarly,

$$\pi_0 \approx \frac{P(\bar{E}|D)P(D)}{P(\bar{E}|\bar{D})}$$

So

$$RR \approx \frac{P(E|D)P(D)P(\bar{E}|\bar{D})P(\bar{D})}{P(E|\bar{D})P(\bar{D})P(\bar{E}|D)P(D)}$$
$$= OR$$

2.3.5 Odds ratios in case–control studies

The odds ratio is particularly useful in a case–control study. Above, we have given the formula for the odds ratio for disease given exposure $OR_{D|E}$. From a case–control study we can calculate the odds ratio for exposure given disease, $OR_{E|D}$ which can be shown to be equal to $OR_{D|E}$. So although we cannot calculate a relative risk for a case–control study, we can estimate its odds ratio, provided that the cases and controls are random samples from the same population.

In addition, for rare diseases, the odds ratio is close to the relative risk as seen previously. So in practice the odds ratio from a case–control study can often be interpreted as if it were the relative risk.

2.4 Standardised mortality ratio (SMR)

When we are investigating differences in risk between populations we can eliminate the effects of, for example, different age structures by looking at age-specific rates. However, this can be clumbersome and it is often easier to make comparisons within a single summary figure. This is achieved by calculating standardised rates that allow for differing rates in specific groups, e.g. age and gender groups and the proportions of the population that are within those groups.

Let Y be the observed number of deaths in the population of interest, and E be the expected number of deaths with respect to some reference population.

$$SMR = \frac{Y}{E}$$

The SMR is a ratio, not a rate or a percentage. An SMR of 1 means that the population of interest has the same number of deaths as we would expect from the reference population. If it is greater than 1, we have more deaths than expected; if it is less than 1 we have fewer deaths than expected.

2.4.1 Rates and expected numbers

A *rate* is defined as the number of events, for example deaths or cases of disease, per unit of population, in a particular time span. To calculate a rate we require: a defined period of time, a defined population, with an accurate estimate of the size of the population during the defined period, the person-years at risk, and the number of events occurring over the period.

$$\text{rate} = \frac{\text{No. events}}{\text{total person-time at risk}}$$

For a fixed time period Δt, an average size of the population at risk during that period $\bar{N}$ and the number of events A the rate is

$$\text{rate} = \frac{A}{\bar{N} \times \Delta t}$$

The *crude mortality rate* is usually calculated as deaths per 1000 population per year. Let D be the number of deaths in a given time period of length Δt, and $\bar{N}$ be the average size of the population at risk during that period (often approximated by the number in the population at the mid-point of the time period). Then the crude mortality rate is given by

$$r = \frac{d}{\bar{N} \times \Delta t} \times 1000$$

Rates may be required for particular sections of a community. Where the populations are *specified* these rates are referred to as *specific rates*. For example, age-specific or age- and sex-specific rates may be used for comparison of different populations. Other common specific rates are area, occupation or social class specific (and combinations of these).

For direct standardisation, we use a standard population structure for reference. We then calculate the overall mortality rate that this reference population would have observed if it had the age-specific mortality rates of the population of interest.

Suppose the reference population has population counts $N'_k; k = 1, \ldots, K$ in each age-group k. We calculate the age-specific mortality rates r_k for the population of interest. The directly standardised rate is then given by

$$\text{directly standardised rate} = \frac{\sum_{k=1}^{K} N'_k r_k}{\sum_{k=1}^{K} N'_k}$$

For indirect standardisation, we take the age-specific rates from the reference population and convert them into the mortality rate we would observe if those reference rates were true for the age-structure of the population of interest. This gives us the expected rate for the population of interest, if age-specific mortality rates were the same as for the reference population.

We calculate the age-specific mortality rates r'_k for the reference population. Suppose the population of interest has population counts $N_k; k = 1, \ldots, K$ in each age-group k.

The expected rate of deaths in the population of interest is

$$\text{expected rate} = \frac{\sum_{k=1}^{K} N_k r'_k}{\sum_{k=1}^{K} N_k}$$

Given the expected rate of deaths in the population of interest, the expected number of deaths is

$$E = \sum_{k=1}^{K} N_k r'_k$$

Example 2.4. *Risk of liver cancer and proximity to incinerators*

In a study of the association between the risk of liver cancer and living in close proximity to a municipal incinerator in the UK (Elliott et al., 1996), the number of people with liver cancer living less than 3km from an incinerator was 152. The expected number of deaths for liver cancer based on national rates applied to the age–sex structure of the population living within 3km of an incinerator was 118.10 and therefore the SMR was 1.29. This indicates there is a 29% increase in risk of liver cancer in those living in close proximity to incinerators compared to what might be expected. Of course, there may be other factors that mean that the observed increase in risk is not wholly due to incinerators. In the next section, we see how the effects of other potential risk factors (covariates) can be used to adjust such risks by modelling SMRs using generalised linear models.

2.5 Generalised linear models

Data from epidemiological studies commonly consist of counts of disease in a defined geographical area and/or time-period or of binary indicators of whether a subject is a case or control. The response variable is therefore going to be non-normal, being non-negative integers in the first case and constrained to be either 0 or 1 in the second. Linear regression is therefore unlikely to be appropriate. Any model used for estimating the association between such outcomes and possible explanatory variables (and subsequently estimates of associated risk) must therefore acknowledge the type of data coming from the study and accommodate suitable distributional assumptions. Examples include the Poisson distribution when predicting deaths and the Binomial when predicting the probabilities of being a case or control.

2.5.1   Likelihood

The likelihood, $L_n(\theta|y) = f(y|\theta)$ is a measure of the *support* provided for a particular set of the parameters, θ, of a probability model by the data, $y = (y_1, ..., y_n)$. It is a relative measure and only determined by the values of the observed data, therefore any constants independent of θ can be omitted. It is often more convenient to consider the log likelihood, $l_n(\theta|y) = \log(L_n(\theta|y))$. If the response vector, y, consists of n independent and identically distributed observations, then the likelihood of θ is $L_n(\theta|y) = \prod_{i=1}^{n} L(\theta|y_i)$ and $l_n(\theta|y) = \sum_{i=1}^{n} l_n(\theta|y_i)$.

The first derivative of the log likelihood is known as the *score function*, U_n, which for a sample of size n is given by

$$U_n(\theta|y) = l_n'(\theta|y) = \frac{dl_n(\theta|by)}{d\theta} \tag{2.1}$$

The maximum likelihood estimate (MLE) is found by solving the equation $U_n(\theta|by) = 0$.

For reasons of clarity, unless stated, the following descriptions consider a univariate unknown parameter, θ. In regular problems analogous results for more than one parameter follow naturally using standard multivariate techniques such as those described in DeGroot (1986) and in Chatfield and Collins (1980).

2.5.2   Quasi-likelihood

Quasi-likelihood only requires that the first two moments of the data generating distribution are specified, i.e. the mean and variance. The integral $Q(\mu|y) = \int_y^\mu \frac{y-t}{V(t)} dt$ acts like a *quasi*-likelihood even if it does not constitute a proper likelihood function and estimates of the parameters of the model can be obtained by maximizing under certain conditions. Further details of the use of quasi-likelihood can be found in Section 2.8.2 (in relation to over-dispersion in Poisson models, where the variance is greater than the mean). Further technical details can be found in McCullagh and Nelder (1989).

2.5.3   Likelihood ratio tests

The likelihood ratio test is a useful way of comparing the fit of two competing nested models (where the set of parameters in one model is a subset of those in another). The models can be constructed to correspond to the null and alternative hypothesis of a test. If Ω is the parameter space of the larger model and Ω_0 is that of the smaller model, then the models are nested if $\Omega_0 \in \Omega$. The null hypothesis corresponds to $H_0 : \theta \in \Omega_0$ and the alternative $H_0 : \theta \in \Omega|\Omega_0$ (the set difference). The likelihood ratio test for comparing the two models, and thus testing the hypotheses, is twice the log of the difference between the maximised likelihoods under the two models.

$$2\left(\frac{\max_{\Omega|\Omega_0} l_n(\theta)}{\max_{\Omega} l_n(\theta)}\right)$$

In the case of regression models, if a particular model is compared to the 'full' or saturated model then the *scaled deviance* can be calculated

$$
\begin{aligned}
\text{Deviance, } D &= -2\log(L_k/L_m) \\
&= -2(l_k - l_m) \qquad\qquad (2.2)
\end{aligned}
$$

where l_k is the log of the likelihood of model M_k and l_m is that for the saturated model, M_m, which contains all n parameters. In what follows, the scaled deviance will be referred to simply as the *deviance, D*.

Large values of the deviance ($L_k << L_m$) suggest that the current model does not fit well. If the model in question has r_k variables then the scaled deviance is asymptotically χ^2 distributed with $n - r_k$ degrees of freedom, if the model is appropriate. If the model fits poorly then the deviance will be large.

Rather than make comparisons with the full model, M_m, it is generally more useful to make comparisons with another model with only one (or a few) extra parameters, i.e. comparing models M_k and M_q, with $r_k + 1$ and $r_q + 1$ parameters respectively. This is because, for all but the Normal linear case, the χ^2 test for change in deviances is a much better asymptotic approximation than the one using comparisons with the full model (McCullagh & Nelder, 1989). If $r_k < r_q$ then M_k is said to be nested within M_q and if $D_k = -2(l_k - l_m)$ and $D_q = -2(l_q - l_m)$ then

$$D_k - D_q = -2(l_k - l_q) \sim \chi^2_{r_q - r_k} \qquad\qquad (2.3)$$

These results assume that the number of parameters is fixed as $n \to \infty$. Further discussion of the selection of variables for use in regression models can be found in Chapter 6.

2.5.4   Link functions and error distributions

Generalised linear models (GLMs) are defined by the distribution of the errors and a link function which relates the linear predictor to the response. They can be used when data is an independent sample from an exponential family probability distribution; which includes the normal, binomial and Poisson distributions.

- The likelihood for a GLM has a general form for distributions in the exponential family of distributions; $f_y(y|\theta,\phi) = \exp\{(y\theta - b(\theta))/a(\phi)\} + c(y,\phi)\}$, where the expectation and variance of Y take the form $E(Y) = b'(\theta)$ and $VAR(Y) = v_i = b''(\theta)a(\phi)$ respectively. The function $a(\phi)$ is often of the form $a(\phi) = \phi/w$, where ϕ is the dispersion parameter and w the prior weights.

- The effects of the explanatory variables are contained in the linear predictor, $\eta_i = \sum_{i=1}^{p} x_{ij}\beta_j$, which simplifies certain aspects of inference, both in terms of computation and properties of the resultant estimates.

- The link function links the mean function $\mu = E(Y)$ to the linear predictor, $g(\mu_i) = \eta_i$

In general, the link and error structure have to be explicitly defined. The parameters, β can be estimated by solving the estimating equations,

$$U_n(\beta) = \sum_{i=1}^{p} \frac{d\mu_i}{d\beta} v_i^{-1}(y_i - \mu_i(\beta)) = 0 \tag{2.4}$$

Linear models are a special case of GLMs with the following set-up:

- $Y_i \sim N(\mu_i, \sigma^2)$, i.i.d. Normal observations
- $g(\mu_i) = \eta_i$, is the identity link function
- $\eta_i = \sum_{i=1}^{p} x_{ij}\beta_j$, a linear predictor

Example 2.5. *Poisson log–linear models*

In the majority of studies examining the effects of air pollution on health, the outcome, e.g. number of daily deaths, is assumed to follow a Poisson distribution. Poisson data, for example, take the form of counts and thus the outcome is in the range $[0, \infty)$. The natural link function is therefore $\log(\mu_i) = \gamma_i$, which restricts the outcome to the required range and assumes an underlying multiplicative relationship. In this case, $f_y(y|\theta) = (\mu, \phi)) = \exp\{y\log\mu - \mu - \log y!\}$, therefore $\log\mu = \theta, b(\theta) = \exp(\theta), c(y, \phi) = -\log y!$ and $a(\phi) = 1$.

2.5.5 Comparing models

Testing between competing nested models is achieved using the difference in deviance, as described in Section 2.5.3. The deviance will be of the form $\sum_{i=1}^{n} 2\{y_i(\tilde{\theta}_i - \hat{\theta}_i) - b(\theta_i) + b(\hat{\theta}_i)\}/\phi = D(y|\hat{\theta})$, where $\hat{\theta}$ is the MLE under the model in question and $\tilde{\theta}$ is the MLE under the saturated model. The deviance for Poisson data is therefore $D = 2\sum_{i=1}^{n}\{y\log(\tilde{\theta}/\hat{\theta}) - (y - \hat{\theta})\}$.

The MLEs in all but the Normal case cannot be found by simply maximizing the log likelihoods numerically. Instead, they can be found using Iteratively Re-weighted Least Squares (IRLS) (McCullagh & Nelder, 1989).

2.6 Generalised additive models

Generalised additive models (GAMs) can be thought of as non-parametric extensions of GLMs. In Section 2.5 the form of the GLM was given as $g(\mu_i) = g(E(y)) = X\beta$. In a GAM, instead of assuming dependence on the sum of linear predictors, the outcome is assumed to be dependent on a sum of smooth functions of the predictors

$$g(E(Y)) = \sum_j S_j(X_i) \tag{2.5}$$

where S_j are smoothing functions. GAMs therefore provide a flexible framework for controlling for non-linear dependence on potential covariates (Hastie & Tibshirani, 1990; Wood, 2006).

GAMs are commonly used in epidemiological studies, especially in time series studies of the short-term effects of air pollution on health where it is now almost the standard approach for studies of this type. For example, in analysing the relationship between PM_{10} and SO_2 and hospital admission in New Haven and Tacoma, U.S., Schwartz (1995) used a GAM to replicate the 19-day moving average used in Kinney and Ozkaynak (1991), modelling the outcome as a Poisson outcome rather than being restricted to the normal distribution as a consequence of using moving averages of the actual outcome variable. GAMs have also been used in this setting to allow for long-term patterns, such as seasonality, in time-series studies, enabling the associations between short-term changes in air pollution and health to be examined.

2.6.1 Smoothers

The class of potential smoothers is very large and only a selection of the most commonly used ones are described here. For a more complete review see Hastie and Tibshirani (1990). Possibly the most intuitive smoother is the moving average or running mean, in which the value of x_i is replaced by the average of that value together with those within a defined period around t

$$S(X_i) = \sum_{k=-p}^{p} w_p X_{i+k} : k = p+1, ..., n-p \tag{2.6}$$

where w_k are the weights and $p > 0$. Here, $2p+1$ is the order of the moving average, i.e. the number of points which are included in calculating the average.

Unless the number of points included in the calculation of the average is large, in which case information will be lost, the resulting line is unlikely to be very smooth due to the sudden drop in weights as points are not included in the calculation. However, if the weights were specified to follow a probability distribution, as in *density estimation* (Silverman, 1986), then a smooth set of weights can be defined. An alternative approach is to calculate a least squares regression line within each set of neighbouring points of the form

$$S(x_i) = \hat{\alpha}_{x_i} + \hat{\beta}_{x_i} x_i \tag{2.7}$$

The fitted value at x_i is then used as the smoothed value. The regression can be fitted using weights, w_i, which can be constructed to be decreasing with distance from x_i, giving a *loess* smoother (Hastie & Tibshirani, 1990).

Another approach is to fit a polynomial to the data, for example

$$S(x_i) = \sum_{k=0}^{p} \beta_k x_i^k \qquad (2.8)$$

where the coefficients, β_k, are estimated from the data. As with the moving average approach, a choice of the extent of the smoothing (the order of the polynomial) has to be made. Simple polynomials are often useful as part of a model for the trend with additional terms describing the seasonality, but polynomials of higher order, whilst theoretically allowing very good fits to part of the data are likely to be too restrictive, especially if the model is to be used on another dataset or for forecasting out of the range of the current set.

2.6.2 Splines

Regression *splines* fit piecewise polynomials separated by *knots*, and are usually reforced to join smoothly at the knots. When the first and second derivatives are continuous, the piecewise polynomial is known as a *cubic spline*. A drawback of this approach is the need to choose the number and position of the knots, although attempts have been made to automate the procedure (Denison, Mallick, & Smith, 1998).

Splines provide a flexible technique for modelling non-linear relationships. They transform possible non-linear relationships into a linear form. They are formed by separating the data, $\mathbf{x}$, into $k+1$ sub-intervals. These intervals are defined by k knots, $k_1, ..., k_K$. The class of precise choice of splines and their basis functions is very large and further details can be found in Wahba (1990); Ruppert, Wand, and Carroll (2003); Wood (2006).

A spline, $f(x)$, is of the form

$$f(x) = \sum_{j=1}^{q} b_j(x)\beta_j \qquad (2.9)$$

where $b_j(x)$ is the jth basis function, and β_j are the basis parameters to be estimated. There are many ways to express the basis function which will represent the spline (Wood, 2006). A popular choice of spline are the cubic splines which have the following properties: (i) within each interval $f(x)$ is a cubic polynomial and (ii) it is twice differentiable and continuous at the knots. Cubic splines comprise cubic polynomials within each interval. Green and Silverman (1994) show a simple way to specify a cubic spline by giving four polynomial coefficients of each interval, i.e.

$$f(x) = d_i(x - x_i)^3 + c_i(x - x_i)^2 + b_i(x - x_i) + a_i, \quad x_i \le x \le x_{i+1}$$

where x_i are the knots and a_i, b_i, c_i and d_i are parameters representing the spline. A cubic spline on an interval $[a, b]$ is called a *natural cubic spline* (NCS) if its second and third derivatives are zero at a and b.

2.6.3   Penalised splines

The shape and smoothness of the spline depend upon the basis parameters and a term that represents smoothness, the latter of which can be represented in a number of ways. The most common of these is to use the number of knots, i.e. to space the knots equally throughout the data, and specify the smoothness by k the number of knots. However, too many knots may lead to oversmoothness, whilst inadequate number of knots leads to rough model fit. Alternatively a penalised approach can be adopted, which uses an overly large number of basis functions and penalises excess curvature in the estimate using a penalty term.

Penalised splines control smoothness by adding a 'wiggliness' penalty to the least squares objective, that is fitting the model by minimizing

$$\|Y - X\beta\|^2 + \lambda \int_a^b [f''(x)]^2 dx$$

where the second part of this formula penalises models that are too 'wiggly'. The trade off between model fit and model smoothness is controlled by the smoothing parameter , λ. Since f is linear with parameters β_i, the penalty term can be written as a quadratic form in β:

$$\int_a^b [f''(x)]^2 dx = \beta^T S \beta$$

where S is a matrix of known coefficients.

Therefore, the penalised regression spline fitting problem is equivalent to minimize:

$$\|Y - X\beta\|^2 + \lambda \beta^T S \beta$$

In this case, the estimation of degree of smoothness becomes the issue of detecting the smoothing parameter λ. If λ is too high then the data will be oversmoothed, and if it is too low then the data will be undersmoothed. In both cases this will mean that the spline estimate $\hat{f}$ will not be close to the true function f. Choosing λ may be done using data-driven criterion, such as cross validation (CV) and generalised cross validation (GCV), details of which can be found in Wahba (1990); Gu (2002); Ruppert et al. (2003); Wood (2006).

Example 2.6. *The use of GAMs in time series epidemiology*

When assessing the short-term effects of an environmental exposure, for example air pollution, on health there will be a need to model underlying long-term trends in the data before shorter-term effects can be assessed. This can be performed within the framework of general additive models (see Section 2) with suitably chosen levels of smoothing over time. This is now the standard approach for epidemiological studies investigating the short-term effects of air pollution on health and we now describe some examples of this approach.

Examples of the use of GAMs to model underlying trends and long-term patterns in time-series studies can be found in Schwartz (1994b), where they were used to analyse the relationship between particulate matter and daily mortality in Birmingham (Alabama, U.S.), between particulate air pollution, ozone and SO_2 and daily hospital admissions for respiratory illness (Philadelphia, U.S.) and between ozone and particulate matter for children in six cities in the eastern U.S. In each of these analyses, Poisson models were fitted using loess smoothers for time, temperature and a linear function of pollution, e.g. $\log(E(Y)) = \alpha + S_1(time) + S_2(temperature) + S_3(dewpoint) + \beta PM_{10} + \text{day}$ of the week dummy variables. A similar model was used in Schwartz (1997), but in this case the effect of weather was modelled using regression splines. One of the major advantages of using non-linear functions could then be exploited, that of being able to include the possibility of thresholds, without forcing the epidemiologist to specify a functional form. In particular there appeared to be a non-linear relationship between temperature and mortality. However, unlike GLMs, there are no fixed degrees of freedom and so although deviances can be calculated, and ways of approximating the degrees have been proposed, there are no formal tests of differences between models (Hastie & Tibshirani, 1990). Despite the possibility of using a non-linear function for the variables of interest, namely air pollution, the final results were always presented for a linear term, in order to produce a single summary measure (relative risk per unit increase) and for comparability with other studies.

2.7 Generalised estimating equations

Even when temporal patterns causing autocorrelation are modelled, for example using GAMs, it is still likely that the residuals will exhibit autocorrelation. One approach to allowing for the possibility of serial correlation (and over-dispersion) that has been used widely in air pollution epidemiology is generalised estimating equations (GEEs). GEEs extend Generalised linear models to the extent that the correlation between individuals is taken into consideration in the formulation of the model, the estimation of the regression coefficients and their standard errors. Here, the concepts behind GEEs are briefly introduced together with examples of how they have

been used in epidemiological studies. Further details of GEE models can be found in
Liang and Zeger (1986), Zeger, Liang, and Albert (1988), Diggle, Heagerty, Liang,
and Zeger (2002) and Zeger and Liang (1992).

Unlike a hierarchical model, where the structure of the covariance model is explicitly defined (see Chapter 6, Section 6.3), a standard GEE is a marginal model. In
a marginal model, the outcome, Y_i, is related to a linear predictor which contains only
fixed effects and the correlation or variance parameters are estimated from the residuals. Comparisons between the uses of GEEs and hierarchical models for analysing
repeated measures data are given in Omar, Wright, Turner, and Thompson (1999)
and Burton, Gurrin, and Sly (1998).

Example 2.7. *The use of GEE in time series epidemiology*

In earlier research, GEEs were widely used in time series analyses of air
pollution and health, to allow for the serial correlation that is likely to be
present in the residuals from a regression model that assumes independence.
Examples can be found in Schwartz (1994a), Schwartz and Dockery (1992a)
and Schwartz (1991), who analysed the effect of particulate matter on daily
mortality in Birmingham, Alabama, Philadelphia, and Detroit, U.S. respectively. Log–linear regression models were used with an autoregressive structure to allow for the autocorrelation between daily mortality counts. An exchangeable covariance structure was used by Schwartz and Dockery (1992b)
and comparison made with an autoregressive structure, however there was no
noticeable difference in the results. It is notable that the model building and
selection in these examples were performed on regular GLMs assuming independence, with the final chosen model fit using GEEs, and it is not clear
whether this might have an effect on the suitability of the model that is used
for estimation.

GEEs were used in a different way by Moolgavkar, Luebeck, Hall, and
Anderson (1995), in an analysis of data from Philadelphia, U.S. for 1973–88.
They performed log–linear regression, but instead of using day as the unit
of analysis in the repeated measures of the GEE, they performed separate
analyses for each season, for example 16 summers, and used the season as
the unit of observation in the GEE with an independence structure. Using
the seasons as replications, more stable estimates could be obtained than if a
single observation were used. In this sense, it is the same as the rationale for
the hierarchical model of Schwartz and Marcus (1990) described in Section
6.3, but achieved through using a marginal model rather than a hierarchical
one.

2.8   Poisson models for count data

Poisson regression may be appropriate when the dependent, Y, variable is count data, for example counts of mortality in a particular area and period of time, as opposed to whether an individual has or does not have a disease (which would arise in a case–control study). Often the interest is in comparing the rates of disease in different areas. Here we consider the estimation of relative risk in the case where observed numbers of disease are available for a number of areas together with the *expected* number of disease counts in those areas. It should be noted that these are not the expected number of cases in the sense of statistical expectation, but instead are what would be expected based on applying national rates of disease to the population structure of those areas. The concept of expected numbers in this case is explained in the following section.

2.8.1   Estimating SMRs

We can model the count Y in an area, i, with expected number E as Poisson, which for fixed λ is

$$\Pr(Y = y|\lambda) = \frac{e^{-E\lambda}(E\lambda)^y}{y!}$$

for $y = 0, 1, \ldots$. Here λ is the relative risk.

For fixed y we have the likelihood function

$$l(\lambda) = \frac{e^{-E\lambda}(E\lambda)^y}{y!} \propto e^{-E\lambda}\lambda^y$$

for $\lambda > 0$.

Example 2.8. *The MLE of a Poisson rate*

The MLE of the relative risk is the SMR $= Y/E$

$$
\begin{aligned}
log(l(\lambda|Y) &\propto -(E\lambda) + Y\log(E) \\
\frac{\delta \log(l(\lambda|Y)}{\delta\lambda} &\propto -(E\lambda) + \frac{Y}{\lambda E} \\
0 &= -(E\hat{\lambda}) + \frac{Y}{\hat{\lambda}E} \\
\hat{\lambda} &= Y/E
\end{aligned}
\tag{2.10}
$$

Estimating the MLE (and thus the SMR) can also be found by expressing the terms of a GLM, when it is assumed that $Y \sim P(E\lambda)$. If the rate is $\mu = E\lambda$, the linear predictor of a Poisson GLM would take the form

$$\log \mu = \log(E) + \beta_0 \qquad (2.11)$$

where $\log E$ is an offset, is a known multiplier in the log–linear mean function, $\mu = \exp(E) \times \beta_0$, and therefore will not have an associated coefficient in the results of the model.

Example 2.9. *Estimating the SMR using a Poisson GLM*

We consider the observed and expected number of cases of respiratory mortality in small areas in the UK from a study examining the long-term effects of air pollution (Elliott, Shaddick, Wakefield, de Hoogh, & Briggs, 2007). Taking a single area (a ward) that had a population of 1601, the observed counts of deaths in people over 65 years old was 29 compared to the expected number which was 19.88. The SMR is therefore 29/19.88 = 1.46.

The R code for finding the MLE of the (log) SMR ($= \beta_0$) and its standard error would be

```
> y <- 29; E <- 19.88
> summary(glm(y ~ offset(log(E)),family=poisson))

Coefficients:
            Estimate Std. Error z value Pr(>|z|)
(Intercept)   0.3776     0.1857   2.033    0.042 *
```

Noting that the linear predictor, and thus the coefficient of interest, β_0, is on the log scale, the estimate of the SMR = $\exp(\beta_0) = \exp(0.3776) = 1.458$. The standard error of the estimate of β_0 can be used to construct a 95% confidence interval: $\hat{\beta}_0 \pm 1.96 \times se(\hat{\beta}_0) = (0.014, 0.741)$. Again this is on the log–scale and so the exponent is taken of both the lower and upper limits; $(\exp(0.014), \exp(0.741))$. The SMR in this case is therefore 1.458 (95% CI; 1.013 – 2.099) meaning that the number of observed cases of disease in the area is significantly greater than that expected based on the age–sex profile of the population.

We now consider the estimation of the SMR over more than a single area; in this case we use the observed counts and expected numbers for $N = 393$ areas. The parameter, β_0, now corresponds to the log of the overall SMR using data from all of the areas, i.e. $\frac{\sum_{i=1}^{393} O_i}{\sum_{i=1}^{393} E_i}$. From the data, this is 8282/7250.2 = 1.142 with the log of this being 0.133.

The R code to find the MLE in this case is

```
summary(glm(Y ~ offset(log(E)),family="poisson",
data=data))
```

```
Coefficients:
             Estimate Std. Error z value Pr(>|z|)
(Intercept)  0.13305     0.01099   12.11   <2e-16 ***
---
Signif. codes:  0 '***' 0.001 '**' 0.01
'*' 0.05 '.' 0.1 ' ' 1
```

We can see that the estimate of the SMR will be $\exp(0.13305) = 1.142$, the overall SMR for all the areas. The 95% CI will be $\exp(0.112) - \exp(0.155) = 1.118 - 1.167$, indicating an overall increased risk of respiratory mortality in these areas compared to what might be expected if they experienced the same mortality rates (by age and sex) as the national population.

2.8.2 Over-dispersion

There is a strong possibility of over-dispersion in the Poisson models (i.e., where the variance is greater than the mean), arising from the presence of unmeasured confounders. These may be operating at the individual level, e.g. smoking, or at the area level, e.g. residual socio-economic confounding. Over-dispersion may also arise because of data anomalies such as errors in the numerators and/or denominators or due to migration.

The choice of a log link and variance that is proportional to the mean is the canonical one for the Poisson distribution and under regular Poisson assumptions, the variance is assumed to be equal to the mean, $VAR(Y_i) = E(Y_i)$, i.e. the dispersion parameter $\phi = 1$. When this is relaxed to proportionality, $VAR(Y_i) = \phi E(Y_i)$, the *dispersion*, or scale, parameter is assumed to be constant over all of the data. The estimating equations for this will generally be different from those obtained by weighted least squares, but solutions can be found using quasi-likelihood techniques.

The likelihood-based approach theoretically allows the different models, and therefore the effect of different exposure scenarios, to be compared, but it makes no allowance for the extra-Poisson variability that may be present. This will lead to confidence intervals for the estimates of risk being too narrow and changes in deviances, used to compare models, being too small. An attempt to correct these effects, and to assess the degree of over-dispersion, can be made using a quasi-likelihood model (see Section 2.5.2), in which the usual Poisson assumption, that the variance is equal to the mean, is relaxed, allowing it to be a function of the mean and an unknown constant, i.e. $Var(Y_i) = cE(Y_i)$.

This *over-dispersion* may arise in a number of different ways, causing the variance to be greater than the mean. One example is when there is inter-subject variability where the Poisson mean parameter is not constant, but is a random variable. If a Gamma distribution, $Ga(a, b)$ were used for the distribution of the random effects, the marginal distribution of Y_{ti} is analytically tractable and takes the form of a negative binomial. Wakefield, Best, and Waller (2000) discusses the effect of different formulations of this Gamma distribution. Two cases are of particular interest, relating to the methods for allowing over-dispersion used here. In the first, the random effects follow a distribution of the form $\beta_{0i} \sim Ga(E_i a_i b, E_i b)$ (where E_i is the expected number of cases in area i), in which case the marginal variance of $Y_i|a_i, b$ is $E(Y_i|a_i, b)(1 + 1/b)$. This linear function of the mean is close to the quasi-likelihood approach described above, with $c = (1 + 1/b)$. If $\beta_{0i} \sim Ga(b, b/a_i)$ then $V(Y_i|a_i, b) = E(Y_i|a_i, b)(1 + E(Y_i|a_i, b))/b$. Here, the variance is a quadratic function of the mean. This quadratic form is also the case when a log–normal model is used for the random effects as described in Chapter 6, Section 6.3. In that case we might have the model $\log(Y_i) = \beta_0 + \beta_{0i} + \beta X_i$ with $\beta_{0i} \sim N(0, \sigma_{\beta_0}^2)$. In this case the marginal distribution of Y_i is intractable but the mean and variance can be found and are

$$V(Y_i|\mu_i, \sigma_{\beta_0}^2) \quad = \quad E(Y_i|\mu_i, \sigma_{\beta_0}^2)\{1 + E(Y_i|\mu_i, \sigma_{\beta_0}^2))(\exp(\sigma_{\beta_0}^2)) \qquad (2.12)$$

where $E(Y_i|\mu_i, \sigma_w^2)E_i \exp(\mu_i + \sigma_{\beta_0}^2/2)$ and $\mu_i = \beta_0 + \beta X_i$

As with the example above, the variance is a quadratic function of the mean and so the results from the log–normal random effects model would not be expected to be the same as those from the quasi-likelihood model, where the variance is a linear function of the mean, although both will result in an increase in the variability of the estimates, and thus wider confidence/credible intervals.

Example 2.10. *Estimating the SMR using quasi-likelihood*

The R code for using quasi-likelihood to find the MLE of the log(SMR)= β_0 and its standard error using the data from the previous example would be

```
summary(glm(Y ~ offset(log(E)),family="quasipoisson",
data=data))
```

```
Coefficients:
            Estimate Std. Error t value Pr(>|t|)
(Intercept)  0.13305    0.02078   6.402 4.39e-10 ***

Dispersion parameter for quasipoisson family taken
to be 3.577536
```

Note that the estimate itself is the same as with the Poisson case but the standard error has increased from 0.01099 to 0.02078, reflecting the over-dispersion which is present, with the dispersion parameter having been estimated to be over 3. The 95% confidence interval will therefore be wider; $\exp(0.092) - \exp(0.133) = 1.096 - 1.190$, with the increase in width reflecting the extra uncertainty that is present.

2.9 Estimating relative risks in relation to exposures

Consider an area i. If Y_i denotes the number of deaths in area i, it is assumed that $Y_i \sim P(E_i\mu_i)$, where E_i represents the pre-calculated age–sex standardised expected number of cases. The rate in each area, μ_i is modelled as a function of the exposures, X_{1i}, together with other area-level covariates, X_{2i},

$$\log \mu_i = \beta_0 + \beta_1 X_{1i} + \beta_d X_{2i} \tag{2.13}$$

where β_l represents the effect of exposure and β_d is the effect of the area-level covariate.

Example 2.11. *Modelling differences in SMRs in relation to differences in exposures*

We now consider the possible effects of air pollution in relation to the SMRs observed in the different areas; the exposure, X_{i1} for each area being the annual average of measurements from monitoring sites located within the health area. In addition, we consider the possible effects of a covariate; in this case the covariate is a measure of deprivation known as the Carstairs score.

Smoking is known to be a major risk factor for respiratory illness and it is known that smoking habits vary with social class (Kleinschmidt, Hills, & Elliott, 1995) and may therefore correlate with pollution levels, and act as a potential confounder. Although routine data on smoking levels at small area level are not available in Great Britain, strong correlations have, however, been demonstrated on several occasions between smoking rates and the Carstairs index of deprivation, which has also been shown to be a strong predictor of disease risk (Carstairs & Morris, 1989). The index is derived from a weighted aggregation of data on four census variables: unemployment, overcrowding, car ownership and social class.

The R code for fitting a model to estimate the relative risk associated with air pollution in this case is as follows:

```
summary(glm(Y ~ offset(log(E))+X1,family="poisson",
data=data))

Coefficients:
            Estimate Std. Error z value Pr(>|z|)
(Intercept) -0.04746    0.02603  -1.823   0.0683 .
X1           0.07972    0.01023   7.797 6.35e-15 ***
```

In this case, the effect of air pollution is highly significant and the associated relative risk will be $\exp(\beta_1) = \exp(0.07972) = 1.082$, indicating an increase in risk of 8.2% associated with every increase of one unit in air pollution (in this case, the units are $10\mu\text{gm}^{-3}$).

Using a quasi-likelihood approach again results in the same estimate but with a larger standard error;

```
summary(glm(Y ~ offset(log(E))+X1,family="quasipoisson",
data=data))

Coefficients:
            Estimate Std. Error t value Pr(>|t|)
(Intercept) -0.04746    0.04835  -0.982    0.327
X1           0.07972    0.01899   4.198 3.34e-05 ***

(Dispersion parameter for quasipoisson family taken
to be 3.449832)
```

The 95% CIs are 1.0615 – 1.1049 for the Poisson case and 1.0434 – 1.1241 when using quasi-likelihood; both indicating that the increase in risk is significant, with the wider intervals in the quasi-likelihood case again reflecting the extra uncertainty associated with the over-dispersion.

Adding the deprivation score, X_2, to the model might be expected to reduce the risk associated with air pollution as areas which are highly polluted are likely to also be deprived and deprived areas, with some exceptions, have higher rates of disease. It is therefore a confounder in the relationship between air pollution and health.

The R code for a model with both air pollution and deprivation is as follows:

```
summary(glm(Y ~ offset(log(E))+X1+X2,family="poisson",
data=data))

Coefficients:
            Estimate  Std. Error z value Pr(>|z|)
(Intercept) -0.073589   0.027086  -2.717  0.00659 **
X1           0.025850   0.011023   2.345  0.01902 *
X2           0.051302   0.002582  19.871  < 2e-16 ***
```

It can be seen that adding deprivation to the model has resulted in a reduction in the size of the effect associated with air pollution for which the RR has changed from 1.083 to 1.026 (95% CI; 1.004 – 1.049). The effect of deprivation is also significant, with an increase in risk of 5.3% (RR = $\exp(0.051302) = 1.053$) associated with a unit increase in Carstairs score.

When using quasi-likelihood the estimates of relative risk are the same, but again they have wider confidence intervals.

```
summary(glm(Y ~ offset(log(E))+X1+X2,family="quasipoisson",
data=data))

Coefficients:
            Estimate  Std. Error t value Pr(>|t|)
(Intercept) -0.073589   0.040756  -1.806   0.0718 .
X1           0.025850   0.016585   1.559   0.1199
```

```
X2              0.051302     0.003885   13.206    <2e-16 ***
```

```
(Dispersion parameter for quasipoisson family taken
to be 2.264041)
```

This gives a RR for air pollution of 1.026 (95% CI; 0.993 – 1.060) and for deprivation a RR of 1.053 (95% CI; 1.045 – 1.061) which in this case leads to the effect of air pollution being non-significant. This suggests that it is deprivation that is playing a large part in the differences in SMRs observed in the different areas. Note the amount of the widening of the intervals is reduced as there is less over-dispersion; some of the extra-Poisson variability has thus been 'explained' by deprivation.

The effect of adding deprivation to the model can be assessed by calculating the change in deviance between two models; (i) with air pollution and (ii) with both air pollution and deprivation. A significant difference in deviance will indicate that deprivation is a significant risk factor.

The R code to perform a test between the deviances of the two models is as follows:

```
anova(glm(Y ~ offset(log(E))+X1,family="quasipoisson",
data=data),
glm(Y ~ offset(log(E))+X1+X2,family="quasipoisson",
data=data),
test="Chisq")
```

```
Analysis of Deviance Table
```

```
Model 1: Y ~ offset(log(E)) + X1
Model 2: Y ~ offset(log(E)) + X1 + X2
  Resid. Df Resid. Dev Df Deviance  Pr(>Chi)
1       391    1218.11
2       390     846.66  1   371.45 < 2.2e-16 ***
---
Signif. codes:  0 '***' 0.001 '**' 0.01
'*' 0.05 '.' 0.1 ' ' 1
```

This shows that deprivation has a highly significant effect on the risk of respiratory mortality. Using this method, the effect of taking air pollution out of the model can also be assessed, which proves to have a non-significant change in deviance; this indicates that when deprivation is included in the model the estimated risk associated with air pollution is non-significant.

```
anova(glm(Y ~ offset(log(E))+X1+X2,family="quasipoisson",
data=data),
glm(Y ~ offset(log(E))+X2,family="quasipoisson",
data=data), test="Chisq")
```

```
Analysis of Deviance Table
```

```
Model 1: Y ~ offset(log(E)) + X1 + X2
Model 2: Y ~ offset(log(E)) + X2
```

```
    Resid. Df Resid. Dev Df Deviance Pr(>Chi)
1          390       846.66
2          391       852.06 -1  -5.4054    0.1223
```

2.10   Modelling the cumulative effects of exposure

In order to assess the effect of environmental hazards on health, models are required that relate risk to the exposure, both in terms of the degree of exposure and the time over which exposure occurred. In cohort studies of individuals, such models need to account for the duration of exposure, time since first exposure, time since exposure ceased and the age at which first exposure occurred (Breslow & Day, 1980; Waternaux, Laird, & Ware, 1989). For the development of carcinogenesis, complex multi-stage models have been developed that use well defined dose–response relationships (Dewanji, Goddard, Krewski, & Moolgavkar, 1999). However, when using aggregated daily mortality counts for a specific day or health period and a specified area, detailed exposure histories and other information are generally not available.

Considering, for ease of illustration, a generic area, if Y_t is the health outcome at time t, e.g. the number of respiratory deaths or on a single day or other period of time, and the true exposure history is $X(u), 0 \leq u \leq t$, then the outcome is modelled as a function of the exposure history.

$$E(Y_t) = f(X(u); \quad 0 \leq u \leq t) \tag{2.14}$$

As true personal exposure to air pollutants is unmeasurable over a lifetime, as it depends on ambient levels and integrated time–activity, the term 'exposure' here relates to cumulative ambient outdoor concentrations of air pollutants, measured at the aggregate area level. The summaries of the exposure history are therefore constructed based on available data, X_t.

If it is assumed that $X(u)$ is piecewise continuous, then the cumulative exposure up to and including time t is

$$\int_0^t X(u)du \tag{2.15}$$

Rather than just considering the effect of the total exposure over a period of time, the contributions from intervals within the period may be of interest, in which case (2.15) can be expressed in the form of weighted integrals (Breslow, Lubin, Marek, & Langholz, 1983; Bandeen-Roche, Hall, Stewart, & Zeger, 1999).

$$C_t = \int_o^{u=t} W(t-u)X(u)du \tag{2.16}$$

where the weights, $W(t-u)$, determine the aspect of the exposure being summarised. For example, if the weights are of the form $W(u) = \min(1, u/b)$, then the exposures are phased in linearly over a period of length b until they reach their maximum. This can allow for delayed, as well as cumulative effects, depending on the form of the weights. In individual studies, the form of the cumulative exposure can be explicitly modelled. For example, in the case of exposure to asbestos fibres, the rate of elimination of the fibres from the lungs, λ, may be incorporated and the model will take the form $W(u) = \{1 - \exp(-\lambda u)\}/\lambda$ (Berry, Gilson, Holmes, Lewinshon, & Roach, 1979).

Since exposure to air pollution starts in infancy, the lower limit of the integral will not be zero; instead, the sum is likely to be over a specified period of time. If the weights are of the form

$$W(u) = \begin{cases} 1/(b-a) & \text{for} & a \leq u < b \\ 0 & \text{otherwise} \end{cases} \tag{2.17}$$

then the summary will represent the average for the period $(t - b, t - a]$, $0 \leq a < b \leq t$. For example, when studying the short-term effects of air pollution, if $a = 0$ and $b = 2$, then $W(t - u)$ would represent a three-day mean. An alternative is to model exposure-time-response relationships based on patterns seen in the data, for example using splines (Hauptmann, Berhane, Langholz, & Lubin, 2001).

When dealing with mortality counts, and exposure measurements, made at discrete times, the integral in Equation 2.16 can be approximated by a summation over a suitable discretisation.

$$C_t = \sum_{k=0}^{t} W_{t-k} X_k \tag{2.18}$$

If the probability of disease given cumulative exposure is assumed to be proportional to $\exp(\gamma C_t)$, i.e. a log–linear model in cumulative exposure, then a Poisson model can be used to estimate the weights, W_{t-k} in Equation 2.18. Assuming that $Y_t \sim P(E_t \mu_t)$, where E_t represents the expected number of cases, then

$$\log \mu_t = \alpha + \gamma \sum_{k=0}^{t} W_{t-k} X_k = \alpha + \sum_{k=0}^{t} \beta_{t-k} X_{t-k} \tag{2.19}$$

Here the parameter, β_{t-k} represents the effect of exposure k time periods ago. Comparison with Equation 2.16 and 2.18 shows that $\beta_{t-k} = \gamma W_{t-k}$.

Example 2.12. *Modelling the risks associated with lagged effects of air pollution*

Following on from the previous example, we might fit the annual averages from the previous three years, $X_{it}, X_{i(t-1)}, X_{i(t-2)}$. The R code to do this is as follows:

```
glm(formula = Y ~ offset(log(E)) + X1 + X1t1 + X1t2,
family = "quasipoisson",  data = data)


Coefficients:
             Estimate Std. Error t value Pr(>|t|)
(Intercept) -0.09740    0.04959  -1.964 0.050233  .
X1           0.02118    0.03223   0.657 0.511514
X1t1        -0.01415    0.03143  -0.450 0.652917
X1t2         0.04271    0.01251   3.414 0.000708 ***

(Dispersion parameter for quasipoisson family taken
to be 3.276761)
```

However, the measurements for areas in the individual years are likely to be highly correlated over this short period of time leading to issues of collinearity. In this case, it has resulted in just one of the covariates being significant and in patterns in the effects over time that may not make much sense, i.e. a decrease in risk associated with the previous year (although this is non-significant). For this reason, where there are high levels of collinearity , more sophisticated approaches may be required, such as those described in Chapter 6.

Developments have been made in specifying the shape of the distributions of the weights, W_{t-k}, within aggregate level studies examining the short-term effects of air pollution on health. Schwartz (2000) describes the use of a distributed lag model, where the weights fit a polynomial function (Harvey, 1993). This requires assumptions to be made, in terms of the polynomial used, on the maximum lags that are likely to have an effect, but has the advantage of increasing the stability of the individual estimates where there is high collinearity between the explanatory variables (Zanobetti, Wand, Schwartz, & Ryan, 2000).

2.11   Logistic models for case–control studies

In case–control studies, the response variable is whether an individual is a case or control and is therefore a binary variable. The most common link function in this case is the logistic link, $\log(y/(1-y))$, which will constrain the results of the linear predictor to produce values between zero and one representing the probability that an individual is a case (i.e. has the disease). This means that the estimates of the parameters in a logistic regression model will be on the log–odds scale.

If Y_i is an indicator of whether the ith individual is a case (one) or a control (zero) then $Y_i \sim Bi(1, \pi_i)$. A logistic regression model with just an intercept will take the form

$$\log(\pi/(1-\pi)) = \beta_0 \qquad (2.20)$$

In this case, the intercept, β_0 represents the overall proportion of cases on the log–odds scale,

$$
\begin{aligned}
\log(\pi/(1-\pi)) &= \beta_0 \\
\pi &= \exp(\beta_0)/(1+\exp(\beta_0))
\end{aligned}
\tag{2.21}
$$

Example 2.13. *Estimating the odds ratio in a case–control study using a logistic model*

In a study of asthma of whether children living near main roads require higher levels of asthma treatment than those who live further away , cases and controls were grouped according to whether or not they lived within 150m of a main road (Livingstone, Shaddick, Grundy, & Elliott, 1996). Of the 1066 cases, 172 lived within 150m of a main road with the corresponding number for controls being 464 (out of 6233).

The MLE of the probability that an individual is a case can be found using R as follows:

```
glm(formula = Y ~ 1, family = "binomial", data = data)

Coefficients:
            Estimate Std. Error z value Pr(>|z|)
(Intercept) -1.76594    0.03314  -53.28   <2e-16 ***
```

The estimate -1.76594 is on the log–odds scale. Using 2.21 this can be converted back to the probability scale as $\frac{\exp(-1.76594)}{(1+\exp(-1.76594))} = 0.146$ which is the same as the proportion of cases (1066/6233).

Including the effect of exposure, X_i in the model gives

$$
\log(\pi_i/(1-\pi_i)) = \beta_0 + \beta_1 X_i
\tag{2.22}
$$

where β_1 is the estimated increase in log–odds per unit increase in X_1. It is the difference between two log–odds, with and without an increase of one unit in X_1 and is $\log(p_0/(1-p_0)) - \log(p_1/(1-p_1)) = \log\{p1/(1-p1)\}$ which is the log of the odds ratio. Therefore taking the exponent of the estimate of β_1 gives us the odds ratio associated with a unit increase in X_1.

Example 2.14. *Estimating the odds ratio of asthma associated with proximity to roads*

We now estimate the effects of living near to a main road on asthma. The R code to do this is as follows;

```
summary(glm(Y~X, family="binomial", data=data))

Coefficients:
            Estimate Std. Error z value Pr(>|z|)
(Intercept) -1.86455    0.03594 -51.875  <2e-16 ***
X            0.87216    0.09623   9.063  <2e-16 ***
```

Here, the odds ratio associated with living close to a main road is $\exp(0.87216) = 2.391$ (95% CI; $1.981 - 2.889$. This indicates that there is a significant increase in risk of asthma in the children under study associated with their living close to a main road. Of course there may be confounders, such as parental smoking, which may affect this. If available, these confounders could be added to the model in the same way as seen in the Poisson example.

2.12 Summary

This chapter contains the basic principles of epidemiological analysis and how estimates of the risks associated with exposures can be obtained. From this chapter, the reader will have gained an understanding of the following topics:

- Methods for expressing risk and their use with different types of epidemiological study.
- Calculating risks based on calculations of the expected number of health counts in an area, allowing for the age–sex structure of the underlying population.
- The use of generalised linear models (GLMS) to model counts of disease and case–control indicators.
- Modelling the effect of exposures on health and allowing for the possible effects of covariates.
- Cumulative exposures to environmental hazards.

Exercises

Exercise 2.1. Show that the odds ratio for exposure given disease, $OR_{E|D}$, can be equal to the odds ratio for disease given exposure $OR_{D|E}$.

Exercise 2.2. Express the probability density for the normal distribution with mean μ and variance σ^2 in the exponential family form

$$f(y) = \exp\left\{ \frac{y\theta - b(\theta)}{a(\phi)} + c(y, \phi) \right\}, y \in S$$

State the general formula for the mean and variance of an exponential family distribution and check that they hold for a normal random variable.

Exercise 2.3. In the case of logistic regression, express $fy(y|\theta)$ in the form of the exponential family and find expressions for $b(\theta), c(y, \phi)$ and $a(\phi)$.

Exercise 2.4. Show that for a Binomial (n, p) variable, the MLE of p is $\sum_{i=1}^{n} Y_i/n$.

Exercise 2.5. The observed, Y_i counts of mortality in N areas all assumed to be Poisson distributed with the mean in each area being $E_i\lambda$ where E_i is the (age–sex adjusted) expected number of deaths and λ_i is the relative risk in that area. Show that MLE of the risk for all areas combined is equal to the overall SMR $= \frac{\sum_{i=1}^{N} O_i}{\sum_{i=1}^{N} E_i}$.

Exercise 2.6. It is thought that there is a relationship between air pollution and respiratory deaths. The table below contains data from two cities, A and B. The first city (A) had much lower pollution measurements than the second (B). A sample of residents from each city was followed and their ages at death were recorded. The number of deaths in 10 year age groups and the number of person-years at risk (PYR) are given for each city.

Table 2.3: Number of deaths and person-years (PYR) for two factories manufacturing rubber under different operating conditions.

Age range	City A Deaths	City A PYR	City B Deaths	City B PYR
20–29	3	6000	1	3000
30–39	3	5000	4	4000
40–49	5	4000	5	3500
50–59	7	4045	8	3701
60–69	27	3571	43	3702
70–79	30	1777	40	1818
80–89	8	381	10	350

It is thought that the death rate in the city B might be higher than that in city A and that this might be due to the differences in air pollution. A regression model was used to assess the effect of city on death rate.

(a) Explain why simple linear regression might not be appropriate in this case, and suggest an alternative model, giving reasons for your choice.

(b) Use the data in the table to calculate the $\log_e$ of the death rates (LDR) by age group for each city. Plot them against age group, labelling the points according to city. What does the plot suggest ?

The results of a Poisson regression model including just the intercept term are as follows.

Model 1

```
                Coefficients (standard error)
Intercept           -5.443109        0.100000
Degrees of Freedom: 14 Total; 13 Residual
Residual Deviance: 120.0000
```

(c) Calculate the overall death rate in the two cities from data in the table, and show that you could have produced the same result using the output from the model.

A city term was then added to Model 1, giving the following results.

Model 2

```
                Coefficient  (standard error)
Intercept         -5.00000    0.100000
City  B            0.400002    0.150000

Degrees of Freedom: 14 Total; 12 Residual
  Residual Deviance: 110.000
```

(d) Calculate a relative risk with a 95% confidence interval comparing city B with city A. What do you conclude about the relative risk of living in city B compared to city A?

(e) Explain how you would assess the effect of adding the city term on the fit of the model. Show that there is a significant effect in this case.

Chapter 3

The importance of uncertainty

3.1 Overview

Uncertainty is a topic that permeates all scientific inquiry and its importance is magnified when the results are applied in decision-making, which in this setting will involve legislation, regulations and designing public policy.

Despite the general importance of the concept of 'uncertainty', its meaning lacks a universally agreed on definition. In fact it shares its general lack of definition with 'information', as described by the late Debabrata Basu (Basu, 1975):

"But what is information? No other concept in statistics is more elusive in its meaning and less amenable to a generally agreed definition."

It has been described in various ways including 'incomplete knowledge in relation to a specified objective' which arises 'due to a lack of knowledge regarding an unknown quantity' (Bernardo & Smith, 2009). However, the lack of a clear cut definition has not stopped people from taxonomizing it! Thus we have for example the distinction between *aleatory* (stochastic) and *epistemic* (subjective) uncertainty (Helton, 1997). It seems generally agreed that some aspects of uncertainty are quantifiable while others are inherently qualitative, that is not subject to quantification. The latter would, for example, include framing the problem to be investigated by defining the system boundaries and explicating the role of values (van der Sluijs et al., 2005). Both qualitative and quantitative aspects of uncertainty need to be taken into account within environmental risk analyses.

There will often be intangible sources of uncertainty which will arise through the subjective judgements that are sometimes required to estimate the nature and magnitude of empirical quantities where other methods are not appropriate. Uncertainty may arise as a result of imprecise language in describing the quantity of interest and disagreement about interpretation of available evidence. There may also be uncertainty about the actual methods being used to assess policy changes, whether the data available to implement them is suitable for the purpose and the extent to which the results can be generalised.

Where models are used to mathematically represent real-life processes, there are a number of potential sources of uncertainty. These may include concerns about the input data, uncertainties in the model description, how well the chosen models represent spatial and other variation and of course the stochastic nature of the physical processes being described. Uncertainty in one part of a model will also propagate to other parts of the model and if multiple models are implemented then uncertainty will propagate through these as well. Often, such systems become so complex that the identification and characterisation of uncertainty will need to be prioritised, looking at the most sensitive and most important parameters for the uncertainty assessment.

This chapter considers various aspects of uncertainty from initial characterisation to methods for quantification where appropriate. Methods for considering and handling uncertainty range from the philosophical and qualitative in nature (Walker et al., 2003; Briggs, Sabel, & Lee, 2009) to more quantitative approaches based on probabilistic or statistical methods (Cullen & Frey, 1999).

3.2   The wider world of uncertainty

Post-normal science (see Chapter 1) called for a search for new approaches to dealing with uncertainty, one that recognized the diversity of stakeholders and evaluators needed to deal with these challenges. New tools were needed to facilitate the conduct of post-normal scientific investigations in an organized way. One such tool is the Numerical–Units–Spread–Assessment–Spread (NUSAP) matrix (Funtowicz & Ravetz, 1990). It was designed to facilitate the analysis and diagnosis of the uncertainties in the knowledge base underlying a complex environmental issue and in particular those expressed in the assumptions made by modellers. The NUSAP matrix consists of five columns, the first three of which comprise the quantitative elements involved in generating the knowledge base. For example, these might be the estimated relative risks of unit changes in air pollution calculated by an environmental epidemiologist, the units of measurement involved and the degree of uncertainty e.g. standard errors (Funtowicz & Ravetz, 1990). In short, the first three columns capture the information that is available from standard statistical approaches.

However this does not capture the underlying uncertainty that may arise from many other sources. In an epidemiological investigation there is typically a hypothesis to be tested that will ideally be based on some preconceived theory or understanding of the aetiological processes involved. However, there will be uncertainty inherent in this understanding (Briggs et al., 2009). In risk and impact assessment, there will be uncertainty arising from the specification of the risks or policy, the question(s) to be analysed, and the conditions, e.g. scenarios and study areas, under which the analyses will be performed. Uncertainty may arise according to the extent, both spatially and temporally, of the analysis, what aspects to include and which to ignore, and, fundamentally, the underlying 'model' of the system under study (Briggs et al., 2009). There will also be uncertainty in the definition of the key relationships of interest and the processes that they represent. The scope for uncertainty is therefore very large and its implications will run throughout the analysis, results and interpretation.

The final two columns of the NUSAP matrix act as an aid to recording this more general uncertainty. This may record the result of a 'pedigree analysis'. This pedigree analysis, summarised in the final column, would be based on an organised discussion amongst experts, possibly from a wide variety of subject areas and interests, charged with determining the uncertainty in the relevant knowledge base related to the overall objectives of the analysis.

Example 3.1. *Concerns about high levels of dioxin*

Controversy arose about the potential human health effects of a waste incinerator located near Antwerp when an unusually high number of children were found to have congenital defects (van der Sluijs et al., 2005). Members of the local population suggested that the cause was dioxin emissions from the incinerator. However the operators, supported by local officials, disagreed and argued that the claims were not supported by scientific evidence. Years of debate ensued due to the uncertainties involved.

A workshop was held in which structured discussions about the uncertainties took place based on the findings from three scientific studies. The first was a spatio–temporal epidemiological analysis to determine if there were increased health risks among children whose parents lived, or had lived, in the region at risk. The second was an exposure assessment of how much dioxin might have been absorbed by people living in that region during the period of the study. The third was a biomonitoring study comparing the region in question with other similar regions which were used as 'controls'. A pedigree matrix (van der Sluijs et al., 2005) was used to evaluate each of the three components of this investigation, each having a specifically designed matrix.

One of the outcomes of the workshop was a recognition that the way the problem had been framed played a strong role in the ensuing debate (van der Sluijs et al., 2005). This pointed to the need for an extension to the pedigree matrix to include factors such as 'problem framing', 'research design' and 'extended review'. The NUSAP process of structured dialogue was shown to lead to insights that can go deeper than relying solely on statistical calculations of uncertainty.

3.3   Quantitative uncertainty

As noted in the introduction, quantitative uncertainty can be dichotomised as that which is unknown, *epistemic* and that which is unknowable, *aleatory uncertainty*. The first refers to a lack of knowledge about the processes which generate estimates of the parameters of interest, while the second refers to intrinsic uncertainty associated with the estimates. Though the distinction between the two may sometimes be blurred, the overall aim is the same; to reduce the uncertainty associated with the underlying processes and information required in order to perform estimation and prediction. In addition, a natural requirement of any estimate or prediction is a characterisation and quantification of the uncertainty associated with them, without which the estimate/prediction has little use in reality.

Example 3.2. *Aleatory and epistemic uncertainty*

As a simple example, let A denote the event that shaking a die results in a six showing. Whether or not A will occur is uncertain; it may or may not. If it were known that the die is perfectly balanced then the uncertainty in this case could be characterised by saying there is one chance in six of its being correct. This uncertainty can never be reduced even if a long string of repeated tosses is observed.

However, if it were not known if the die is balanced and that the likelihood of a six in this case is θ, one might say the chances of correctly predicting that A occurs is θ. But if θ is unknown we have an example of epistemic uncertainty. This uncertainty can be eliminated by observing repeated tosses so that eventually uncertainty about θ would be eliminated.

3.3.1   Data uncertainty

Uncertainty will arise when attempting to obtain information about empirical quantities which have some true value that could, in principle at least, be measured. Uncertainty arises when these quantities are inaccurately measured. For example, data may be missing or incomplete, either spatially or temporally. These missing values may be ignored completely, inducing uncertainty as data are not available. Alternatively values of the missing data may be imputed, for example by using an average of existing values. This will introduce uncertainty due to (unknown) differences between the true unknown value and the imputed value. In some cases no data may be available, in which case we may make use of surrogate or proxy variables. In the case of air pollution for example, the full ambient pollution field cannot be completely measured but instead monitors are located at selected points in time and space. Interpolating to other locations, in either time, space or both, will introduce uncertainty due to assumptions that will be required in order to fit statistical models. Interpolating will be further hampered by missing values at the monitoring sitess and the non-random placement of monitors. Monitoring stations may be situated in response to suspected 'hot-spots' of pollution whereas methods of interpolation will assume that the sites are located randomly or will require some information on how the non-randomness operates. Even apparently appropriate data can be subject to inherent randomness. This may be subject to measurement error arising as a result of variability arising from measurement instruments and methods. If continuous variables are discretised this often requires some subjective assessment of the boundary, for example between hot and cold temperatures.

3.3.2 *Model uncertainty*

The basis of a statistical analysis is the formation of a statistical model that is intended to represent the associations between the variables of interest. However, a model will always be a simplified approximation of underlying causal structure. Different choices of the model could include different parameters or structural form, as well as statistical choices such as frequentist versus Bayesian methods and choice of Bayesian priors. There will be many possible choices of model and many methods are available for selecting between candidate models as described in Chapter 6. In addition to statistically based approaches, the choice will to some extent be pragmatic and will be based on the availability of data and what is computationally feasible. Uncertainty about the model can arise through uncertain data as discussed in Section 3.3.1 and factors such as measurement error or missing values can influence the choice of model. Mathematical models, even those involving physical laws, as described in Chapter 14, will have uncertainty associated with them through the use of approximations and dependence on parameters that must be estimated using experimental data. The latter will be subject to all the sources of data uncertainty described in Section 3.3.1 and these will be propagated through the process and manifest themselves in the output.

3.4 Methods for assessing uncertainty

There are many methods for assessing the effects of quantitative uncertainty and we describe a selection of them here. An underlying concept behind many of them is the idea of changing the inputs to a system and observing the effects on the outputs and often this is done in order to identify the components that are likely to have the greatest effect on the overall uncertainty. Other methods are based on mathematical approximations to the uncertainty characterisation.

3.4.1 *Sensitivity analysis*

Sensitivity analysis aims to assess how changes in model inputs, either one at a time or in combination, can affect outputs from a model. The model is run a number of times with different input values in order to observe the resulting change in the output. It allows parameters that have large (or small) effects to be identified. Possible procedures for such sensitivity analysis include the following:

- Parameters varied one at a time—local sensitivity analysis which sees how changes in one parameter affect the result.
- Parameters varied in combinations—global sensitivity analysis, where several parameters are varied together.
- Extreme values—all parameters are set at their maximum/minimum values.

The identification of variables that may have a substantial effect on model output is an important step in any analysis of uncertainty as it is these variables which are liable to contribute most to the final uncertainty. This is especially important when the models are complex and computationally expensive to run and it may be infeasible to fully incorporate full uncertainty modelling for every variable. Sensitivity analysis can therefore be used to inform decisions as to which input variables might be suitable candidates for the application of more complex or computationally demanding methods, such as the Taylor expansion or Monte Carlo simulation.

3.4.2   *Taylor series expansion*

This is a mathematical technique that approximates the underlying distribution which characterizes uncertainty in a process (MacLeod, Fraser, & Mackay, 2002; Oden & Benkovitz, 1990; Morgan & Small, 1992). Once such an approximation is found, it can provide a computationally inexpensive way of characterising the uncertainty in model output and so is useful when dealing with large and complex models for which simulation-based methods may prove to be infeasible. Firstly, a set of important input variables have to be identified using sensitivity analysis by performing model runs with a set of pre-defined values of all the candidate variables and computing a coefficient indicating their relative sensitivity (RS). The final distribution of uncertainty, based on a log–normal distribution, is constructed by combining the uncertainties from each of the variables weighted by their RS. It is important to note that in identifying dominant sources of uncertainty for a model output, it is the combination of the sensitivity and uncertainty of the parameters that is important and they cannot be considered in isolation, except in the unlikely case that they can all be considered independent.

3.4.3   *Monte Carlo sampling*

This replaces single values of input variables for a model with repeated samples from probability distributions (Metropolis & Ulam, 1949; Sobol, 1994; Rubinstein & Kroese, 2011). This results in uncertainty being propagated through the model resulting in a distribution for the output. This distribution gives a value for the most likely value (the mean or median) together with a measure of its uncertainty, often represented by quantiles of the distribution. This is a very useful technique where models are highly non-linear, with many input variables, although it can be computationally expensive especially in cases where models are large and incorporate large amounts of data.

The classic form of Monte Carlo simulation implies the use of the simple random sample technique; however it is possible to apply other sampling methods to improve the coverage and efficiency of the Monte Carlo methods. Latin Hypercube Sampling (LHS) is an example of such a method (McKay, Beckman, & Conover, 2000; Stein,

1987). Instead of randomly generating realisations of the prior distributions this approach divides the distribution up into areas of equal probability, for example using percentile bands of interest and selecting from the median of these bands only, known as 'Median LHS', or alternatively selecting randomly from within the specified probability range. The order of the sampling is usually random. LHS can be applied to reduce the number of realisations, particularly when accumulated distributions are to be assessed. For a general overview see e.g. Cullen and Frey (1999). LHS has been used in a variety of applications where correlation between parameters exists, for example Iman and Conover (1982); Pebesma and Heuvelink (1999). This is a particularly important point when assessing the uncertainty between two different scenarios where there can be a large degree of correlation.

3.4.4   Bayesian modelling

Bayesian analysis is the subject of Chapters 4 and 5 but we briefly discuss it here in the context of a framework for quantifying uncertainty. The Bayesian philosophy considers that uncertainty can be described by means of probability distributions, and is thus highly parametric. The basic concept of Bayesian philosophy is that of the conditional distribution (Bernardo & Smith, 2009). Data, if available, are actively used and are assigned a probability distribution that will be conditional on a set of parameters. For example, data may be assumed to be Gaussian, conditional on the values of the mean and the variance. Moreover each parameter has a probability distribution, known as a prior, that can incorporate differing degrees of subjectivity . If expert opinion or information from the literature are available, the prior distribution will reflect this giving higher probability to suggested values. The prior distribution and the data are combined according to Bayes' theorem leading to the posterior distribution of the parameters of interest . The uncertainty associated with the parameters is included in the model via the prior distributions and is propagated through to the posterior distribution. The output is not a single number but an entire probability distribution from which both measures of central tendency and uncertainty can be obtained. Often however the posterior distribution is not a known probability distribution or it is difficult to treat it analytically and so simulation or approximations are often used. This is the subject of Chapter 5.

3.5   Quantifying uncertainty

Although uncertainty is a fundamental concept it is somewhat absent from the mainstream of statistical science and the best ways to define and quantify it remain unclear (Chen, van Eeden, & Zidek, 2010; Chen, 2011). The importance of random variability has led statisticians to model it, seek its causes and its relative importance. Following from this, statistical scientists have traditionally seen the identification of sources of uncertainty, its quantification and reduction through measurement and modelling as hallmarks of their discipline.

As described in Section 3.4.4 and more fully in Chapter 4, Bayesian theory represents uncertainty through probability (Oh & Li, 2004). That seems satisfactory for the uncertainty associated with the occurrence of a chance event such as $A = Heads$ on the toss of a coin. In this case, $p = P(A) = 0.9$ would imply a high degree of certainty in that case whereas $p = 0.5$ would reflect a state of complete uncertainty. However in more complex situations where uncertain quantities T are involved, probabilities would need to be expressed as distributions to adequately describe the uncertainties (Frey & Rhodes, 1996). In other words, $p = P(T \leq t)$ would work for a single t but it would not characterise the degree of overall uncertainty about T. For that, a metric calculated from that distribution is required. The topic is an important one as any contemplated action or decision must take into account uncertainty associated with the result. For example, deciding which smelters to close because of the high levels of arsenic they were generating could be based on spatial predictions of soil deposits produced through the methods seen in Chapter 9. Such decisions would then depend critically on the degree of confidence that could be placed on predictions made in the vicinities of the various smelters.

Various metrics have been proposed, notably the variance of and entropy of T's distribution, and these will be the topic for this Section. Notably, both of these are additive when independent random variables are combined appropriately but little is known about their behaviour when additional information about T itself becomes available. One would heuristically expect uncertainty about it to go down, but does it? Starting with the simplest case above where probability is the index of uncertainty, we see that such heuristics can often be too simplistic.

Example 3.3. *Uncertainty as probability*

Let $P(Y \in C)$ represent our uncertainty about the event, $A = \{Y \in C\}$. Now we learn that in fact $Y \in A$, $A \cap C \neq \phi$ Will our uncertainty about the event decrease as our heuristics suggest?

The answer can be 'no' when the new information conflicts with our prior beliefs. In fact $P(Y \in C | Y \in A)$ may be closer to $1/2$ than $P(Y \in C)$ as shown by Zidek and van Eeden (2003). In their example, if $Y \sim U[0, 1]$, C=[0,c], and A=[a,1] with $0 < a < c < 1$, then $P(Y \in C | Y \in A) = (c - a)/(1 - a) = 1/2$ when a = $2c - 1$. If c = 7/8 and a = 3/4 we would move from a state of near certainty about the event to one of complete uncertainty (conditional probability 1/2) about whether or not it has occurred. So new information need not reduce our uncertainty with this simple measure of uncertainty.

3.5.1 Variance

Suppose T denotes an uncertain quantity, for example an estimate of the long-run annual average temperature, μ_T, or a prediction of an unmeasured pollutant at a spatial location. How might the degree of our uncertainty about T be indexed? A common approach is to use the variance $Var(T)$ and it, or rather the standard deviation, σ_T, is used for example by The U.S. National Institute of Standards and Technology, to express measurement uncertainty. The institute's web page (http://physics.nist.gov/cuu/Uncertainty/basic.html) states that "Each component of uncertainty, however evaluated, is represented by an estimated standard deviation, termed *standard uncertainty* ?". Perhaps a preferable alternative would be the coefficient of variation $CV_T = \sigma_T / \mu_T$ which does not depend on the scale on which T is measured, making this a more intrinsic property. Expressed as a percentage (which will also be unitless), CV_t , it is even used as a metric for assessing quality of the data from which the estimate was produced, assuming the latter is obtained in a principled way (Bergdahl et al., 2007). Unless that metric is under a specified percentage, statistical agencies may not publish the estimates meaning that for example national estimates may be published while those for small areas may not since the samples are too small. In the authors' experience 20% is a challenging target to reach in practice with samples of a realistic size.

When T is in fact an estimate, σ_T is referred to as the standard error se_T. The se_T has been so well accepted as an appropriate measure of an estimate's uncertainty that it (or an estimate of it) is routinely quoted in most all scientific literature together with the estimate.

Example 3.4. *Uncertainty as variance*

Here, interest is in making inference about $T \sim N(\mu_T, \sigma_T^2)$. The variance $\sigma_T^2 = 1$ indexes the uncertainty about T. Some feedback reveals to the investigator that $|T| < C$ for a known constant C. Thus the uncertainty index has to be recalculated as $Var(T \,||\, T \,|< C)$. Heuristics suggest that the new information should reduce uncertainty about T. But does it? The answer is that it does and in fact $Var(T \,||\, T \,|< C) < 1$ for any μ_T and $0 < C$. Moreover the new index is a monotone increasing function of C. In other words, the result is quite striking!

The result seems plausible when $\mu_T = 0$ for then the condition $|T| < C$ tells us at least if C is small that $T \approx \mu_T$. However it does not seem plausible when $\mu_T \neq 0$ since the new information puts T in a place very different from where the distribution is centred.

The results in Example 3.4 are true in great generality (Chen, 2011), at least when new information comes in a relatively simple form. However nothing seems to be known beyond these simple cases. Simple counterexamples show that they do not always hold as in Exercise 3.4. Apart from the fact that as an estimate or predictor, T would need to be unbiased for the variance to play the role of an uncertainty index, it is not even clear whether or not it will generally pass the simple test of downward revision under increasing information.

These considerations, and other doubts about the suitability of the variance as an index of uncertainty, lead us to consider another index of uncertainty in the following Section (13.7).

3.5.2 Entropy

Entropy's role as an index of aleatory uncertainty has a long history but it has relevance in many modern applications including that encountered in Chapter 13 where it is used in monitoring network design to select sites where uncertainty (entropy) is highest.

To describe this index we begin by letting E be a random event and $\overline{E}$ its complement. Furthermore $T = E$ if E occurs and $T = \overline{E}$ if not. Next $p = Pr(E)$ denotes the probability that the event E occurs so that $1 - p$ is the probability that $\overline{E}$ occurs. Concentrating on E for the moment, we are uncertain about whether it does or does not occur. Letting $\phi(.)$ represent as reduction-in-uncertainty function to be determined below, we get a reduction of $\phi(p)$ in that uncertainty if E does occur. In the same way we get a reduction of $\phi(1 - p)$ if $\overline{E}$ occurs. The idea is that if $p = 0.9$ we would get a smaller reduction $\phi(p)$ in our uncertainty if E does occur than we would get $\phi(1 - p)$ if $\overline{E}$ occurs, the latter being quite a surprise. Thus the expected reduction in uncertainty for observing $\mathbf{T}$ is

$$H(T) = p\phi(p) + (1 - p)\phi(1 - p)$$

To specify ϕ we need to impose some requirements and these are: $\phi \in [0, 1]$; $H(T_1, T_2) = H(T_1) + H(T_2)$ when T_1 and T_2 are independent. Then it turns out that $\phi(p) = -log(p)$ in this simple case.

We now turn to the more general case of discrete random variables T where $T \in \{t_1, \ldots, t_n\}$. In that case with $p_i = P(T = t_i)$, we obtain

$$H(T) = \sum_{i=1}^{n} -p_i \log p_i \tag{3.1}$$

as the average reduction of uncertainty for observing T.

In the case of a continuous random variable T, things are a bit more complicated. A naive approach would be to first partition the continuous domain of T into n equiprobable discrete subsections, S_i, with $n^{-1} = p_i = P(T \in S_i) = \int_{S_i} f_T(u)du$ where f_T denotes the pdf of T. With this approximation, we get the result in Equation 3.1 namely

$$H(T) \approx = -\sum_{i=1}^{n} n^{-1} \log n^{-1} = \log n$$

as an approximation. Letting the width of those subsections go to zero and $n \rightarrow \infty$ we get ∞ in the limit. Obviously this approach does not work.

The next approach is by analogy:

$$H(T) \equiv -E[\log f(T)] = -\int \log f(t) f(t) dx.$$

However this proves unsatisfactory in more subtle ways. Firstly the pdf is not invariant under transformations from say Celsius to Fahrenheit if T were temperature. Uncertainty should ideally be an intrinsic property of an uncertain quantity and in fact should be invariant under any 1:1 transformation of the scale of measurement. Recognising this problem, Jaynes (1963) introduces the idea of a reference measure with PDF $h(x)$:

$$H(T) \equiv -E\left[\log \frac{f(T)}{h(T)}\right] \tag{3.2}$$

In fact Jaynes gives an ingenious argument, which shows that if h is defined correctly Equation 3.2 can be obtained from Equation 3.1 by an approximation argument similar in spirit to the failed naive argument above. However commonly in practice h is taken to be $1(T\text{units})$ so that it cannot be explicitly seen. However it does mean that the units of measurement cancel in f_T/h, making this a unitless quantity as it must be if its logarithm is to be taken (see Section 9.3).

3.5.3   Information and uncertainty

It is not clear precisely how information and uncertainty are related. At one time uncertainty seems to have been thought of exclusively as aleatory uncertainty and then entropy was seen as a way of quantifying the uncertainty in the probability distribution of a random variable (Harris, 1982). When the variable was observed, all uncertainty was gone so the entropy could then be thought of as the amount of information that had been gained by observing the variable (Shannon, 2001; Renyi, 1961).

That was all well and good until it was recognised that not all quantitative uncertainty could be considered to be due to chance (aleatory) and that some of it had to be ascribed to a lack of knowledge (epistemic) (Helton, 1997). In this case, the latter is knowable, meaning that with enough data that uncertainty would disappear.

The approach in Section 13.7 fails when epistemic uncertainty is introduced meaning that aleatory uncertainty (corresponding to the amount of information) forms just one component of a decomposition. The other corresponds to epistemic uncertainty when the uncertainty about the model is expressed as a probability distribution.

Example 3.5. *Aleatory and epistemic uncertainty*

Recall that epistemic uncertainty is that component of uncertainty related to the truth of a proposition and that it is reduced or eliminated given enough data. Let the indicator function $Z = I\{ace\}$ represent the truth of the proposition "the next toss of this die will be an ace" when the probability of an ace p is uncertain. Repeated tosses of the die would resolve this uncertainty. However even if eventually $p = 1/6$ came to be known for certain, the value of Z would not be known. This uncertainty due to chance variation is aleatory uncertainty. To get a better understanding of what is going on, let's consider the example within a Bayesian setting and assign a beta prior distribution on p with prior density proportional to $p^{\alpha-1}(1-p)^{\beta-1}$. After n repeated tosses we would have a posterior density proportional to $p^{\alpha+r-1}(1-p)^{\beta+n-r-1}$ where r is the number of times a six turned up in these independent tosses.

Although it is not clear that the variance is the correct way to quantify uncertainty (Chen et al., 2010), we consider it here because it is commonly used. The posterior variance of Z is approximately

$$Var[Z \mid r] \quad = \quad Var\{E[Z \mid p,r]\} + E\{Var[Z \mid p,r]\} \qquad (3.3)$$

$$\approx \quad \frac{\hat{p}(1-\hat{p})}{n}\} + \hat{p}(1-\hat{p}) \qquad (3.4)$$

where $\hat{p} = r/n$ (Chen et al., 2010). The first term represents the epistemic uncertainty, which declines to zero as $n \to \infty$ while the second represents the aleatory uncertainty.

We may ask "what happens if a 7 turns up on the second toss of the die?" Unfortunately we are stuck since all our calculations were done conditional on a six-sided die. In this case, our uncertainty would increase. Bayesian theory can allow for the 'known unknowns' to be accommodated by acknowledging model uncertainty. Here, if we had allowed a die with an unknown number of faces into our model, we could have dealt with the '7'. An insurmountable problem arises from the 'unknown unknowns' as there is no formal theory that allows us to represent the increase in uncertainty that arises due to a surprise outcome. Nevertheless, it is important to recognise that information can both increase or decrease our uncertainty.

So what lessons does Example 3.5 teach us about spatio–temporal modelling? First it tells us that additional data may increase rather than decrease our uncertainty when the initial class of models is insufficiently rich. The analyst should try to ensure a class of models that is big enough to admit even seemingly unrealistic scenarios to the maximum feasible extent leading to a restatement of the late Dennis Lindley's version of Cromwell's rule: "don't put a probability of zero on anything". Secondly we learn that spatio–temporal models must be able to retain the aleatory component of uncertainty in a dynamic system. A meteorologist with sufficient data would be able to estimate the correlation between temporally stationary responses seen at two sites say Y_{s_1t} and Y_{s_2t}, due to chance variation in things like variable wind direction. There is no model here. A naive Bayesian approach might model the process like this (Chen et al., 2010): $Y_{st} = \beta_{st} + \varepsilon_{st}$, or in vector form $\mathbf{Y}_s = \beta_s + \epsilon_s$ with $\beta_s \sim N_p(\mu_0 \mathbf{1}_{T+1}, \sigma_\beta^2 \mathbf{I}_{T+1})$ and $\mu_0 \sim N(\mu^*, \sigma_\mu^2)$. It is easily shown that conditional on the *data*

$$Corr(Y_{s_1(T+1)}, Y_{s_2(T+1)}) = [\sigma_\mu^{-2} + pT\sigma^{-2}]^{-1} \to 0, \text{ as } T \to \infty \qquad (3.5)$$

in contradiction to the meteorologist's calculation. This is because the Bayesian failed to distinguish between the two forms of uncertainty and built the model accordingly. Even in less extreme cases, uncertainties will shift between components of a Bayesian posterior distribution when the marginal distribution of the process responses are fixed and reflect aleatory uncertainty.

3.5.4 Decomposing uncertainty with entropy

In this section we return to the problem of decomposing uncertainty into the part that is aleatory and the part that is epistemic with entropy playing the role of the uncertainty index. The latter means in this section that uncertainty about the model parameters is represented by a probability distribution as in Chapter 4. To begin, assume $T \sim f(. \mid \theta)$. Given data D

$$H(T, \theta) = H(T \mid \theta) + H(\theta) \qquad (3.6)$$

where

$$
\begin{aligned}
H(T \mid \theta) &= E[-\log(f(T \mid \tilde{\theta}, D)/h_1(T)) \mid D] \\
H(\theta) &= E[-\log(f(\tilde{\theta} \mid D)/h_2(\theta)) \mid D] \\
h(T, \theta) &= h_1(T)h_2(\theta)
\end{aligned}
$$

with $\tilde{\theta}$ meaning that the expectation is being computed over a random θ (Caselton, Kan, & Zidek, 1992; Le & Zidek, 1994). An important example of this decomposition, for an application detailed in Chapter 13 is given in Appendix A.2.6.

3.6 Summary

This chapter contains a discussion of uncertainty, both in terms of statistical modelling and quantification but also in the wider setting of sources of uncertainty outside those normally encountered in statistics. The reader will have gained an understanding of the following topics:

- Uncertainty can be dichotomised as either qualitative or quantitative, with the former allowing consideration of a wide variety of sources of uncertainty that would be difficult, if not impossible, to quantify mathematically.

- Quantitative uncertainty can be thought of as comprising both aleatory and epistemic components, the former representing stochastic uncertainty and the latter subjective uncertainty.

- Methods for assessing uncertainty including eliciting prior information from experts and sensitivity analysis.
- Indexing quantitative uncertainty using the variance and entropy of the distribution of a random quantity.
- Uncertainty in post-normal science derives from a wide variety of issues and can lead to high levels of that uncertainty with serious consequences. Understanding uncertainty is therefore a vital feature of modern environmental epidemiology.

Exercises

Exercise 3.1. Referring to Example 3.2, give a reasonable expression of your uncertainty about the prediction of a six when θ is unknown. How is this case different from the case when θ was known to be $1/6$?

Exercise 3.2. Referring to Example 3.4,

(i) determine in explicit form both $E(T \mid\mid T \mid < C)$ as well as $Var(T \mid\mid T \mid < C)$

(ii) numerically evaluate and plot both functions of C in (i).

Exercise 3.3. Repeat Exercise 3.2, but this time for the uniform distribution where $T \sim U[-1,1]$

Exercise 3.4. The results in Exercise 3.2 do not always hold. Give two counterexamples.

Exercise 3.5. You are to predict the outcome T of the toss of a die but you do not know if the die is fair.

(i) Model both the aleatory as well as the epistemic uncertainty in this situation and show how the total uncertainty can be decomposed using your model.

(ii) You are to bet on the event $T = 1$ occurring and asked to state the best odds you can offer me, below which you will refuse to bet. What are those odds?

(iii) The die is tossed and a 7 turns up. What are the best odds you would be willing to give me now?

Exercise 3.6. Using the notation of Appendix 14.5, $Z \sim N_p(\mu, \Sigma)$,

(i) find the entropy of the distribution of Z.

(ii) interpret your result from (i) when $p = 2$.

Exercise 3.7. RESEARCH QUESTION: How does the entropy of the distribution of T $H(T)$ change if it learned that $\mid T \mid < C$ for a known C?

Chapter 4

Embracing uncertainty: the Bayesian approach

4.1 Overview

In Bayesian statistics, probability describes all types of uncertainty, both through unpredictability and through imperfect knowledge. Uncertainty is described by means of probability distributions that are assigned to uncertain quantities. When estimating the values of unknown parameters, which are often properties of the population, these values will be uncertain due to lack of knowledge rather than due to random variation. In the Bayesian setting, parameters are treated in a similar fashion to all other uncertain quantities; they are treated as random variables and assigned probability distributions. As such, probability statements can be made about them, such as "an interval that has 95% probability of containing the true value" or "the probability that a null hypothesis is true is...". This is in contrast to the frequentist approach under which statements about probability are based on the idea of repetition, e.g. "95% of all confidence intervals calculated under repeated sampling will cover the true parameter value" or "the p–value for a hypothesis test is the probability that we would observe this, or something more extreme (under the assumption that the null hypothesis is true)". Bayesian statistics may be thought of as *subjective* in the sense that prior knowledge and beliefs about what we expect to see can be incorporated into the inferential process. Frequentist statistics on the other hand is based solely on observed data and is therefore referred to as *objective*.

This chapter introduces the Bayesian approach, upon which the majority of the methodology introduced in this book is based, while making comparisons to the likelihood approach. The choice of prior distributions, central to the idea of Bayesian statistics, is considered together with the ways in which prior distributions then combine with information from data to obtain posterior distributions. A posterior distribution captures our beliefs about a particular quantity and will be the basis of interpretation and possible further inference. In the following chapter, Chapter 5 we consider methods for implementing these ideas in practice.

4.2   Introduction to Bayesian inference

When considering the effects of a set of covariates on a response variable, interest often lies in drawing conclusions about the unknown parameters, θ. In the case of a regression model unknown parameters might include the set of coefficients, β and the variance, σ^2. Interest might lie in predicting values, Z, of the response variable for particular values of the explanatory variables. When data, y, are observed, such inference will be expressed as $p(\theta|y)$ and $p(Z|y)$ for the cases of parameter estimation and prediction respectively.

In the frequentist setting, in order to estimate the values of p unknown quantities, $\theta = (\theta_1, \theta_2, \ldots, \theta_p)^T$, data would be collected $\mathbf{y} = (y_1, \ldots, y_n)^T$ and a model proprosed for how the data depend on θ; $p(\mathbf{y}|\theta)$. This is then used to find maximum likelihood estimates $\widehat{\theta}$.

In the Bayesian setting there are three key elements:

(i) $p(\theta)$ is the prior distribution for the parameters.

(ii) $p(\mathbf{y}|\theta)$ is the likelihood of the data given the parameters.

(iii) $p(\theta|\mathbf{y})$ is the posterior distribution of the parameters given the data.

Before we collect any data we formulate our *a priori* beliefs about the values of θ. These may be based on previous studies that suggest, for example, a range of values we expect θ to take. These beliefs are expressed in terms of a probability density function, $p(\theta)$. After data, $\mathbf{y}$, have been observed we again specify a model $p(\mathbf{y}|\theta)$ for how the data depend on θ. Now the question is how our beliefs about θ can take into account both the prior beliefs and the information from the data. These *posterior* beliefs are expressed in terms of the *posterior* distribution $p(\theta|\mathbf{y})$. This posterior distribution can be found using Bayes theorem:

$$p(\theta|\mathbf{y}) = \frac{p(\mathbf{y}|\theta)p(\theta)}{p(\mathbf{y})} \tag{4.1}$$

When $p(\theta)$ and $p(\mathbf{y}|\theta)$ are proper densities, $p(\theta|\mathbf{y})$ will also be a probability density function.

The distribution of the data, $p(\mathbf{y}|\theta)$, is related to the frequentist likelihood, since $l(\theta|\mathbf{y}) \propto p(\mathbf{y}|\theta)$ and this is often used to obtain the density $p(\theta|\mathbf{y})$ up to proportionality:

$$
\begin{aligned}
p(\theta|\mathbf{y}) &\propto & p(\mathbf{y}|\theta)p(\theta) \\
&\propto & l(\theta|\mathbf{y})p(\theta) \\
\text{posterior} &\propto & \text{likelihood} \times \text{prior} \\
p(\theta|\mathbf{y}) &= & c \times l(\theta|\mathbf{y})p(\theta)
\end{aligned}
$$

The constant c is known as the *normalising constant*. We will see in Chapter 5 that it is not always needed in practice but where it is needed, such as when using Bayes factors (Chapter 6, Section 6.8), it can be found by using the fact that the posterior distribution must integrate to one, i.e. $\int p(\theta|\mathbf{y}) \, d\theta = 1$ and therefore $c = [\int l(\theta|\mathbf{y})p(\theta) \, d\theta]^{-1}$.

Although the prior distribution allows knowledge from previous studies or experiments to be incorporated it can introduce an increased level of complexity into the calculations. There may be cases where there are actual data available with which to construct priors, but decisions often have to be based on knowledge of the literature and other sources. In such cases, there may be doubts as to the accuracy of the prior distributions and it is important to assess the sensitivity of the posterior to the choice of priors.

4.3 Exchangeability

Given n observations, $\mathbf{y} = (y_1, y_2, \ldots, y_n)^T$ and a joint distribution $p(\mathbf{y}|\theta) = p(y_1, y_2, \ldots, y_n|\theta)$, in frequentist statistics *independence* is often assumed, i.e. $p(\mathbf{y}) = \prod_{i=1}^{n} p(y_i)$. However, this assumption does not really seem so reasonable in a Bayesian setting. Independence implies that $p(y_{m+1}, \ldots, y_n|y_1, \ldots, y_m) = p(y_{m+1}, \ldots, y_n), 1 \le m < n$ and hence there is no learning experience. In this case, the first m values tell us nothing about what might happen next which is in contrast to the ideas underpinning Bayesian statistics, those of incorporating prior information. In this setting, a more reasonable assumption may be that of *exchangeability* which means that for a finite set of random variables $(Y_1, \ldots, Y_n)^T$ every permutation $(Y_{\pi(1)}, \ldots, Y_{\pi(n)})^T$ has the same joint distribution as every other permutation. Similarly, an infinite collection of random variables is exchangeable if every finite subcollection is exchangeable.

If $Y_i; i = 1, \ldots, n$ are independent and identically (IID) distributed, then they are exchangeable.

$$
\begin{aligned}
p(y_1, \ldots, y_n) &= p(y_1) \ldots p(y_n) \\
&= p(y_{\pi(1)}) \ldots p(y_{\pi(n)}) \\
&= p(y_{\pi(1)}, \ldots, y_{\pi(n)})
\end{aligned}
$$

However being exchangeable does not necessarily imply IID. Exchangeable Y_i have the same marginal distribution but they need not be independent; exchangeability is a weaker assumption than IID.

This can cause a problem in specifying the joint distribution $p(\mathbf{y})$. We can use de Finetti's Representation Theorem to help this specification (Bernardo & Smith, 2009). If $Y_1, \ldots, Y_n$ is an exchangeable sequence of 0-1 random variables with joint mass $p(y_1, \ldots, y_n)$ then there exists a distribution function $q(\cdot)$ such that we can write

$p(y_1,\ldots,y_n) = \int_0^1 \prod_{i=1}^n \theta^{y_i}(1-\theta)^{1-y_i} q(\theta)\, d\theta$ where $\theta = \lim_{n\to\infty}\left(\frac{1}{n}\sum_{i=1}^n y_i\right)$ is the 'long-run average number of 1s'. The theorem provides justification for the Bayesian approach of conditioning on prior information. The idea is that Y_i are independent Bernoulli random variables and are conditional on the value of some random quantity θ where θ represents the limiting relative frequency of 1s and $q(\theta)$ represents beliefs about θ. If Y_i are exchangeable Bernoulli variables then they can be treated as if they were conditionally independent Bernoulli trials, conditional on θ. If θ has prior distribution $q(\theta)$, then $p(y_1,\ldots,y_n|\theta) = \prod_{i=1}^n p(y_i|\theta) = \prod_{i=1}^n \theta^{y_i}(1-\theta)^{1-y_i}$.

Analogous results are available for situations other than Bernoulli and provide justification for conditional independence in these cases (Bernardo & Smith, 2009).

Example 4.1. *Normal distribution with unknown mean and known variance*

Suppose $Y_1,\ldots,Y_n$ are exchangeable observations that are normally distributed with unknown mean θ, and known variance σ^2, $[Y_i|\theta \sim$ *i.i.d.* $N(\theta,\sigma^2)$.

Suppose our prior beliefs about θ can also be expressed in terms of a Normal distribution, $\theta \sim N(\theta_0,\sigma_0^2)$ with known constants θ_0,σ_0^2. In this case,

$$
\begin{aligned}
p(\theta) &= (2\pi\sigma_0^2)^{-1/2}\exp\left\{-\frac{1}{2\sigma_0^2}(\theta-\theta_0)^2\right\} \\
&\propto \exp\left\{-\frac{1}{2\sigma_0^2}(\theta^2-2\theta\theta_0)\right\}
\end{aligned}
$$

The distribution of the data is $p(\mathbf{y}|\theta) = \prod_i^n p(y_i|\theta)$. Since Y_i are treated as conditionally independent,

$$
\begin{aligned}
p(\mathbf{y}|\theta) &= \prod_{i=1}^n\left[(2\pi\sigma^2)^{-1/2}\exp\left\{-\frac{1}{2\sigma^2}(y_i-\theta)^2\right\}\right] \\
&= (2\pi\sigma^2)^{-n/2}\exp\left\{-\frac{1}{2\sigma^2}\sum_{i=1}^n(y_i-\theta)^2\right\} \\
l(\theta|\mathbf{y}) &\propto p(\mathbf{y}|\theta) \\
&\propto \exp\left\{-\frac{n}{2\sigma^2}(\theta^2-2\bar{y}\theta)\right\}
\end{aligned}
$$

The posterior is then given by:

$$
\begin{aligned}
p(\theta|\mathbf{y}) &\propto p(\theta)l(\theta|\mathbf{y}) \\
&\propto \exp\left\{-\frac{1}{2\sigma_0^2}(\theta^2-2\theta\theta_0)-\frac{n}{2\sigma^2}(\theta^2-2\bar{y}\theta)\right\} \\
&\propto \exp\left\{-\frac{n}{2w\sigma^2}\left(\theta^2-2\theta(w\bar{y}+(1-w)\theta_0)\right)\right\}
\end{aligned}
$$

It can be seen that the posterior distribution is also Normal, $\theta|\mathbf{y} \sim N(w\bar{y} + (1-w)\theta_0, \frac{w\sigma^2}{n})$ with mean $w\bar{y} + (1-w)\theta_0$ where the weights are $w = \frac{\sigma_0^2}{\sigma_0^2 + \sigma^2/n}$. This is a weighted combination of the *prior mean* θ_0 and the mean of the data $\bar{y}$. Note that $n = 0$ corresponds to having no data in which case the posterior is the same as the prior distribution and when $n \to \infty$, $E[\theta|\mathbf{y}] \to \bar{y}$; i.e. all the information comes from the data.

Example 4.2. *Over-dispersion in a Poisson model*

Here we revisit over-dispersion as seen in Chapter 2, Section 2.8.2 in a Bayesian setting by using a Gamma prior for the rate of disease. Suppose we are interested in studying the incidence of a disease. As discussed in Chapter 2, a Poisson may be a plausible model for such count data,

$$Y_i|\theta \sim Po(E_i\lambda), i = 1, ..., N \qquad (4.2)$$

where Y_i is the observed number of cases of disease, λ is the disease rate and E_i is the expected number of cases based on the age–sex structure of the underlying population.

Using a Gamma(a, b) prior distribution, $p(\lambda) \propto \theta^{a-1} \exp^{-b\lambda}$, the posterior distribution will be a Gamma$(\sum y_i + a, \sum n_i + b)$ distribution.

Therefore the marginal mean and variance are as follows.

$$
\begin{aligned}
E[Y_i|a,b] &= E_\lambda[E[Y_i|\lambda]|a,b] \\
&= E_\lambda[E_i\lambda|a,b] \\
&= E_i E_\lambda[\lambda|a,b] \\
&= = \frac{aE_i}{b} \qquad (4.3) \\
var(Y_i|a,b) &= E_\lambda[var(Y_i|\lambda)|a,b] + var_\lambda(E[Y_i|\lambda]|a,b) \qquad (4.4) \\
&= E_\lambda[E_i\lambda|a,b] + var_\lambda(E_i\lambda|a,b) \\
&= E_i E_\lambda[\lambda|a,b] + E_i^2 var_\lambda(\lambda)|a,b) \\
&= \frac{aE_i}{b} + \frac{aE_i^2}{b^2} \\
&= \frac{aE_i}{b}\left(1 + \frac{E_i}{b}\right) \\
&= E[Y_i|a,b]\left(1 + \frac{E_i}{b}\right) \qquad (4.5)
\end{aligned}
$$

So the over-dispersion is of the form $(1 + \frac{E_i}{b})$, which is always positive; and, for fixed E_i, the smaller the value of b in the prior distribution, the larger the over-dispersion parameter.

4.4   Using the posterior for inference

Strictly speaking there is no need for point estimates within a Bayesian framework as a parameter is considered a random variable and a lot more information is available from examining the entire (posterior) distribution. However, point estimates can be useful in comparison and discussion; if a point estimate from the posterior distribution of the unknown parameters is required, then the posterior mean, $\bar{\theta} = E(\theta|y) = \int \theta p(\theta|y)d\theta$, for example, can be found, although this is not necessarily an easy task. Alternatively, if samples can be taken from the posterior distribution, using, for example Gibbs sampling (see Chapter 5), then the mean and median together with credible intervals can be obtained.

Posterior beliefs are captured by a whole distribution and typically we want to summarise this distribution. This can be achieved in a variety of ways including the following:

- Using graphs and plots of the shape of the distribution.
- Calculating summaries such as the mean, median, variance and intervals that contain most of the values.
- Calculating posterior probabilities that θ is greater than some value.

Credible intervals are in some sense the Bayesian analogue of confidence intervals, or rather the way in which confidence intervals are commonly interpreted. They comprise an interval that captures the most likely values of θ. A $(1-\alpha)100\%$ credible interval (θ_l, θ_u) is an interval within which $(1-\alpha)100\%$ of the posterior distribution lies, i.e. $P(\theta_L < \theta < \theta_U|y) = 1 - \alpha$. Of course, this could define many possible intervals and typically the $\alpha/2$ and $1 - \alpha/2$ percentiles of the posterior distribution are used such that $P(\theta < \theta_L|y) = \alpha/2, P(\theta > \theta_U|y) = \alpha/2$. Hence there is a $(1-\alpha)100\%$ probability that θ is between θ_L and θ_U.

4.5   Predictions

Let Z be the value of an observation that has not been measured. Since Z is unknown, within the Bayesian framework it is considered a random variable with an associated distribution. In performing prediction there are two cases to consider, both before and after data have been observed. In the case of the former, the *marginal* or *prior predictive distribution*, $p(y)$, is a distribution that does not depend on any previous observations.

After the data have been observed, a predicted value, Z, is obtained,

$$p(z|\mathbf{y}) = \int p(z|\theta)p(\theta|\mathbf{y}) \, d\theta$$

where $p(z|\theta)$ is the distribution of the data given the value of θ, the same distribution we assumed for the rest of the data $p(y_i|\theta)$ and $p(\theta|\mathbf{y})$ is the posterior distribution, reflecting the uncertainty about the true value of θ. The predictive distribution is averaged over all possible values of θ.

Example 4.3. *A predictive distribution for normally distributed data*

Given normally distributed data $Y_1, \ldots, Y_n$ with unknown mean θ, and known variance σ^2 and prior distribution $\theta \sim N(\theta_0, \sigma_0^2)$ where θ_0, σ_0^2 are known constants, the posterior distribution is $\theta | \mathbf{y} \sim N(w\bar{y} + (1-w)\theta_0, \frac{w\sigma^2}{n})$. In this case, the distribution of predictions of a new observation, Z, will be normal,

$$Z | \mathbf{y} \sim N\left(w\bar{y} + (1-w)\theta_0, \sigma^2 + \frac{w\sigma^2}{n} \right)$$

4.6 Transformations of parameters

Sometimes we may be more interested in a function of the parameter rather than in the parameter itself. For example, with Binomial data the interest may be in the odds of success, $\theta/(1-\theta)$ rather than in the probability θ directly. In this situation we need to transform the prior and/or posterior distribution from one parameterisation to another.

4.6.1 Prior distributions

Given a function g and prior distribution for θ, in order to find the corresponding prior beliefs, $p_\phi(\phi)$ we use change of variables. For discrete θ we have $p_\phi(\phi) = p_\theta\left(g^{-1}(\phi)\right)$ and for the continuous case, $[p_\phi(\phi) = |J| p_\theta\left(g^{-1}(\phi)\right)$ where J is the Jacobian matrix with $(i,j)^{th}$ element $J_{ij} = \frac{\partial \theta_i}{\partial \phi_j}$ the partial derivative of $\theta = g^{-1}(\phi)$ with respect to ϕ_j.

4.6.2 Likelihood

The likelihood is *invariant* to transformation of the parameters; and so $g^{-1}(\phi)$ may be substituted for θ, i.e. $p(\mathbf{y}|\phi) = p\left(\mathbf{y}|\theta = g^{-1}(\phi)\right)$. The likelihood will be $l_\phi(\phi|\mathbf{y}) \propto p(\mathbf{y}|\phi) = p\left(\mathbf{y}|\theta = g^{-1}(\phi)\right) \propto l_\theta\left(g^{-1}(\phi)|\mathbf{y}\right)$ and the maximum of $l(\theta|\mathbf{y})$ occurs at the same point as the maximum of $l(\phi|\mathbf{y})$. The maximum likelihood estimator for ϕ is $\hat{\phi} = g(\hat{\theta})$.

4.6.3 Posterior distributions

If the posterior distribution of θ is denoted $p_\theta(\theta|\mathbf{y})$ the corresponding prior beliefs of ϕ will be expressed as $p_\phi(\phi|\mathbf{y})$. For continuous θ we have:

$$
\begin{aligned}
p_\phi(\phi|\mathbf{y}) &\propto l_\phi(\phi|\mathbf{y}) p_\phi(\phi) \\
&\propto l_\theta\left(g^{-1}(\phi)|\mathbf{y}\right) p_\theta\left(g^{-1}(\phi)\right) |J| \\
&= |J| p_\theta\left(g^{-1}(\phi)|\mathbf{y}\right)
\end{aligned}
$$

In general, posterior moments are not invariant under reparameterisation unlike maximum likelihood estimators. The posterior mean for θ is $E_{\theta|y};[\theta|\mathbf{y}]$ but, unless $g(\theta)$ is a linear transformation, $E_{\phi|y}[\phi|\mathbf{y}] = E_{\theta|y}[g(\theta)|\mathbf{y}] \neq g(E[\theta|\mathbf{y}])$.

Posterior quantiles are invariant to transformations. In particular, this means that the median and credible intervals based on quantiles are also invariant to transformations. If m is the median of the posterior distribution, $P(\theta \leq m|\mathbf{y}) = 0.5$ then $0.5 = P(\theta| \leq m|\mathbf{y}) = P(g(\theta) \leq g(m)|\mathbf{y}) = P(\phi \leq g(m)|\mathbf{y})$. Therefore, $g(m)$ is the median of the posterior distribution expressed in terms of ϕ. More generally, if $s(\theta,q)$ is the quantile of the posterior distribution such that $P(\theta \leq s(\theta,q)|\mathbf{y}) = q$ then $P(\phi \leq g(s(\theta,q))|\mathbf{y}) = q$.

The mode, $\tilde{\theta}$, occurs at the maximum of the posterior distribution and so maximises the log posterior: $\log p_{\theta|y}(\theta|\mathbf{y}) \propto \log p(\mathbf{y}|\theta) + \log p_\theta(\theta)$ which takes the form $\log p_{\theta|y}(\theta|\mathbf{y}) \propto \log p(\mathbf{y}|\theta = g^{-1}(\phi) + \log p_\theta(g^{-1}(\phi)) + \log |J|$. In this case $\tilde{\phi} = g(\tilde{\theta})$ only if $J = \frac{\partial \theta}{\partial \phi}$ does not depend on ϕ. In general, posterior modes are not invariant to transformations.

4.7 Prior formulation

4.7.1 Conjugate priors

When the prior and the likelihood are from the same family of distributions, they are called *conjugate prior distributions*.

As seen in Chapter 2, the exponential family of distributions has the form, $p(y_i|\theta) = f(y_i)g(\theta)\exp\{\phi(\theta)^T\mathbf{u}(y_i)\}$. Given a sufficient statistic for θ, the likelihood can be written $l(\theta|\mathbf{y}) \propto g(\mathbf{t},\theta)$ and the posterior can be written as $p(\theta|\mathbf{y}) = p(\theta|\mathbf{t}) \propto g(\mathbf{t},\theta)p(\theta)$. If Y_i, $i = 1,\ldots,n$ are exchangeable then the likelihood is $l(\theta|\mathbf{y}) = g(\theta)^n \exp\{\phi(\theta)^T\mathbf{t}(\mathbf{y})\}$ where $\mathbf{t}(\mathbf{y})$ is the sufficient statistic $\mathbf{t}(\mathbf{y}) = \sum_{i=1}^n u(y_i)$.

The conjugate prior will have the same form as the likelihood, $p(\theta) \propto g(\theta)^\eta \exp\{\phi(\theta)^T v\}$ for some η and v. This general form specifies the distributional family whilst specific choices of η and v characterise the moments of the prior.

The posterior can be found as follows:

$$
\begin{aligned}
p(\theta|\mathbf{y}) &\propto g(\theta)^n \exp\{\phi(\theta)^T\mathbf{t}(\mathbf{y})\} g(\theta)^\eta \exp\{\phi(\theta)^T v\} \\
&= g(\theta)^{\eta'} \exp\{\phi(\theta)^T v'\}
\end{aligned}
$$

with $\eta' = \eta + n$ and $v' = v + \mathbf{t}(\mathbf{y})$. This has the same form as $p(\theta)$ so the prior is conjugate.

4.7.2 Reference priors

Although one of the central premises of the Bayesian approach is the ability to capture subjective beliefs about parameters, through *informative priors* there may be times where *non-informative* or *reference* priors may be useful. When using non-informative priors, all the information comes from the data.

Example 4.4. *Reference priors*

Suppose we have $Y|\theta \sim \text{Bin}(n, \theta)$, with a uniform prior $p(\theta) = 1$, $0 \leq \theta \leq 1$. Therefore

$$l(\theta|y) \propto \theta^y (1 - \theta)^{n-y}$$

The posterior is

$$p(\theta|y) \propto \theta^y (1 - \theta)^{n-y} \propto l(\theta|y)$$

and $\theta|y \sim \text{Beta}(y + 1, n - y + 1)$.

4.7.3 Transformations

As seen in Section 4.6, prior and posterior distributions are not invariant to transformations of parameters. If we use the transformation $\phi = \theta^{-1}$ then a uniform prior for θ is equivalent to using $p_\phi(\phi) = \frac{1}{\phi^2}$, $1 < \phi < \infty$. Here, $p_\theta(\theta)$ represents the situation where there is no knowledge about θ; in this case it seems reasonable to assume that there is equally no knowledge about ϕ. However, the prior for ϕ is informative.

In practice lack of knowledge about θ might not be too much of a problem as long as there is sufficient data to give greater weight to the likelihood. This will mean there is less influence of the prior in the posterior meaning that posterior $p_\theta(g^{-1}(\phi)|\mathbf{y})$ will not differ greatly from $p_\phi(\phi|\mathbf{y})$.

4.7.4 Jeffreys' prior

Jeffreys' prior is a choice of prior based on the amount of information contained in the data, as captured by the Fisher Information matrix $I(\theta|\mathbf{y}) = -E_{y|\theta}\left[\frac{\partial^2}{\partial\theta^2}\log l(\theta|\mathbf{y})\right] = -E_{y|\theta}\left[\left(\frac{\partial}{\partial\theta}\log l(\theta|\mathbf{y})\right)^2\right]$, where the expectation is with respect to the data $\mathbf{y}|\theta$. It takes the form $p_\theta(\theta) \propto [I(\theta|\mathbf{y})]^{1/2}$ and is often used as a reference prior as it has the property of invariance under reparameterisation.

4.7.5 Improper priors

A density is *proper* if it integrates to one. Priors that do not not have a finite integral, i.e. priors that are not real densities, are known as *improper* priors. Reference priors often lead to improper priors but even when this is the case, the posterior may still be proper. A posterior density will always be proper if the prior density is proper.

Example 4.5. *A reference prior for the Binomial distribution*

Using the improper prior, $p(\theta) \propto \theta^{-1}(1-\theta)^{-1}$, as a reference prior for Binomial data leads to a posterior of the form $p(\theta|\mathbf{y}) \propto \theta^{y-1}(1-\theta)^{n-y-1}$. This is a Beta distribution; $\theta|\mathbf{y} \sim \text{Beta}(y, n-y)$; but in the case of either $y = 0$ or $y = n$, this will be an improper posterior.

4.7.6   Joint priors

In all but the simplest examples there will be more than one unknown quantity, which means that a *joint prior distribution* will have to be specified leading to a *joint posterior distribution*. Considering now a vector of unknown parameters, θ, then a joint distribution $p(\theta)$ is combined with the data to obtain the joint posterior distribution: $p(\theta|\mathbf{y}) \propto l(\theta|\mathbf{y})p(\theta)$. Conjugate priors are more difficult to work with for more than one parameter as they are not generally independent.

Example 4.6. *Normal data with unknown mean and variance*

If data Y_i; $i = 1, \ldots, n$, is normally distributed with $Y_i|\mu, \phi \sim N(\mu, \phi)$ then the unknown parameters will be $\theta = (\mu, \phi)^T$. A joint distribution is required for both μ and ϕ and choosing an appropriate joint distribution may not be entirely straightforward. Here we consider the non-informative (improper) prior, $p(\mu, \phi) \propto \frac{1}{\phi}$, which assigns reference priors for both μ and ϕ and assumes independence between the two.

The likelihood is $l(\mu, \phi|\mathbf{y}) \propto \phi^{-n/2}\exp\left\{-\frac{1}{2\phi}\left(S + n(\bar{y}-\mu)^2\right)\right\}$ where $\bar{y} = \frac{1}{n}\sum_{i=1}^{n} y_i$ or $S = \sum_{i=1}^{n}(y_i - \bar{y})^2$ are sufficient statistics. The posterior will be $p(\mu, \phi|\mathbf{y}) \propto \phi^{-n/2-1}\exp\left\{-\frac{1}{2\phi}\left(S + n(\bar{y}-\mu)^2\right)\right\}$, which is sometimes referred to as the Normal-Inverse-Gamma distribution.

4.7.7   Nuisance parameters

The set of unknown parameters in a model may include some that are not of primary interest and could be considered nuisance parameters. When this is the case the joint posterior distribution need only include the parameters of interest and not the nuisance parameters. This will be the *marginal posterior distribution* for the parameters of interest.

Consider a parameter vector $\theta = (\theta_1, \theta_2)^T$ containing the parameter of interest, θ_1, and a nuisance parameter, θ_2. The marginal distribution for θ_1 will be $p(\theta_1|\mathbf{y}) = \int p(\theta|\mathbf{y})\, d\theta_2$.

Example 4.7. *Marginal distribution for normal data with unknown mean and variance*

If data Y_i; $i = 1, \ldots, n$, are normally distributed with $Y_i|\mu, \phi \sim N(\mu, \phi)$, and prior distribution $p(\mu, \phi) \propto \frac{1}{\phi}$ then, as seen in Example 4.6, the full joint posterior is $p(\mu, \phi|\mathbf{y}) \propto \phi^{-n/2-1} \exp\left\{-\frac{1}{2\phi}\left(S + n(\bar{y} - \mu)^2\right)\right\}$. If interest is only in μ, then ϕ is integrated out of the joint posterior distribution, $p(\mu|\mathbf{y}) \propto \int_0^\infty \phi^{-n/2-1} \exp\left\{-\frac{1}{2\phi}\left(S + n(\bar{y} - \mu)^2\right)\right\} d\phi$.

4.8 Summary

This chapter introduces the Bayesian approach which provides a natural framework for dealing with uncertainty and also for fitting the models that will be encountered later in the book. The reader will have gained an understanding of the following topics:

- The use of prior distributions to capture beliefs before data are observed.
- The combination of prior beliefs and information from data to obtain posterior beliefs.
- The manipulation of prior distributions with likelihoods to formulate posterior distributions and why conjugate priors are useful in this regard.
- The difference between informative and non-informative priors.
- The use of the posterior distribution for inference and methods for calculating summary measures.

Exercises

Exercise 4.1. Suppose a probability density function for Y is given by $p(y|\theta) = c(\theta)k(\theta, y)$, for some functions $c(\cdot)$ and $k(\cdot)$, where $c(\theta)$ does not depend on Y. Then $k(\theta, y)$ is said to be the *kernel* of the distribution, and $p(y|\theta) \propto k(\theta, y)$.

For each of the following distributions, write down the probability density function and find the kernel:

(i) $Y|\theta \sim \text{Po}(\theta)$

(ii) $X|\theta, b \sim \text{Beta}(b\theta, b)$

(iii) $\theta|a, b, y \sim \text{Gamma}(a + y + 1, b - 3y)$

(iv) $\phi|\mu, \bar{x}, \tau \sim N(\tau\mu + (1 - \tau)\bar{x}, \bar{x}^2\tau^{-2})$

Exercise 4.2. Let $Y_i, \ldots, Y_n$ be a sample of data. For each of the following distributions for Y_i find the conjugate prior distribution and the corresponding posterior distribution.

(i) $Y_i | \theta \sim \text{Bern}(\theta)$

(ii) $Y_i | \theta \sim \text{Po}(\theta)$

(iii) $Y_i | \theta \sim N(\theta, \sigma^2)$, with σ^2 known

(iv) $Y_i | \phi \sim N(\mu, \phi)$, with μ known

Exercise 4.3. Suppose that Y_i are n exchangeable Normal random variables, with mean θ and known variance σ^2. We are interested in inference about the mean θ. Consider the prior distribution $\theta \sim N(\theta_0, \sigma_0^2)$ for known constants θ_0, σ_0^2.

Suppose we wish to make predictions about a future observation Z. Show that the predictive distribution is

$$z | \mathbf{y} \sim N\left(w\bar{y} + (1 - w)\theta_0, \sigma^2 + \frac{w\sigma^2}{n} \right)$$

Exercise 4.4. Suppose $X | \mu, \phi \sim N(\mu, \phi)$ and $Y | \mu, \delta \sim N(\mu + \delta, \phi)$, where ϕ is known, and X and Y are conditionally independent.

(i) Find the joint distribution of X and Y.

(ii) Consider the improper non-informative joint prior distribution:

$$p(\mu, \delta) \propto 1$$

Find the joint posterior distribution. Are $\mu | x, y$ and $\delta | x, y$ independent?

(iii) Find the marginal posterior distribution $p(\delta | x, y)$.

(iv) Suppose a future observation z is given by $z | \mu, \delta \sim N(\mu - \delta, \phi)$. Find the predictive distribution $p(z | x, y)$.

Jeffreys' prior takes the form $p_\theta(\theta) \propto [I(\theta | \mathbf{y})]^{1/2}$ where I is the Fisher's information. Considering the reparameterisation $\phi = g(\theta)$,

(i) show that the information contained in the data, $I(\phi | y)$ about ϕ is $\left(\frac{\partial \theta}{\partial \phi} \right)^2 \times I(\theta | \mathbf{y})$

(ii) show that the Jeffreys' prior is therefore invariant to reparameterisation, i.e. the Jeffreys' prior for ϕ is the same as that for θ.

Exercise 4.5. Let Y_i, $i = 1, \ldots, n$, be exchangeable Poisson random variables, with mean θ. Consider a prior distribution $\theta \sim \text{Gamma}(a, b)$.

(i) Find the posterior distribution $p(\theta | \mathbf{y})$.

(ii) Show that the posterior mean can be written as a weighted average of the prior mean, θ_0, and the maximum likelihood estimator $\hat{\theta} = S_n/n$, where $S_n = Y_1 + \cdots + Y_n$ is the sum.

(iii) Let z be a future (unobserved) observation. Find the mean and variance of the predictive distribution $z | \mathbf{y}$.

(iv) The data in the table below are the number of cases of a disease in a specified area between 2001 and 2010.

2001	2002	2003	2004	2005	2006	2007	2008	2009	2010
213	189	222	231	199	245	220	299	289	267

Suppose our prior beliefs about cases of the disease in this area can be expressed as $\theta \sim$ Gamma$(375, 1.5)$. Let z be the number of cases in 2011.
Assuming

$$p(z|\mathbf{y}) \sim_{\text{approx}} N\left(E[z|\mathbf{y}], \text{var}(z|\mathbf{y})\right)$$

find an approximate 95% predictive interval for the number of cases in 2011.

Exercise 4.6. Suppose that $Y_i|\phi \sim N(\mu, \phi)$, $i = 1, \ldots, n$, where μ is known, and the aim is to estimate ϕ.

(i) Show that $S = \sum_i (y_i - \mu)^2$ is a sufficient statistic for ϕ.

(ii) Consider an inverse-gamma prior distribution for ϕ:

$$p(\phi) = \frac{b^a}{\Gamma(a)} \phi^{-(a+1)} \exp(-b/\phi)$$

Show that this corresponds to a gamma distribution for τ, where $\tau = 1/\phi$, the *precision* of y.

(iii) Find the posterior distribution of $\phi|\mathbf{y}$.

Chapter 5

The Bayesian approach in practice

5.1 Overview

In stylised problems, conjugate prior distributions are available that allow for analytical solutions to finding posterior distributions (see for example Gelman et al. (2013)). However, in most practical situations such conjugacy is either not available or overly restrictive and in order to evaluate the posterior model probabilities there are several computational matters that have to be considered. There is therefore a need to compute, or estimate, integrals of the form $p(y) = \int p(y|\theta)p(\theta)d\theta$.

5.2 Analytical approximations

One approach is to use an analytic approximation such as Laplace's, which is based on the Taylor series expansion of a real valued function $f(u)$

$$\int e^{f(u)}du \approx (2p)^{r/2}|H|^{1/2}\exp\{f(u^*)\} \tag{5.1}$$

where r is the dimension of the vector u, u^* is the value of u at which f attains its maximum and H is minus the inverse Hessian information of f evaluated at u^*. We return to the subject of Laplace approximations in more detail when we discuss Integrated Nested Laplace approximations (INLA) in Section 5.6.

Example 5.1. *Using a normal approximation for a posterior distribution*

If $p(\theta|\mathbf{y})$ is unimodal and roughly symmetric we may approximate it by a multivariate normal distribution. In general cases where the posterior is not in a tractable form, finding the mean and variance of the posterior may be non-trivial. In such cases we may consider a Taylor expansion of $\log p(\theta|\mathbf{y})$ about the mode $\tilde{\theta}$.

We begin by recalling the one-dimensional Taylor series expansion for a function $g(x)$ about x_0 expressed in terms of its derivatives $g^{(j)}$, $j = 0, 1, \ldots$

$$g(x) = g(x_0) + (x - x_0)g'(x_0) + \ldots + \frac{(x - x_0)^r}{r!} g^{(r)}(x_0) + \ldots$$

Then we apply this expansion by letting $x = \theta$, $x_0 = \widetilde{\theta}$ and $g(x) = \log p(\theta|\mathbf{y})$, so that

$$
\begin{aligned}
\log p(\theta|\mathbf{y}) &\approx \log p(\widetilde{\theta}|\mathbf{y}) + (\theta - \widetilde{\theta}) \left[\frac{d}{d\theta} \log p(\theta|\mathbf{y}) \right]_{\theta = \widetilde{\theta}} \\
&\quad + \frac{1}{2}(\theta - \widetilde{\theta})^2 \left[\frac{d^2}{d\theta^2} \log p(\theta|\mathbf{y}) \right]_{\theta = \widetilde{\theta}} \\
&= \log p(\widetilde{\theta}|\mathbf{y}) + \frac{1}{2}(\theta - \widetilde{\theta})^2 \left[\frac{d^2}{d\theta^2} \log p(\theta|\mathbf{y}) \right]_{\theta = \widetilde{\theta}} \\
&\propto \frac{1}{2}(\theta - \widetilde{\theta})^2 \left[\frac{d^2}{d\theta^2} \log p(\theta|\mathbf{y}) \right]_{\theta = \widetilde{\theta}} \\
p(\theta|\mathbf{y}) &\approx \exp\left\{ -\frac{1}{2I^{-1}(\widetilde{\theta})}(\theta - \widetilde{\theta})^2 \right\}
\end{aligned}
$$

The remaining terms of the Taylor series will be small when θ is close to $\widetilde{\theta}$ and n is large. Under these conditions, we obtain the useful approximation

$$\theta|\mathbf{y} \sim_{\text{approx}} N\left(\widetilde{\theta}, I^{-1}(\widetilde{\theta}) \right)$$

where $I(\theta)$ is the *observed information*:

$$I(\theta) = -\frac{d^2}{d\theta^2} \log p(\theta|\mathbf{y})$$

More generally, we can use a multivariate Taylor series expansion to get

$$\theta|\mathbf{y} \sim_{\text{approx}} N\left(\widetilde{\theta}, I^{-1}(\widetilde{\theta}) \right)$$

5.3 Markov Chain Monte Carlo (MCMC)

Often posterior densities may be difficult or impossible to integrate explicitly, particularly when we have a complex model with many parameters. An alternative approach, which avoids the problems of integration, is to use simulation techniques where a random sample from the posterior distribution is used to estimate $\int g(\theta)p(\theta|y)d\theta$ for some function g. In this situation a large number of samples are drawn from $p(\theta|y)$, which can be used to estimate quantities of interest such as the posterior mean. These samples can be generated using a number of techniques including *rejection sampling, importance sampling and adaptive rejection sampling*.

For further details on these direct sampling methods see Ripley and Corporation (1987); Gilks, Richardson, and Spiegelhalter (1996).

Markov Chain Monte Carlo (MCMC) provides another approach to obtaining samples from the required posterior distributions and is based on the premise that it is possible to construct a Markov chain , $\theta^1, \theta^2, \theta^3...$, whose stationary distribution is the joint posterior $p(\theta|y)$ that is of interest. This again avoids the problems of integration and may be used in situations where direct simulation is not feasible. MCMC is used extensively to perform inference in Bayesian analyses. In this section we give a brief review of MCMC techniques and their use. For more comprehensive treatments see Gilks et al. (1996); Gelman et al. (2013); Gamerman and Lopes (2006).

The chain is initialised by a starting value θ^0, and run until it has converged to its target distribution. Convergence can be assessed using objective criteria such as those suggested by Gelman and Rubin (1992) or 'by eye'. The initial period of non-convergence is known as 'burn-in'. After this the Markov chain will produce samples from the posterior distribution. It is important that the Markov chain has covered the entire posterior distribution, but this may be difficult to check. However, this can be partially alleviated by using multiple Markov chains initialised at different points in the sample space, which reduces the likelihood that areas will be missed.

We now give a brief review of two methods for obtaining samples from the posterior distribution using MCMC: (i) Metropolis–Hastings algorithm (Metropolis, Rosenbluth, Rosenbluth, Teller, & Teller, 1953; Hastings, 1970) and (ii) the Gibbs sampler (Smith & Roberts, 1993).

5.3.1 Metropolis–Hastings algorithm

Starting with an initial value θ^0, a sequence, $\theta^{(k)}$, $k = 1, ..., K$ is generated. At each stage, a 'candidate' value, θ^*, is drawn from a *proposal distribution*, and the ratio of the (unnormalised) posterior probabilities under each of the values calculated. The current candidate is then either accepted or rejected where in the former case it becomes the next value in the chain. If it is rejected, another candidate value is drawn.

1. Arbitrarily draw a staring point θ^0 for the Markov chain ensuring that its posterior probability $p(\theta^0|y)$ is positive.

2. At each iteration k, for $k = 1, ..., K$, generate a candidate θ^* from a proposal distribution $p(\theta^*)$, that is based on the current value of the Markov chain. The candidate value is then accepted with probability r, given by

$$r = min\left\{ \frac{p(y|\theta^*)p(\theta^*)}{p(y|\theta^0)p(\theta^0)}, 1 \right\}$$

If the candidate value is accepted, then $\theta_{k+1} = \theta^*$, otherwise the chain does not move.

For illustration, consider an initial value θ^0, with 'candidate' value, θ^*, drawn from a normal (proposal) distribution with mean θ^0 and variance b^2. The ratio of the (unnormalised) posterior probabilities under each of the values is calculated. The minimum of that ratio and 1 is then taken. If a random sample is then taken from a Uniform distribution, $U \sim (0,1)$ and $U < r$, $\theta^1 = \theta^*$, the value sampled from the Normal distribution, otherwise let $\theta^1 = \theta^0$, the original value. When this step is repeated N times, by drawing each of the candidate variables from $N(\theta^1, b^2), N(\theta^2, b^2), \dots$ etc. then the resulting sample $\theta^0, \theta^1, \dots, \theta^N$ will approximate a random sample from the posterior distribution.

3. Repeat step 2 until the sequence of drawn samples reaches convergence.

Since the proposal density need only be proportional to the interested density rather than exactly equal to it, the Metropolis–Hastings algorithm presents a relatively straightforward and efficient way of simulating from any unspecified probability distribution.

MCMC can be complex to implement, and the results can be affected by the choice of starting values. Although the starting point should not impact the stationary distribution when chains are *mixing* well, i.e. covering all possibilities in the parameter space, it may need to be chosen carefully for slowly mixing chains which can often stick in a small area of the parameter space for a long time.

The choice of proposal distribution can have a large impact on the convergence and acceptance rate of the Metropolis–Hastings algorithm. Acceptance rates are significantly influenced by the variance of proposal distribution. A 'cautious' proposal distribution with relatively small variance will generate small steps, i.e. candidate values which are close to the current (accepted) value and therefore the acceptance rate will be high. Proposal distributions with large variance will result in much more variation in the candidate values and therefore may explore more of the parameter space in a shorter time but the probability of acceptance at each step is likely to be smaller and thus more iterations of the algorithm will be required in order to reach convergence. Typically, 20–30% of acceptance rate is considered reasonable (Gelman et al., 2013).

5.3.2 Gibbs sampling

The Gibbs sampler is a special case of the Metropolis–Hasting algorithm where the proposal distribution is the full conditional distribution of the parameter in question. This gives an acceptance probability of one which effectively removes the accept or reject stage; hence the movement of the Markov chain is ensured.

For a vector random variable θ with joint density $p(\theta = p(\theta_1, \theta_2, \dots, \theta_n)$ where the full conditional distribution $p(\theta_i | \theta_{-i})$ for each parameter that is known, then the Gibbs sampler algorithm works as follows:

1. Arbitrarily choose a starting point $\theta^{(0)} = (\theta_1^0, \ldots, \theta_n^0)$ with $f(\theta^0|y) > 0$

2. At iteration k, draw θ_i^k from the full conditional distribution $p(\theta_1|\theta_{-i}^{p-1})$

3. Then repeat step 2 until the sequence converges.

5.3.3 Block updating

In simple implementations, parameters are updated independently as it is often easier to sample from univariate distributions and they are more likely to be available in closed form. However this single updating method can have very poor convergence properties, poor mixing and can be computationally slow. Knorr-Held (1999) suggests that using block updating is potentially faster with better mixing properties.

Example 5.2. *Gibbs sampling with a Poisson-Gamma model*

In this example we show the full conditional distributions for a Poisson model where prior distribution on the rates is Gamma. If $\mathbf{y} = y_1, \ldots, y_N$ is a set of observed counts of a health outcome in a set of areas i then $y_i \sim Poi(\theta_i E_i)$ where E_i is the age–sex standardised expected number of health outcomes in area l as described in Chapter 2, Section 2.4.

If we assign a Gamma prior to the random effects θ_l, i.e. $\theta_l \sim Gamm(a,b)$ for $l = 1, \ldots N$ and independent exponential distributions to the hyperparameters a and b then we can find the full conditional distributions required for Gibbs sampling easily as the Gamma is conjugate to the Poisson and so they will be of standard form.

The joint posterior will be

$$p(a,b,\theta) \propto \prod_{i=1}^{N} \frac{\theta_l E_l)^{y_i}}{y_i!} \exp(-\theta_l E_i)$$
$$\times \prod_{i=1}^{N} \frac{b^a}{\Gamma(a)} \theta_i^{a-1} e^{-b\theta_i}$$
$$\times \lambda_a \exp(-\lambda_a)\lambda_b \exp(-\lambda_b) \tag{5.2}$$

and the full conditionals will be

$$p(\theta_i|\theta_{-i}, a, b, \mathbf{y}) \propto \theta_i^{y_i+a-1} \exp[-(E_i+b)\theta_i]$$
$$p(a|\theta, b, \mathbf{y}) \propto \frac{(b^N \prod_{i=1}^{N} \theta_i)^{a-1}}{\Gamma(a)^N}$$
$$p(b|\theta, a, \mathbf{y}) \propto b^{Na} \exp[-(\sum_{i=1}^{N} \theta_i + \lambda_b)b]$$

These can be used as the proposal distributions in Gibbs sampling.

5.4   Using samples for inference

If the Markov chain is ergodic, meaning that all of its states can be revisited with positive probability in a finite mean number of steps, then samples can be used to perform inference on the posterior distribution with assurance that the estimates will converge to the correct values as the length of the chain tends to infinity (Gamerman & Lopes, 2006). Starting at an initial value θ_0, after a period of time the Markov chain will converge. The period before convergence is known as the *burn–in* period. The length of that period will depend on the rate of convergence. Estimates of the rate of convergence are in general difficult to obtain and determining the length of burn-in required for a specific problem may not be feasible. The most commonly used method for determining burn-in is to produce a time series plot of the values over time; this will also highlight any obvious problems such as slow mixing, and running multiple chains from different starting points.

Samples can be used to summarise the posterior distribution in terms of modes, percentiles and probabilities based on the sample. In addition, we can compute quantities derived from the posterior such as a marginal mean that can be estimated by

$$E[\theta_i|\mathbf{y}] \approx \frac{1}{K-m} \sum_{k=m+1}^{K} \theta_i^{(k)} = \bar{\theta}_1$$

where N is the number of samples omitting m, the number within the burn-in period.

We can also transform between different parameterisations. For example, if the Markov chain gives samples from $p(\theta|\mathbf{y})$ but we are interested in $\phi|\mathbf{y}$, where $\theta = g(\phi)$ we can transform each value of the chain

$$\phi_k = g^{-1}(\theta_k)$$

to produce a Markov chain with samples (ϕ_k) from the distribution $p(\phi|\mathbf{y})$. These can then be used to estimate quantities of the posterior distribution of $\phi|\mathbf{y}$.

To determine how many samples we need to obtain adequate precision in the estimator $\bar{\theta}_1$ we need to consider the variance of $\bar{\theta}_1$. This is complicated since the samples produced by MCMC techniques will be dependent samples. If the chains display slow mixing, which may arise due to high dependence in the Gibbs sampler, or low acceptance rates when using the Metropolis–Hastings algorithm, we will require larger runs. In such cases reparameterisation may help (Gamerman & Lopes, 2006). While formal methods to estimate the variance have been proposed, a simple informal method is to run multiple chains of length K, and compare the estimates $\bar{\theta}_1$. If they do not agree adequately, then K should be increased.

5.5   WinBUGS

We now provide a short introduction to the BUGS language which can be used for specifying Bayesian models. WinBUGS is a powerful tool that allows the user to perform Markov chain Monte Carlo (MCMC) using Gibbs sampling. It currently has two forms; WinBUGS and OpenBUGS (an open source version) . Details of how to obtain them are included in the online resources. From here on, unless specified, we will refer to WinBUGS although all the material is equally applicable to OpenBUGS.

As described in Section 5.3, in many situations there is no exact form for the posterior distribution, and we must resort to analytical ways of finding summaries such as means, credible intervals etc. If we could draw a sample of values from the posterior distribution of interest, we could use properties of the sample, e.g. mean, variance, quantiles, to estimate the properties of the posterior distribution itself. So if we had a way of generating samples from the posterior, then we could get the posterior summaries we require. Markov Chain Monte Carlo is a way of obtaining such samples without having to write down a direct form for the posterior distribution. The basic idea is to start from any point and generate a sequences of values, each based on the last. These form a Markov chain. After a suitably long time, the chain converges to the posterior distribution and, from this point, on the sequence of values are all samples from the posterior.

WinBUGS requires the user to define their model but not to have to code the actual MCMC. Given a model, WinBUGS will derive the required full conditionals for the Gibbs sampling and offer visual representations of the chains and the facilities to summarise the resulting samples from the posterior distributions of the parameters of interest.

The input to WinBUGS is comprised of

- a model describing the prior and likelihood distributions, contained within model{};
- a list containing the observed data, contained within data{};
- a list containing starting values for all the unknown parameters contained within inits{}.

The output is

- samples of each of the specified parameters;
- summaries and plots of these samples;
- a list of initial containing starting values for all the unknown parameters.

The BUGS language looks very similar to R and many features, such as data structures, are handled in the same way. Notably, both can now be called from other software including R. The main addition is that of specifying prior distributions for all the unknown parameters. For example, if Y is distributed $\sim N(\mu, \sigma^2)$ this would be represented in WinBUGS as y $\sim$ dnorm(mu, tau), noting that WinBUGS expresses the variance in terms of precision, $\tau = 1/\sigma^2$ with a prior specified for μ and τ as follows:

$$\text{mu} \sim \text{dnorm(a, b), tau} \sim \text{gamma(c, d)}$$

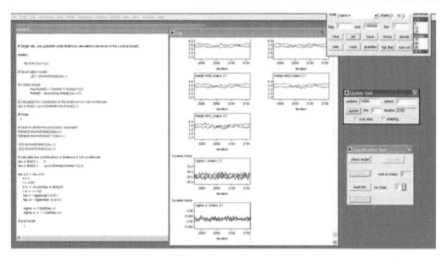

Figure 5.1: A screenshot showing a WinBUGS session. The model description can be seen together with dynamic trace plots which show the updating of the chains associated with the parameters being sampled.

It is then possible to obtain summaries of samples from the posterior distributions of the parameters, including means, standard deviations, medians and quantiles. Figure 5.1 shows an example of a WinBUGS session including a selection of dynamic trace plots which show the updating of the chains associated with the parameters being sampled. This screenshot shows the model presented in Chapter 10, Section 10.8 being fitted.

Further details of WinBUGS can be found in Lunn, Jackson, Best, Thomas, and Spiegelhalter (2012). In Chapters 8, 9 and 10 we show examples of its use in fitting complex Bayesian hierarchical for spatial, temporal and spatio–temporal models.

Example 5.3. *Fitting a Poisson regression model in WinBUGS*

In this example we consider the Poisson log–linear model seen in Chapter 2, Section 2.8;

$$\log \mu_l = \beta_0 + \beta_1 X_l + \beta_d X_l \qquad (5.3)$$

where β_1 represents the effect of exposure and β_d is the effect of the area-level covariate.

The WinBUGS code to fit this model is as follows:

```
model
{
    for (i in 1 : N) {
```

```
        Y[i]    ~ dpois(mu[i])
        log(mu[i]) <- E[i]+exp(beta0+ beta1*X1[i]  + betad*X2[i
            ]X2)
    }
# Priors
    beta0   ~ dnorm(0,0.001)
    beta1   ~ dnorm(0,0.001)
    betad   ~ dnorm(0,0.001)
# Functions of interest:
    base <- exp(beta0)
    RR <- exp(beta1)
}
list(N = 393,
    Y = c( 29 ,27   ,9 ,18 ,24, 29 ,14 , ..., 16, 26, 25), E =
        c(19.883, 13.525, 7.712, 16.99, 19.635, 11.227, ...,
        19.287,
        20.839), X1 = c(1.4, 2.3, 3.4, 2.3, 2, 2.9, ..., 2.2,
            2.8, 1.8), X2 = c(1.599, 8.123, 5.413, 6.857, 7.772,
            ..., 10.283, 6.04))
```

Initial estimates (for 2 chains)

```
list(beta0 = 0, beta1=0, betad=0)
list(beta0 = 1, beta1=1, betad=1)
```

The result is that samples will be drawn from the posterior distributions of the parameters of interest, β_0, β_1 and β_d and also of the functions of those parameters, the underlying mortality rate over all of the areas (the exponent of the intercept), $base = \exp(\beta_0)$, and the relative risk associated with differences in air pollution, the exponent of the estimated coefficient associated with X, $RR = \exp(\beta_1)$.

5.6   INLA

INLA uses a Laplace approximation to the posterior distribution of the parameters, θ, given measurements of the response, Y. For clarity of exposition, we drop the $Y^{()}, Z^{()}, \theta^{()}$ notation of Section 1.5 in what follows. Following the hierarchical setup introduced in Chapter 1, Section 1.5, when the process model is Gaussian we have a latent Gaussian model:

- Observation model; $y_i|z_i \sim p(y_i|z_i, \theta_1)$
- Process model; $z|\theta_2 \sim N(\mu, \Sigma_{\theta_2})$
- Parameter model; $\theta = (\theta_1, \theta_2) \sim p(\theta)$

Therefore $p(z, \theta|y) \propto p(\theta)p(z|\theta)\prod_{i=1}^{N} p(y_i|z_i, \theta)$. In the applications considered here, the response consists of measurements of an environmental hazard that are assumed to depend stochastically on a latent process, Z which will be indexed by spatial–temporal locations.

If the observation model is Gaussian, then the resulting joint distribution will be a Gaussian Markov Random Field (GMRF) (Rue & Held, 2005). For a GMRF, the precision matrix, $Q_\theta = \Sigma_\theta^{-1}$ will be sparse, allowing efficient computation.

The aim is to obtain posterior marginal quantities such as $p(\theta_i|\mathbf{y})$ and $p(z_i|\mathbf{y})$ where for example,

$$p(\theta_i|\mathbf{y}) = \int p(\theta|\mathbf{y})d\theta_{-i} \qquad (5.4)$$

$$\text{and } p(z_i|\mathbf{y}) = \int p(\theta|\mathbf{y})p(z_i|\theta,\mathbf{y})d\theta \qquad (5.5)$$

In order to achieve this, approximations are required for $\tilde{p}(\theta|\mathbf{y})$ and $\tilde{p}(z_i|\theta,\mathbf{y})$.

The Laplace approximation (see Section 9.13 for more detail) to the posterior $\tilde{p}(\theta|\mathbf{y})$ is given by

$$\tilde{p}(\theta|\mathbf{y}) =\propto \left.\frac{p(\mathbf{z},\theta,\mathbf{y})}{\tilde{p}_G(\mathbf{z}|\theta,\mathbf{y})}\right|_{z=z^*(\theta)}$$

where $\tilde{p}_G$ is a Gaussian approximation at the mode $z^*(\theta)$ of the conditional distribution of $\mathbf{z}$ given θ. Given such an approximation, numerical integration can be used to evaluate the required integral.

5.6.1   R–INLA

The R–INLA package provides a practical implementation of Integrated Nested Laplace Approximations (INLA). It can be used with hierarchical GMRF models of the form, for areas for example

$$y_i|z_l, \theta_1 \sim p(y_i|z_i, \theta_1), \quad l = 1,...,N \qquad (5.6)$$

The likelihood of the observed data, $p(y_i|z_i, \theta_1)$ is assumed to be conditionally independent given the latent parameters $\mathbf{z}$ and possibly additional parameters θ_1. The latent variable z_i enters the likelihood through a known link function, for example a log link in a Poisson log–linear model. A latent process, z_l is represented in the following form,

$$z_j = \beta_0 + \mathbf{x}'_j\beta + \sum_{k=0}^{N_f-1} f_k(c_{kj}) + v_l, \quad j = 0,...,J \qquad (5.7)$$

where $\mathbf{v}$ is a vector of unstructured random effects of length J with i.i.d Gaussian priors having precision τ_v: $v|\tau_z \sim N(0, \tau_z I)$. It is noted that I is a subset of J, i.e. not all latent parameters z are necessarily observed through the data y.

In Equation 5.7, β_0 is a intercept term and $\mathbf{x}'_i\beta$ is a linear effect of covariates. The terms $\{f_k(.)\}$ are non-linear, or smooth effects, of covariates c_k which may include continuous covariates, time trends and seasonal effects, two-dimensional surfaces, i.i.d. random intercepts and slopes and spatial random effects. These functions may be weighted by adding weights to Equation 5.7, e.g. $w_{kj}f_k(c_{kj})$, where $\mathbf{w}_k$ are known weights defined for each observed data point. An offset can be added to the linear predictor in Equation 5.7 if required, for example when using expected numbers of counts in a Poisson model as shown in Chapter 2, Example 2.8.

The key point here is that the set of parameters $(\beta_0, \beta, \mathbf{v})$ and the latent processes, f_k, will together have a Gaussian prior.

The class of models that can be expressed in this form and thus can be used with R–INLA is very large and includes, amongst others, the following:

- Dynamic linear models.
- Stochastic volatility models.
- Generalised linear (mixed) models.
- Generalised additive (mixed) models.
- Spline smoothing.
- Semi–parametric regression.
- Disease mapping.
- Log–Gaussian Cox–processes.
- Model–based geostatistics.
- Spatio–temporal models.
- Survival analysis.

The syntax of R–INLA

There are three main parts to fitting a model using R–INLA:

1. The data.
2. Defining the model formula.
3. The call to the INLA program.

The basic syntax of running models in R–INLA is very similar in appearance to that of glm and takes the general form formula, data, family but with the addition of the specification of the nature of the random effects, f(). For the latter component, common examples include f(i, model="iid") (independent), f(i, model="rw") (random walk of order one) and f(i, model="ar") (autoregressive of order p).

Example 5.4. *Fitting a Poisson regression model in R–INLA*

The Poisson log–linear model seen in Chapter 2, Example 2.9 can be extended to incorporate random effects (see Chapter 6, Section 6.3 for further details). An extension of the standard Poisson model shown in Equation 2.11 to include log–normal random effects in the linear predictor will allow for over-dispersion as described in Chapter 2, Section 2.8.2. Using log–normal random effects gives the model,

$$\log \mu_l = \beta_0 + \beta_{0i} + \beta_1 X_l + \beta_d X_l + \varepsilon_l \tag{5.8}$$

where β_l represents the effect of exposure, β_d is the effect of the area-level covariate and β_{0i} denotes the random effect for area i. The syntax of the R–INLA code to fit this model is very similar to that of a standard glm in R:

```
> formula = Y ~ X1+X2 + f(i, model="iid")
> model   = inla(formula, family="poisson", data=data)

Call:
"inla(formula = formula, family = "poisson", data = data)"

Time used:
  Pre-processing    Running inla Post-processing        Total
     0.278389          0.286911       0.125699         0.690999

Integration Strategy: Central Composite Design

Model contains 1 hyperparameters
The model contains 3 fixed effect (including a possible
    intercept)

Likelihood model: poisson

The model has 1 random effects:
1.'i' is a IID model
```

The result is similar to that obtained when using a standard glm although now there are additional details related to the random effects:

```
> summary(model)

Call:
"inla(formula = formula, family = "poisson", data = data)"

Time used:
  Pre-processing    Running inla Post-processing        Total
     0.2784            0.2869         0.1257           0.6910

Fixed effects:
            mean     sd 0.025quant 0.5quant 0.975quant
```

```
(Intercept) 2.4960 0.0713    2.3553   2.4962    2.6355
X1          0.1187 0.0310    0.0578   0.1186    0.1796
X2          0.0578 0.0074    0.0433   0.0578    0.0722

Random effects:
Name        Model
  i    IID model

Model hyperparameters:
                mean      sd 0.025quant 0.5quant 0.975quant
Precision for i 3.784 0.3548      3.131    3.769      4.525

Expected number of effective parameters(std dev):
   321.42(3.926)
Number of equivalent replicates : 1.223

Marginal Likelihood:  -1513.92
```

Future details on R–INLA, including the latent process models that can be accommodated, can be found on the R–INLA webpage: http://www.R-INLA.org. In Chapters 8, 9 and 10 we show how R–INLA can be used to provide a computationally efficient way of implementing complex Bayesian hierarchical models.

5.7 Summary

This chapter describes methods for implementing Bayesian models when their complexity means that simple, analytic solutions may not be available. The reader will have gained an understanding of the following topics:

- Analytical approximations to the posterior distribution.
- Using samples from a posterior distribution for inference and Monte Carlo integration.
- Methods for direct sampling such as importance and rejection sampling.
- Markov Chain Monte Carlo (MCMC) and methods for obtaining samples from the required posterior distribution including Metropolis–Hastings and Gibbs algorithms.
- Using WinBUGS to fit Bayesian models using Gibbs sampling.
- Integrated Nested Laplace Approximations (INLA) as a method for performing efficient Bayesian inference including the use of R–INLA to implement a wide variety of latent process models.

Exercises

Exercise 5.1. Suppose $X|\mu \sim N(\mu, \phi)$ and $Y|\mu, \delta \sim N(\mu + \delta, \phi)$ from Question 4.4 in Chapter 4, where ϕ is known, and consider the improper noninformative joint prior distribution, $p(\mu, \delta) \propto 1$.

(i) Describe how the Gibbs sampler may be used to sample from the posterior distribution, deriving all required conditional distributions.

(ii) Suppose we have samples from the Gibbs sampler $\left\{\mu^{(t)}, \delta^{(t)}\right\}, t = 0, \ldots N$, where N is large. Explain how these samples may be used to estimate the marginal mean, $E[\delta|x,y]$.

Exercise 5.2. The *independence sampler* is the Metropolis–Hastings algorithm with proposal distribution

$$q(\theta^*|\theta_{t-1}) = q(\theta^*)$$

(i) Describe the Metropolis–Hastings algorithm for this transition probability.

(ii) Show that if $q(\theta)$ is proportional to the required posterior distribution $p(\theta|\mathbf{y})$ then the Metropolis algorithm reduces to simple Monte Carlo sampling.

(iii) Suppose we take $q(\theta) = p(\theta)$, the prior distribution. Show that the acceptance probability depends only on the ratio of likelihoods. Under what circumstances will using the prior distribution as the proposal distribution be a good choice?

Exercise 5.3. Derive the full conditional distributions for the Poisson–Gamma model shown in Example 5.2.

Exercise 5.4. The exercise involves performing Gibbs sampling where the full conditionals are from a standard distribution.

(i) Based on the full conditional distributions for the Poisson–Gamma model shown in Example 5.2 and your own derivations from Exercise 5.3 write a Gibbs sampler in R to perform inference using this model.

(ii) Test your code using simulated data and then use it to fit the model to the data on COPD hospital admissions which are included in the online resources and are the basis of Figure 1.1.

(iii) Fit the same model in WinBUGS and compare the results with those obtained when using your Gibbs sampler. Are the results the same?

Exercise 5.5. This exercise involves performing MCMC where the full conditionals are from a standard distribution. Considering an extension to the Poisson–lognormal model shown in Example 5.4 with log–normal random effects assigned to each of the areas, e.g.

$$\begin{aligned} Y_l &\sim Poisson(\mu_t) for\ l = 1, \ldots, N_L \\ \beta_0 &\sim N(0, \sigma_{\beta_0}^2) \end{aligned}$$

In this case the Gaussian prior is not conjugate to the Poisson data, which results in a non-standard full conditional distribution. In such cases a Metropolis–Hastings algorithm can be used. This model can be fit in WinBUGS which will automatically choose a suitable sampling scheme.

(i) Fit the Poisson model shown in Example 5.4 (without random effects) in WinBUGS to the black smoke data which are included in the online resources.

(ii) Fit the model with the area-level random effects shown above and compare the results to those you obtained in part (i).

(iii) Produce a histogram of the area-level random effects. What does this tell you about the heterogeneity of baseline risks in this example?

Exercise 5.6. Repeat the analysis in Exercise 5.5 using R–INLA. Compare your results to those obtained using WinBUGS.

Chapter 6

Strategies for modelling

6.1 Overview

The most common reason for performing a regression analysis in epidemiology is to obtain estimates of the coefficients associated with the variables of interest, for example, the effect of an increase in particulate matter air pollution on the risk of respiratory death. In order to perform such analyses there will be a need for accurate estimates of exposures on which to base the associations with health. Often there will be locations and periods of time for which such data will not be available. This may be due to a fault in monitoring equipment or may be due to design. In many epidemiological studies, the locations and times of exposure measurements and health assessments do not match, in part because the health and exposure data will have arisen from completely different data sources and not as the result of a carefully designed study. This is termed the 'change of support problem' (Gelfand, Zhu, & Carlin, 2001). In such cases, a direct comparison of the exposure and health outcome is often not possible without an underlying model to align the two in the spatial and temporal domains. In this chapter we consider how predictions from exposure models, which are covered in detail in Chapters 9, 10 and 11 can be used in models for estimating health risks.

In developing a model intended for estimating parameters, e.g. risks to health, the choice of which variables are included or eliminated is of great importance. If several of the variables are highly correlated, which is likely to be the case when using several different pollutants and particularly so when dealing with different lagged values of variables, including a variable which is highly correlated with the one of interest, may dramatically alter the estimate of its effect or possibly even lead to the variable of interest being excluded from the model. In terms of interpretation and drawing conclusions, there is an obvious desire to have a relatively simple model that nevertheless includes all the important variables, but there are also important theoretical reasons for eliminating irrelevant variables from the model. The effect that this can have on selection procedures, both for explanatory variables but also for models themselves is explored. In this chapter we consider the effects of covariates and model selection including those within the Bayesian setting in which the choice of the models themselves is part of the overall inferential process.

6.2   Contrasts

Variability in a process, Y, may be partially accounted for by a set of covariates, X. Covariates may be

- spatial contrasts - constant over time but vary over space;
- temporal contrasts - constant over space but vary over time;
- spatio–temporal contrasts - vary over both space and time.

The variability that creates uncertainty in a process is also responsible for creating the 'contrasts', which lie at the heart of statistical inference. In order to best determine relations between a response and a predictor it is important to ensure strong contrasts in the levels of the predictors. In spatial sampling, this might involve putting 1/2 the observations in a (quasi-) control region to measure background levels where exposures are small with the other 1/2 being located where exposures are large.

Example 6.1. *Contrasts in a linear model*

Suppose we have a covariate, $X_s \in [a,b]$, which varies over space and

$$Y_s = \alpha + \beta X_s + v_s, s = 1,2,\ldots,N_S$$

with the X's being fixed and $v_s \sim$ i.i.d.$N(0,\sigma_v^2)$. How should the x's be chosen to best estimate β?

The solution is simple in this case. Given observations y_s of Y_s, we have the standard formula $\hat{\beta} = \sum_s^{N_S} y_s[x_s - \bar{x}]/\sum_s[x_s - \bar{x}]^2$.

$$Var(\hat{\beta}) = \frac{\sigma^2}{\sum_i[x_i - \bar{x}]^2} \qquad (6.1)$$

Therefore putting $N/2$ of the observation's x's at a and the other $N/2$ at b minimizes the standard error of estimation.

Of course this solution would not be robust against model misspecification. If, for example, the model were actually parabolic in shape this design would fail.

Example 6.2. *Spatial network design and organisms on the sea floor*

The principle of obtaining strong contrasts underlies the spatial sampling design proposed to determine changes in the concentrations of benthic organisms in the sea floor off the north slopes of Alaska before and after the start up of exploratory drilling for petroleum (Schumacher & Zidek, 1993). There was concern about the effects of toxic trace elements in the drilling mud that was to be discharged into the sea when drilling commenced. That concern was more about human welfare rather than health; the benthic organisms lay at the bottom of a food chain, the top end of which was the bowhead whale, a staple food of the Innuit People who lived there. There is an extensive literature on the optimal design of experiments for model fitting (Müller, 2007; Zidek & Zimmerman, 2010) which is discussed in Chapter 13.

6.3   Hierarchical models

Many epidemiological studies involve data that inherently have a hierarchical structure, for example when assessing the differences in mortality rates across hospitals for a particular procedure, data may be measured both at the hospital 'level' (e.g size, speciality) and on individual patients within the hospitals (e.g. age, gender). An introduction to the need for hierarchical models and the implementation of the linear case are given in Goldstein (1987), Sullivan, Dukes, and Losina (1999) and Burton et al. (1998). In this section, a simple regression example is presented leading to a discussion of *Generalised Linear Mixed Models* (GLMM) (Breslow & Clayton, 1993), which are an extension of GLMs that allows the inclusion of random effects. Hierarchical models are also known as multi-level models and may be fitted using likelihood methods, but are naturally viewed from a Bayesian perspective.

Hierarchical models offer a convenient way of handling the different levels of correlation that may be present between outcomes in longitudinal studies where repeated measurements, j, may be made on a set of subjects, i. They provide a framework for incorporating the correlation that may be present between measurements made at different times but on the same individual. A comprehensive introduction to the subject of analysing longitudinal data of this type can be found in Diggle et al. (2002). For example, the first, or top level of the model might relate the outcome on individual i to a set of level 1 covariates

$$Y_{ij} = \beta_0 + \beta X_{ij} + \varepsilon_{ij} \tag{6.2}$$

In a naive pooled analysis, the ε_{ij}'s would be considered *i.i.d* $N(0, \sigma^2)$. However, as each of the i sets of j readings made on each individual are likely to be correlated outcomes, this assumption is unlikely to be tenable. One approach would be to fit a different intercept term for each individual (a fixed effect) but this requires the ratio of J to I to be suitably large.

Alternatively, a random effect can be fitted, referring to the fact that the regression coefficients can vary from individual to individual. This alternative to fitting a separate intercept for each individual, i, assumes that there exists an overall intercept (β_0) for the whole population and that a separate intercept β_{0i} for each individual comes from a distribution with expectation β_0 and a suitable variance. For example $\beta_{0i} \sim N(\beta_0, \sigma_{\beta_0}^2)$ or $u_i = \beta_{0i} - \beta_0 \sim N(0, \sigma_{\beta_0}^2)$. Model (6.2) can therefore be expressed as

$$Y_{ij} = \alpha_0 + u_i + \beta X_{ij} + \varepsilon_{ij} \tag{6.3}$$

with $E[\varepsilon_{ij}] = 0, Var(\varepsilon) = \sigma_e^2, Cov(\varepsilon_{ij}, \varepsilon_{ik}) = 0$ and $E[u_i] = 0, Var(u_i) = \sigma_u^2$, $Cov(u_i, u_j) = 0$. The variance of Y_{ij} is thus $\sigma_e^2 + \sigma_u^2$, where the variance between Y_{ij} within a single individual is σ_u^2. The intra-subject correlation coefficient is therefore $\sigma_u^2/(\sigma_u^2 + \sigma_e^2)$. Given estimates of σ_u^2 and σ_e^2, the *exchangeable* correlation structure means that every observation within an individual is equally correlated with every other observation in that individual, i.e. $\rho_{jj} = 1, \rho_{jk} = \rho$ $j \neq k$, the intra-class correlation coefficient.

In non-hierarchical models, tests of statistical significance are associated with the addition (or deletion) of one or more parameters and the likelihood ratio test (see Chapter 2, Section 2.5.3). However when drawing inference about random parameters, the usual null hypothesis, that a single variance is zero, will violate one of the standard regularity conditions needed to ensure that the asymptotic distribution of the test statistic is χ^2. In such cases, a standard χ^2 test will underestimate the significance of an observed departure from the null hypothesis (Laird & Ware, 1982).

6.3.1   Cluster effects

Often in environmental epidemiology, samples are collected using a multi-stage sampling design where the first stage is the selection of a random subset of large areas and the second stage the selection of smaller areas within them. The sample will therefore contain cluster effects; items from the same areas (clusters) will be more alike than items from different clusters. There is then the possibility of inducing correlation across space and hence reducing the effective sample size. In the extreme case where all the items in a cluster are identical, the information will be the same as the information obtained from a single item. If this effect is ignored and the assumption is made that the items are drawn independently, then inferences will be flawed and the uncertainty associated with estimates will be under-represented resulting in the error bands on estimates being too narrow. An expected $\alpha = 0.05$ level test of significance will in reality be much larger.

Example 6.3. *The effect of clusters on modelling*

Let Y_s be the measured response with realisations indexed over space, $s = 1, \ldots, N_S$. Interest lies in the association between a variable of interest, X_s, i.e. an exposure and Y_s. The data are collected in spatial clusters $i, 1, \ldots I$ with random cluster effects β_{0i}. The following model relates the Y to X:

$$\begin{aligned} Y_{si} &= \beta_0 + \beta_{0i} + \beta X_{si} + \varepsilon_{si} & (6.4) \\ &= \beta_0 + \beta x_{si} + \varepsilon_{si}^* & (6.5) \end{aligned}$$

where $\varepsilon_{si} \sim$ i.i.d.$N(0, \sigma^2)$ and $\beta_{0i} \sim$ i.i.d. $N(0, \sigma_{\beta_0}^2)$.

It can be seen in Equation 6.5 that $\varepsilon_{si}^* \sim$ i.i.d. $N(0, \sigma_{\beta_0}^2 + \sigma^2)$. The random cluster effects are now hidden in the so-called error term ε_{si}^* with the result that the residual variance is inflated and the effect of X will be masked. We can see that if the cluster-effect exceeds the effect of X, then the relative lack of variation in the X_s would mean a lack of power to detect a relationship between X and Y. Equation 6.5 shows the naive model that ignores the cluster effect which will be 'hidden' in ε_{si}^*. This will inflate the residual sum of squares and the significance level of the test will be increased.

Cluster effects can also induce temporal autocorrelation and if such an effect were strong there would be little incremental information in the data obtained over time. In such a case, little information would be lost by taking averages of the process and the measurements generated over time, $\bar{Y}_{s.} = \beta_{0s.}^* + \bar{\varepsilon}_{s.}^*$, where Y_{st} and ε_{st} are now indexed by space and time and β_{0s}^* are the associated cluster (random) effects. In this situation it will only be the spatial variation in $\bar{X}$ that provides the contrasts in X and this would have to be substantial for a significant effect to be seen.

Problems with this approach arise when β_{0i} and X_s are aligned and correlated over space, causing confounding. If their variations were roughly equal then the resulting co-linearity would lead to the information in the sample being split between the two effects, leading to the possibility that neither would prove to be significant, meaning important risk factors may be masked. This may occur even if β were large.

Overall, the confounding of the effects of the variable of interest by cluster effects generally leads to a set of undesirable effects. These concerns may suggest that an alternative to averaging over time may be preferable, for example based on similar principles to the paired t–test, we may use $Y_s' = Y_{s'} - Y_s = \beta(x_{s'} - x_s) + (\varepsilon_{s'} - \varepsilon_s)$, which eliminates cluster effects and thus autocorrelation.

6.4 Generalised linear mixed models

Generalised linear mixed models (GLMMs) are essentially GLMs (see Chapter 2, Section 2.5) that include one or more random effects. Conditional on random effects, b_i, the Y_i are assumed independent, with expectation, $E(Y_i|b) = \mu_i = g^{-1}(\gamma_i) = g^{-1}(X_i\beta + z_ib_i)$ and variance, $VAR(Y_i|b) = \sigma^2 = a(\phi)v(\mu_i)$. The likelihood for N subjects, each with N_j measurements will thus be

$$L(\beta, b, \sigma^2|Y) \propto \prod_{i=1}^{N}\prod_{j=1}^{n_j} p(Y_{ij}|\beta, b_i, \sigma^2) \tag{6.6}$$

If the random effects are assumed to be normally distributed, $b \sim N(0, D)$, then within subject correlation can be taken into account whilst allowing each subject to have a unique covariance structure. The correlation structure can be defined using the co-variance, D (rather than simply using $\sigma^2 I$). In a Bayesian setting, where the covariance matrix will be assigned a prior distribution, D^{-1} is often assumed to have a Wishart prior, this being conjugate to the multivariate normal distribution.

Frequentist approach

The parameters of the distribution of the random effect are known as *hyperparameters*, θ and must also be estimated from the data. In the case of linear regression, the parameters, β, ϕ and θ can be estimated by maximum likelihood using iterative generalised least squares (Goldstein, 1987). In the more general case, the usual strategy is to attempt to estimate the parameters by using the integrated likelihood, details of which can be found in Laird and Ware (1982) and Diggle et al. (2002),

$$p(Y|\beta, \phi, \theta) \propto \int p(Y|\beta, b, \phi) p(b|\theta) db \qquad (6.7)$$

In general, $\hat{b} = E(b|\hat{\theta}, \hat{\beta}, \hat{\phi})$, using the MLEs obtained from Equation 6.7. Whilst this approach is feasible for Normal models, the computations are intractable in other cases, although they can be approximated. Breslow and Clayton (1993) review approaches to approximate these calculations, including penalised quasi-likelihood which is estimated using a variation of the iteratively reweighted least squares calculations used for GLMs (see Chapter 2, Section 2.5).

The problem with this approach however, is that the underlying uncertainty in $\hat{\beta}, \hat{\phi}$ and $\hat{\theta}$ is not acknowledged.

Bayesian approach

In assigning the prior distributions of the hyperparameters, θ, the Bayesian approach provides a natural framework for dealing with hierarchical models, incorporating the uncertainty in the estimates of the parameters. For illustration, the set of level one (non-hyper) parameters are denoted simply by β (omitting σ^2/ϕ for clarity), in which case the appropriate Bayesian posterior distribution is the vector (θ, β), with joint prior distribution, $p(\theta, \beta) = p(\theta) p(\beta|\theta)$. The joint posterior distribution is then $p(\theta, \beta|y) \propto p(y|\beta) p(\theta, \beta)$. The hyperparameters can be estimated by obtaining the marginal posterior distribution , $p(\theta|y)$, which involves integrating the joint posterior distribution, $p(\theta|y) = \int p(\beta, \theta|y) d\beta$.

Example 6.4. *A Bayesian hierarchical model for air pollution and health*

Dominici, Samet, and Zeger (2000a) used a Bayesian hierarchical model to combine estimates of the effect of PM_{10} and SO_2 on daily mortality from the twenty largest cities in the U.S. A GAM (see Chapter 2, Section 2.6) was fitted separately for each city, modelling deaths as a Poisson regression with smoothed functions of time and lagged temperature. The relative risks from each city, β^c (representing short-term changes in air pollution), were then assumed to come from an underlying Normal distribution $\beta^c | \alpha, \Sigma \sim N(z^c \alpha, \Sigma)$, where z^c are city-specific covariates. The trend and weather terms are treated as nuisance parameters, η^c and are assumed to be independent across cities. A fully Bayesian analysis of this would involve assigning probability distributions (and associated hyperparameters) to all the parameters at the second level (the first level being the regression equation itself), whether of interest or nuisance. Integrating the joint posterior distribution of all the parameters with respect to the nuisance parameters, would then give the marginal posterior distributions of the β^c's, i.e. $p(\beta^c | \text{data, hyperparameters}) = \int p(\beta^c, \eta^c | \text{data, hyperparameters}) d\eta^c$. This is difficult in practice as the regression was performed using GAMs, and although the pollutant was used as a linear term, the other explanatory variables were modelled non-parametrically. Instead, after the first stage, the regression parameters are taken as given and essentially transferred to a separate analysis to describe the variability between cities. In this analysis, the likelihood estimates of β^c and η^c were assumed to be multivariate normal with means and covariance matrices equal to the corresponding MLEs. The marginal likelihood of β^c then has a Normal distribution. An alternative approach would be to use MCMC to sample from the posterior. This was done for a sample of the cities, but it was said to be computationally prohibitive for the whole analysis. The covariance matrix, Σ, was assumed to have a Wishart prior, for which the off-diagonals (the correlation between cities) took into account the distance between cities by using an inverse exponential distance function, although it seems unlikely that there would be any meaningful spatial structure at this resolution, as the cities were spread across the United States, with very large distances between them, and so the idea that correlations between the cities might be based on their geographical proximity seems unlikely.

6.5   Linking exposure and health models

In order to perform health analysis there will be a need for accurate estimates of exposures during periods and in locations where there is missing data, either by design, for example where a monitor is not located or in operation, or due to shorter periods where measurements are not available. In addition, in many epidemiological studies, the locations and times of exposure measurements and health assessments do not match, in part because the health and exposure data will have arisen from completely different data sources and not as the result of a carefully designed study; termed the 'change of support problem' by Gelfand et al. (2001). Hence a direct comparison of the exposure and health outcome is often not possible without an underlying model to align the two in the spatial and temporal domains (Gryparis, Paciorek, Zeka, Schwartz, & Coull, 2009; Peng & Bell, 2010).

A few studies have used spatio–temporal modelling within such health studies, largely due to the health data being available at a lower geographical temporal resolution than the exposure data (Zidek, White, Sun, Burnett, & Le, 1998; Zhu, Carlin, & Gelfand, 2003; Fuentes, Song, Ghosh, Holland, & Davis, 2006; Lee & Shaddick, 2010) meaning pollution concentrations were not available in each spatial unit. These studies are ecological in nature, being based on spatially aggregated health and exposure data modelled at the same resolution. As such, there is the potential for ecological bias; assuming that associations observed at the level of the area hold for the individuals within the areas can lead to the so-called ecological fallacy. Ecological bias can manifest itself in a variety of ways and in this case bias in the resulting health risks may occur due to the aggregation of a non-linear model. For more details of the issues related to ecological studies and ecological bias see Chapter 12.

More commonly, simple methods for handling missing values are used including simply discarding them from the analysis or replacing them by a specific single value, for example the overall mean. By discarding missing values, we may lose useful information and may in fact introduce bias. By replacing missing values by a single value, for example the posterior mean from an exposure model, important features of the data and the intrinsic variability in using a summary value may be ignored.

Environmental exposure data are generally obtained from N_S fixed site monitors, S located within the spatial domain, $\mathscr{S}$. The set of monitoring sites are collectively denoted by $S = \{s_1, \ldots, s_{N_S}\}$, where $s_l = (a_l, b_l) \in \mathscr{R}^2$. However, health data are commonly available only at aggregated level for administrative areas, $A_l, l = 1, \ldots, N_L$ and therefore a suitable summary of the concentrations in an area for a particular time period is required. Using the notation described in Chapter 1, Section 1.5, the true values are denoted by $Z^{(2)}$ where the $^{(2)}$ indicates that they are measurements of exposures rather than $^{(1)}$ which would indicate health outcomes. The true mean exposure in a health area, A_l is given by

$$\bar{Z}_l^{(2)} = \int_{s \in A_l} M_s Z_s^{(2)} ds \tag{6.8}$$

where $Z_s^{(2)}$ is true level of exposure at all possible locations s in A_l and M_s is the population density such that $\int_{s \in A_l} M_s ds = 1$. However the information required to compute the integral will be unavailable. Therefore there is a need to approximate this, with the simplest and most commonly used approach being to take the average of the observed measurements from actual monitoring sites located within the health area,

$$\bar{y}_l^{(2)} = \frac{1}{N_{A_l}} \sum_{s \in A_l} y_s^{(2)} \tag{6.9}$$

where N_{A_l} is the number of monitoring sites located within area A_l.

There may be situations where the assumption that the average of measured values will be a suitable representation of the (average of the) true concentrations may not be tenable, i.e. $\bar{y}_l^{(2)} \neq \bar{Z}_l^{(2)}$. If this is the case, then bias may potentially occur in the estimation of the effects of pollution with the magnitude being largely dependent on two characteristics of the underlying pollution surface:

- **Spatial variation**

  If the underlying surface exhibits substantial spatial variation, the measurements at the monitor locations may not be a representative sample of the pollution concentrations throughout $\mathscr{S}$. This is because the number of monitor locations is likely to be small, unequally spaced throughout the study region and may be located for specific reasons (for example at a well-known pollution hot spot), meaning that averaging the values at these sites may not produce a good estimate of Z.

- **Measurement error**

  The ambient monitors are known to measure with error (DETR, 1998), meaning that $Y^{(2)}$ may be a biased estimate of $Z^{(2)}$. Details of the effects of measurement error can be seen in Chapter 12, Section 7.3.

If both these factors are negligible then the pollution surface across $\mathscr{R}$ will be relatively flat and the observed monitoring data is likely to provide an adequate representation of the pollution surface meaning that the simple average $Y^{(2)}$ will be an unbiased estimate of the true (unobserved) average over Z. However if that is not the case then this naive approach can induce bias and underestimate uncertainty (Director & Bornn, 2015).

6.5.1 Two-stage approaches

Much of the material in this book advocates a Bayesian approach to modelling environmental exposures in space and time. An introduction to the Bayesian approach is given in Chapter 4 and methods of implementation in Chapter 5. In a fully Bayesian framework estimation of health and exposure models, including prediction at locations where data is not available, is performed simultaneously. The uncertainty in estimating the coefficients of the exposure model is therefore acknowledged and 'fed through' the model to the predictions and further to the estimation of the coefficients in the health model. Often the exposure models are fit separately from the health model, removing the dependence between $Y^{(2)}$ and $Y^{(1)}$, in order to ease the computational burden in running a combined model, an approach that has also been adopted in Carlin, Xia, Devine, Tolbert, and Mulholland (1999) and Zhu et al. (2003).

There are likely to be computational considerations associated with jointly fitting the health and exposure models, especially if the latter uses large amounts of data over space and time. When the exposure model is complicated or when one is interested in running multiple candidate epidemiological models with different sets of covariates either for a single outcome or multiple outcomes, a single model is not going to provide an efficient method of investigation.

Often the exposure models are fit separately from the health model, removing the dependence of $Z^{(2)}$ on $Y^{(1)}$. The joint model is therefore decomposed into separate health and exposure components. The exposure component is of the form:

$$[Z^{(2)}|\theta^{(2)},Y^{(2)}] \propto [Y^{(2)}|Z^{(2)},\theta^{(2)}][\theta^{(2)}]$$

and the health component of the form

$$[\theta^{(1)}|Z^{(2)},Y^{(1)}] \propto [Y^{(1)}|Z^{(2)},\theta^{(1)}][\theta^{(1)}]$$

noting that the first-term exposure model is different from $[Z^{(2)}|\theta^{(2)},Y^{(2)},Y^{(1)}]$ which would be used in a fully Bayesian analysis. This is often done in order to ease the computational burden in running a combined model, and that has been adopted in a number of cases, for example, Carlin et al. (1999), Zhu et al. (2003), Lee and Shaddick (2010), Chang, Peng, and Dominici (2011) and Peng and Bell (2010). This *two-stage approach* has the advantage that the exposure model, which is likely to be the most computationally demanding, does not have to be refit when running multiple health effect analyses. Two stage approaches separate the exposure and health components whilst still allowing uncertainty from the exposure modelling to be incorporated into the health model (Chang et al., 2011; Peng & Bell, 2010; Lee & Shaddick, 2010).

There are other reasons why fitting a joint model may be unappealing; it is not intended that the health counts should inform the estimation of the exposures which should be based on data from the monitored concentrations. It is possible to 'cut' feedback between the stages within MCMC, for example in WinBUGS (Lunn et al., 2000), however the result is that the posteriors may not be proper probability distributions (Plummer, 2014).

6.5.2 Multiple imputation

One approach to performing a two-stage analysis is to use multiple imputation (Little & Rubin, 2014). This allows the uncertainty in predictions to be represented by using a set of plausible values for the exposures, which comprise samples from the posterior distributions of the predictions at the required locations in space and time. Taking D multiple (joint) samples from the posteriors results in D multiple datasets that are repeatedly used in the health model, $(Y^{(1)}|Z^{(1)})$.

Repeatedly running the health model results in an estimate of the log relative risk, β_1, and associated standard error for each dataset. These are then combined to give an overall estimate of relative risk together with a combined standard error that can be used to calculate confidence intervals (Little & Rubin, 2014). Assume β_{1d} is the estimate obtained from data set $d = 1, 2, ..., D$ and $\sigma_{\beta d}$ is the standard error associated with β_d. The overall estimate is the average of the individual estimates,

$$\bar{\beta} = \frac{1}{D} \sum_{d=1}^{D} \beta_d \tag{6.10}$$

The overall estimate of the standard error will be a function of a combination of within-imputation variance and between-imputation variance. The first of these is given as

$$\sigma_{w\beta}^2 = \frac{1}{D} \sum_{d=1}^{D} \sigma_{\beta d}^2$$

and the between-imputation variance by

$$\sigma_{b\beta}^2 = \frac{1}{D-1} \sum_{d=1}^{D} (\beta_d - \bar{\beta})^2$$

The total variance is therefore

$$\sigma^2 = \sigma_{w\beta}^2 + (1 + \frac{1}{D})\sigma_{b\beta}^2$$

Confidence intervals are obtained using quantiles of the t-distribution with degrees of freedom

$$df = (D-1)\left(1 + \frac{D\sigma_{w\beta}^2}{(D+1)\sigma_{b\beta}^2}\right)^2$$

This requires the ability to draw joint samples from the posterior distributions of the predictions from the exposure model which is straightforward when using MCMC. It is also possible in the R–INLA package using the function inla.posterior.sample. In computing the approximation to the required distributions, $\tilde{p}(\theta|\mathbf{y})$ $\tilde{p}(z_i|\theta,\mathbf{y})$, R–INLA uses numerical integration based on interpolation between a number of chosen 'integration points' (Rue et al., 2009). Taking $\tilde{p}(z_i|\theta,\mathbf{y})$ as an example, the integration points are selected from a set of candidate points on a grid. After exploring $\log(\tilde{p}(z_i|\theta,\mathbf{y}))$ to find the mode, a point is selected if the difference between $\log(\tilde{p}(z_i|\theta,\mathbf{y}))$ evaluated at that point and the value evaluated at the mode is greater than a prespecified constant. Apart from the integration based on this procedure for finding approximations to the marginal distributions as described in Chapter 5, Section 5.6, the information stored about the distribution at these integration points can be kept. This allows the function inla.posterior.sample to be used after the main INLA run. Joint samples from the posteriors can be obtained by sampling from Gaussian approximations at the integration points for all of the parameters, including predictions from the exposure model. A combined analysis of these datasets is then performed. This results in valid statistical inferences that properly reflect the uncertainty due to missing values.

6.6   Model selection and comparison

In constructing a regression model, decisions will have to be made on which variables should be included, and what form they should take, i.e. should they be transformed or categorised. The choice of variables and the methods used may, in part, be determined by the reasoning behind the regression analysis, whether for estimation or prediction.

6.6.1   Effect of selection on properties of estimators

This section concentrates on the important subject of subset selection, the choice of which variables to include in the regression equation, and the effect this has on the inferential process.

In choosing which models should be considered, it is important that they have some plausibility outside the realms of statistical significance. Careful consideration of prior information is advocated by Rothman and Greenland (1998), although it is acknowledged that in most cases the prior knowledge available is likely to be too limited to provide much guidance in model selection. Chapter 4 describes how, in a Bayesian setting, if such information were available, it could be directly incorporated into the analysis. They suggest the selection procedure should start with a *credible model*, which would be considered 'compatible with available information' and give the example of modelling cancer rates, where such a model would include age and sex. Variables are then added to the starting model, their importance and thus whether they are retained, being assessed, based on their contribution to the fit of the model to the data. There are many techniques for automatically selecting variables to be included in a model, and a selection of these are described in Section 6.6.2, but the danger in using these so-called 'black box' routines is that a known confounder may be excluded from the model, possibly in favour of a variable with no biological plausibility. In addition to a credible starting model, the final model should also not conflict with prior information. The choice of which other variables are included can have a significant effect on the interpretation of the effect of the variables of interest, although if the model is to be used only for prediction there is more tolerance to the choice of variables, a point emphasised by Clayton and Hills (1993).

In this section, the effect of variable selection on the results of the regression equation are explored. It should be noted that the theoretical effects described hold for cases when the selection is not based on information from the current data, which is often not the case. In order to completely eliminate the effect of bias caused by the selection of variables, completely independent data sets would be required for (i) the selection of the model and for (ii) the estimation of the regression coefficients. However, this is unlikely to be an efficient approach in practice as less information will be available for the actual estimation of the parameters.

For illustration, the effects of variable selection are demonstrated using linear regression, although the concepts generalise to other forms which may be better suited to an epidemiological investigation, such as Poisson regression.

Given a response variable, Y, with a single possible explanatory variable, X, to which a linear regression of the form $E(Y|X) = \alpha + \beta X$ is fitted, then an unbiased estimate, $\hat{\beta}$ of β can be found and then tested to see whether it is significantly different from zero. If it is then β is included in the equation, if not then it is dropped. Using this procedure means the estimate $\hat{\beta}$ will no longer be unbiased for β, as it depends on decisions made based on the data. If β is included in the model, rather than just fitting the intercept term, then the expectation of $\hat{\beta}$ should be written

$$E(\hat{\beta}|\hat{\beta} \text{ is significantly different from 0})$$

which will not be equal to β. Chatfield (1995) describes a *pre-test estimator* which also considers *not* fitting the line, when $\beta = 0$, in order to give the unconditional estimate of $\hat{\beta}_{PT}$

$$\hat{\beta}_{PT} = \begin{cases} \hat{\beta}, & \text{if } \hat{\beta} \text{ is significant,} \\ 0, & \text{otherwise,} \end{cases}$$

from which it is obvious that $E(\hat{\beta}_{PT})$ is not generally equal to the unconditional estimator, $E(\hat{\beta})$. The effect of variable selection is investigated more fully later in this section, but briefly, Chatfield advises that (i) least squares theory does not apply when the same data are used to formulate and fit a model, and (ii) the analyst must always be clear exactly what any inference is conditioned on.

In order to investigate further the effects of variable selection on the estimates of the parameters and prediction, consider a linear model with n observations, each with a response variable Y and a set of p possible covariates $X_1, ..., X_p$. The full model, but not necessarily the 'true' underlying model, is obtained by fitting all the available variables,

$$Y_l = \beta_0 + \sum_{j=1}^{p} \beta_j X_{lj} + \varepsilon_l \qquad (6.11)$$

Here, the residual terms, ε_l, are assumed independently and identically distributed, and in the case of Normal linear regression are such that $\varepsilon_l \sim N(0, \sigma^2)$.

The full model is expressed in matrix notation as

$$\mathbf{Y} = \mathbf{X}\beta + \varepsilon \qquad (6.12)$$

where $\mathbf{Y}$ is the vector (of length n) of observed responses, $\mathbf{X}$ the full design matrix dimension $n \times (p+1)$, containing all of the p variables for which information is available together with a column representing the intercept term (which is assumed to be included in all of the models), β is the $p+1$ vector of unknown regression coefficients and ε the n vector representing the random component.

With p potential covariates, by considering all the possible models that comprise combinations of each of them being included/excluded there will be $m = 2^p$ possible models, M_k $(k = 1,...,m)$. Assume that model, M_k, contains r_k regressors, so that $p + 1 - r_k$ have been omitted. For model, M_k, let $\mathbf{X}_k$ and β^k represent the design matrix and parameter vector respectively and $\mathbf{X}'_k$ and $\beta^{k'}$ the complement, so that $\mathbf{X}_k = (\mathbf{X}_k, \mathbf{X}'_k)$ and $\beta^T = \{(\beta^k)', (\beta^{k'})'\}$.

A natural ordering is used, so that $k = 1$ represents the null model containing just the intercept term with design matrix $\mathbf{X}_1 = \mathbf{1}_n$, an n vector of one's. The full model is represented by $\mathbf{X}_m = \mathbf{X}$, the complete $(n \times (p+1))$ design matrix containing all the variables.

Example 6.5. *Possible design matrices when using two covariates*

To clarify this notation, an example with two possible covariates, X_1 and X_2 is considered. Here, $p = 2$ and therefore the total number of possible models is $m = 2^2 = 4$. The full design matrix, X can be written as $[\mathbf{1}_n, X_1, X_2]$, which is equivalent to X_4. The four possible models are shown in Table 6.1.

Table 6.1: Possible design matrices and corresponding regression coefficients when using combinations of two covariates.

Model, k	$\mathbf{X}_k$	r_k	β^k	$\mathbf{X}_{k'}$	$p+1-r_k$	$\beta^{k'}$
1	$[\mathbf{1}_n]$	1	β_0^1	$[X_1, X_2]$	2	$(\beta_1^1, \beta_2^1)'$
2	$[\mathbf{1}_n, X_1]$	2	$(\beta_0^2, \beta_1^2)'$	X_2	1	β_2^2
3	$[\mathbf{1}_n, X_2]$	2	$(\beta_0^3, \beta_2^3)'$	X_1	1	β_1^3
4	$[\mathbf{1}_n, X_1, X_2]$	3	$(\beta_0^4, \beta_1^4, \beta_2^4)'$	-	0	-

As discussed above, from maximum likelihood theory, if the full model is correct then the expected value of $\hat{\beta}$ is equal to the true value of β, however, if the model includes variables that are extraneous then this is no longer true. If the true model is $Y = \mathbf{X}_k \beta^k + \varepsilon$ but the full model is fitted then the expected value and variance of $\hat{\beta}$ will be as follows:

$$\begin{aligned} E(\hat{\beta}_k) &= \hat{\beta}_k + (\mathbf{X}'_k\mathbf{X}_k)^{-1}(\mathbf{X}'_k\mathbf{X}_{k'})\hat{\beta}_{k'} \\ &= \beta^k + A\beta^{k'} \quad\quad\quad\quad\quad (6.13) \\ &= \textit{true value} + \textit{bias term} \\ VAR(\hat{\beta}^k) &= (\mathbf{X}'_k\mathbf{X}_{k'})^{-1}\sigma^2 \quad\quad\quad (6.14) \end{aligned}$$

where $A = (\mathbf{X}'_k\mathbf{X}_k)^{-1}(\mathbf{X}'_k\mathbf{X}_{k'})$.

This shows the bias in the regression coefficients that can be expected if the extraneous variables contained in $\mathbf{X}_{k'}$ are included in the model. This bias will exist unless (i) $\beta_k = 0$, the true coefficients of the additional variables are actually zero and thus the variable has no effect on the response, or (ii) $\mathbf{X}'_k \mathbf{X}_{k'} = 0$, there is no correlation between the two sets of variables meaning that the inclusion of non-important variables will not have an effect on the explanatory power (and the estimates) of the variables of interest. This is often a product of design in a controlled experiment, as in the case of randomised trials where randomisation is used to allow the investigator to separate the effects of different covariates, but is unlikely to be the case in observational studies.

It can be shown that the components of $\hat{\beta}^k$ will generally have less variation than those from the full model, $\hat{\beta}$. This means that the precision of the estimates can be reduced by using a subset model, even if the terms excluded are not completely extraneous, i.e. do not have a true coefficient of zero. It can also be shown that the estimate of the variance, $\hat{\sigma}_k$, will generally be biased upwards.

The reason that this result is not taken to the extreme case, where just the intercept would be fit and thus predict the same value irrespective of the values of the covariates, is because of the bias that is introduced when important variables are ignored. As variables are added to the model, the usual trade-off between reducing the bias and increasing the variance has to be considered. This bias term can also be expressed in terms of the mean square error (MSE) of the estimates $\hat{\beta}$, where

$$
\begin{aligned}
MSE(\hat{\beta}) &= E(\hat{\beta}^k - \beta^k)(\hat{\beta}^k - \beta^k)' \\
&= (\mathbf{X}'_k \mathbf{X}_k)^{-1}\sigma^2 + A\beta_{k'}\beta'_{k'}\mathbf{A}' \qquad (6.15) \\
&= variance + (bias)^2
\end{aligned}
$$

If a variable is included that has no effect on the response then the variance will be increased without a decrease in the bias. If a variable has a small effect on the response, then the reduction in bias has to be weighed against the increase in variance.

The bias in the regression estimates and the change in the variability associated with variable selection will have an associated effect on any predicted values generated from the regression equation. If a set of explanatory variables is considered $x' = (\mathbf{X}'_k \mathbf{x}^*_{k'})$ (comprising values for both the important and extraneous variables) then the full model will produce response, $\hat{Y} = \mathbf{x}'\beta$, with mean $x'\hat{\beta}$ and prediction variance,

$$
\text{VAR}(\hat{Y}) = \sigma^2(\mathbf{x}'(\mathbf{X}'\mathbf{X})^{-1}\mathbf{x}) \qquad (6.16)
$$

As the point predictions are essentially based on the individual estimates, the variances decrease as terms are excluded from the model.

6.6.2   Selection procedures

Chapter 6, Section 2.5.3 describes how changes in deviances between nested models can be tested to assess differences in model fit. Procedures have been developed based on such tests, which aim to identify good (though not necessarily the best) subset models. Generally they aim to use less computation than would be necessary if all the possible regressions were performed and examined. These methods are often referred to as *stepwise regression procedures*. Dependent on the method, subset models are identified sequentially by adding or deleting variables, Three commonly used stepwise methods are described here. These methods are in no way guaranteed to find the 'best' or even the same subset of variables.

Forward selection of variables adds one variable at a time to the previously chosen subset. At the beginning of the process the first variable chosen is that which accounts for the largest amount of variation in the dependent variable, i.e. the variable having the highest correlation with Y. At each successive step, each of the remaining variables are added one at a time to the model and that which results in the largest reduction in the residual sum of squares is retained. This will be the variable with the highest correlation with the *residuals* from the current model. A stopping rule is needed to terminate the process, or it would continue until all the variables have been included.

If RSS_k denotes the residual sum of squares with r_k variables in the model and if the smallest RSS which can be obtained by adding another variable is RSS_q, for new model, M_q, then

$$R = \frac{RSS_k - RSS_q}{RSS_q/(n - r_k - 2)} \tag{6.17}$$

is calculated and compared to a 'F-to-enter' value, F_e, i.e. if R is greater than F_e, then the variable is added to the selected set.

Backward elimination is similar but starts with the full model and then chooses a variable to delete at each step according to which would result in the smallest increase in the residual sum of squares (the variable with the smallest partial sum of squares). Again, a stopping rule is needed or the process will continue until all the variables have been deleted.

Let RSS_k denote the residual sum of squares with r_k variables in the model and let RSS_t be the smallest that can be obtained by deleting any variable resulting in model M_t . Then let

$$R = \frac{RSS_t - RSS_k}{RSS_k/(n - r_k - 1)} \tag{6.18}$$

is calculated and compared to the 'F-to-stay' value, F_s. If R is less than F_s, then the variable is deleted from the selected set.

It should be noted that although the terms 'F-to-enter' and 'F-to-stay' suggest that the ratios, R, follow an F-distribution, this is not the case. If, after r_k variables have been entered and the next variable is chosen at random, then the assumptions required to use an F-test (that the residuals are independently, identically and normally distributed) are met. However, if the next variable is chosen so as to maximise R then the distribution will not be an F-distribution. Using approximations to the distribution of R to obtain nominal 5% points for the 'F-to-enter' taken from the F-distribution have been shown to give true significance levels in excess of 50% (Draper, Guttman, & Lapczak, 1979).

Stepwise selection is a combination of the previous two approaches, neither of which takes into account the effect that adding or deleting a variable can have on the contributions of the other variables previously included in the model. A variable that has been added to the model early on may become unimportant after another, possibly highly correlated, variable has been added. This is certainly going to be the case when considering the effects of pollutants and temperature. For this reason, stepwise selection is an extension of forward selection that re-checks at each step the importance of the variables currently included in the model. If the partial sum of squares for any previously included variable(s) does not meet some predetermined criteria, then those variables are excluded from the model before the next addition step is attempted.

The stopping rules, referred to as 'F-to-stay' and 'F-to-go' are both required in stepwise selection. However, there is no reason why they should be set at the same level, indeed it might be preferable to have a more relaxed criterion for inclusion as this would force the procedure to consider larger numbers of subset models.

One criticism of these selection methods is that they imply an order of importance to the variables. This can be misleading since, for example, in forward selection, the first variable that was included is often deemed unimportant when other variables have been included. Indeed, it could be the first variable to be excluded when backwards elimination is applied. It should also be remembered that the choice of variables in the final model can be determined to a varying extent by the particular data under consideration the selection method used and the stopping rule. Unless an independent dataset is used for the selection process, the bias in the individual parameter estimates will also be affected by these factors. The more extensive the search for the model, the more chance there is that the final model might be an extremely good fit to the data in question but possibly with little other general use. By basing the inference on the results of a single model, the possibility of extreme or non-typical values in the data affecting the parameter estimates will be increased.

A danger in using these methods is that they might appear to replace the need for the investigator to obtain a genuine feel for the data, with respect to the problem under consideration, and may mask the possible effects that model assumptions and variable selection might have on the inferential process.

There are alternatives to the forward and backward selection criteria, including the Akaike Information Criterion (AIC) and the Bayesian Information Criterion (BIC). The AIC chooses the model which maximises

$$l(\hat{\theta}) - r_k \qquad (6.19)$$

where $l(\hat{\theta})$ is the log likelihood evaluated at the MLE and r_k is the number of parameters in the model. The penalty term, $-r_k$, aims to guard against models being chosen simply because they contain more variables and thus will have greater explanatory power, however limited the increase.

The BIC chooses the model which maximises

$$l(\hat{\theta}) - (r_k/2) \log n \qquad (6.20)$$

This has an additional penalty term which considers the sample size, guarding against accepting variables into the model which might be considered 'significant', just because they are based on large numbers. The BIC will tend to favour simpler models than the AIC and also has the benefit of being closely related to Bayes Factors (see Section 6.8).

Example 6.6. *Model choice in studies of air pollution and health*

In studies of air pollution and health it is common for regression models to contain a number of highly correlated pollutants in the same regression and to attempt to draw inferences about which variables were causal, or at least had the strongest relationship, from that model. Many studies have used stepwise procedures, including many highly correlated pollutants and meteorological variables in the candidate list. For example, Kinney and Ozkaynak (1991) performed multiple regression using backward selection on 5 pollutants and 3 temperature variables in Los Angeles County. After pre-whitening both the outcome (daily deaths) and the explanatory variables using moving averages, they found significant correlations only between deaths and same-day levels of the pollutants and temperature, except for ozone, for which a 1-day lag was significant. These variables were then included in the multiple regression on which backwards selection was performed, with the resulting model including same-day NO_2 and temperature, and 1-day lag O_x. The difficulties in using an automatic procedure to select between highly correlated variables have been discussed, and here there was the additional complication of the use of filtering on the data. Although this was an attempt to eliminate the problems of autocorrelation and cross correlation between the variables, the choice of which lagged variables to include in the initial model could be determined to a large extent by the exact filtering mechanism. For example, if a particular lag did not show significant correlation with the outcome after the filtering, it was not a candidate for inclusion in the initial model, even before the selection procedure was performed.

Given the problems discussed with automatic procedures of this type, the high levels of correlation between the explanatory variables and the generally low explanatory power of air pollution for mortality or morbidity, such an approach is unlikely to produce stable conclusions. The APHEA study (Katsouyanni et al., 1995) took a different approach. Initially, single pollutants models were fitted and if one or more pollutant was associated with the outcome then attempts were made to separate the effects by examining associations with one pollutant stratified by the other. There was also the advantage of having a multi-city study, in which the relative levels of the pollutants were unlikely to be the same in each city. This could help in identifying interactions that may mask the effects of some of the pollutants. However, the model building was performed using a log transformed outcome with the final model using Poisson regression (Katsouyanni et al., 1995).

It is not clear what effect this strategy of basing the model building and the estimation on different distributional assumptions will have. This may be done to benefit from the simplicity associated with using the Normal distribution, even if the Poisson has been stated to be more appropriate. Given the relative ease with which more appropriate models can often be used, there seems little reason to introduce this additional complication to the analysis.

6.7 What about the p–value?

As noted in Chapter 3, reproducibility has been a hallmark of science. Statistics came to play a key role in compensating for the inevitable measurement error that arises even in well-designed lab experiments and statistics provided a solution; results from repetitions of the experiment could be considered in agreement with a hypothesis if they were not *significantly* different from what one would expect to see under that hypothesis. To ensure objectivity, that meant extracting the evidence provided by the experiment and testing the (null) hypothesis (H_0) of no difference.

For reasons that are not entirely clear, the p–value (p) has become the tool of choice, even in model selection. It represents the chance of seeing evidence of a difference, i.e. against H_0, at least as strong as that seen in the experiment if H_0 were true. This will include for example hypotheses about model parameters. In other words given the sampling distribution for a test statistic $U = U(Y)$ calculated from the random sample Y and its sampling distribution, when H_0 is true, compute

$$p = P[U(Y) > U(Y_{obs})]$$

Reject H_0 if $p < 0.05$, a widely accepted 'significance level'. This simple, seemingly natural, criterion came to be universally accepted as the perfect tool for implementing the scientific method as it was seen as representing evidence in support of H_0.

Fisher originally proposed the p–value and the method described above as a 'test of significance', although he did not intend it to be used in the way it is now for confirmatory analysis. Instead he meant it to be a descriptive tool for exploratory analysis (Nuzzo, 2014). In particular, he did not see it controlling the probability of a false positive. That came with Neyman–Pearson theory which formalised testing. If H_0 is true, the chances are no more than 5% of falsely rejecting H_0. While it seems plausible that the test criterion $p < 0.05$ will ensure the same result, it is not true. Notably, an alternative hypothesis does not need to be specified, unlike in Neyman–Pearson theory.

Berger (2012) finds this use of the p–value as one of the reasons published scientific studies are not reproducible. Citing Ioannidis (2005), Berger states that "of the forty-nine most famous medical publications from 1990–2003 resulting from randomized trials; 45 claimed successful intervention". Out of those 45 studies, 7 (16%) were contradicted by subsequent studies, 7 (16%) had found effects that were stronger than those of subsequent studies, 20 (44%) were replicated but with issues and 11 (24%) remained largely unchallenged. Of course, not all the replication failures were due to the use of the p–value, but their use was a factor in some cases and Berger elaborates on this point noting that few non-statisticians understand p–values.

The question of whether there is a sensible alternative to the p–value has been extensively explored (Bayarri & Berger, 2000; Sellke, Bayarri, & Berger, 2001). One approach, based on interpreting the p–value within a Bayesian testing framework, is to quote

$$B(p) = -eplog_e(p) \qquad (6.21)$$

instead of the p–value as a measure of the evidence in the sample against the null hypothesis. This is a lower bound for the appropriate Bayes factor giving the posterior odds in favour of H_0 over H_1 at least when a priori the two hypotheses are equally likely (the conservatively objective case) (Sellke et al., 2001). As we will see in more detail in Section 6.8, these factors are a ratio of the posterior probabilities of the models specified under H_0 over H_1, respectively where for each model a prior has been put on the model parameters. The lower bound $B(p)$ is obtained by finding the minimum value over all reasonable choices of those priors. If $B(p)$ were substantially larger than p, then the evidence against H_0 would not be as strong as the p would suggest.

Example 6.7. *When a p–value might not be all it seems*

This is an example based on Bayarri and Berger (1999). Let $Y_i \sim_{iid} N(\mu, 1)$, $i = 1, \ldots, N$ be a random sample and $U(\mathbf{Y}) = \bar{Y}$ the test statistic you would use to test $H_0 : \mu = 0$ if the alternative hypothesis were $H_1 : \mu \neq 0$. If $U = u$ is observed then the p–value is

$$p = 2[1 - \Phi(\sqrt{N}u\,|)]$$

If the prior distribution were $p_1(\mu)$ under H_1, the Bayes factor would be

$$B = \frac{(2p/N)^{-1/2}\exp\{-Nu^2/2\}}{\int (2p/N)^{-1/2}\exp\{-N(u-\mu)^2/2\}p_1(\mu)d\mu} \tag{6.22}$$

Thus given any given p, we can find the observed value, u_p, of the test statistic that would yield that p. In turn we can find the Bayes factor and $B(p)$; the minimum value over the priors, p_1. Table 6.2 shows the result, a B–value for any p–value. The table also contains the ratio for three specific priors under the alternative. For example, a value of $p = 0.01$, which would seemingly give strong evidence against H_0, yields a B–value of 0.13; which indicates much less evidence against it.

p	0.10	0.05	0.01	0.001
B	0.62	0.41	0.13	0.02
Normal	0.70	0.47	0.15	0.02
Unimodal–symmetric	0.64	0.41	0.12	0.02
Symmetric	0.52	0.29	0.07	0.01

Table 6.2: A comparison of the p and B values for Example 6.7, the latter being a lower bound for the Bayes factor in favour of H_0 over H_1. Here three different classes of priors under the alternative are considered. From the Bayesian perspective, the p–value substantially misrepresents the evidence in the data against H_0.

In conclusion, although the p–value has been reported routinely by almost every statistical package for coefficients in a model for several decades and is routinely used by countless researchers to decide on the significance model parameters its value is the subject of controversy as this book is being written. There is the real possibility that it has led to the false rejection of null hypotheses in scientific inquiries and has contributed to the non-reproducibility of scientific findings.

6.8   Comparison of models - Bayes factors

In this section, an alternative to the likelihood ratio (LR) based methods of model comparison described in Chapter 2, Section (2.5.3) is introduced. In common with those methods, Bayes factors give an indication of how much more likely the data are to have arisen under one model than another, but unlike the LR tests, the models do not have to be nested. Further details on Bayes factors, their interpretation and calculation can be found in Kass and Raftery (1995) and O'Hagan (1995).

For illustration, the simplest case is considered, where there are just two possible models, M_0 and M_1. If the two models are assigned prior probabilities $p(M_0)$ and $p(M_1)$ then these are compared with the probabilities of the data under each of the models, $P(y|M), i = 0, 1$, the *prior predictives*, to obtain *posterior model probabilities*, $p(M_0|y)$ and $p(M_1|y)$. Contained in each model is the uncertainly associated with its parameters. For each of the models, M_k,

$$p(y|M_k) = \int p(y|\theta_k, M_k) p(\theta_k|M_k) d\theta_k \qquad (6.23)$$

From Bayes' theorem the probability that the data was generated under the first model can be obtained,

$$p(M_0|y) = \frac{p(y|M_0)p(M_0)}{p(y)} = \frac{p(y|M_0)p(M_0)}{p(y|M_0)p(M_0) + p(y|M_1)p(M_1)} \qquad (6.24)$$

As there are only two models under question the posterior odds of model 1 in favour of model 2 can be expressed as

$$\frac{p(M_0|y)}{p(M_1|y)} = \frac{p(y|M_0)}{p(y|M_1)} \cdot \frac{p(M_0)}{p(M_1)} \qquad (6.25)$$

The term $p(y|M_0)/p(y|M_1)$ is known as the *Bayes' factor*, B_{01} in favour of model 0 over model 1. Therefore,

$$\text{Posterior Odds} = \text{Bayes factor} \times \text{Prior Odds} \qquad (6.26)$$

As described in Chapter 4, there are often difficulties associated with the, often complex, integrals involved in obtaining the posteriors. One approach, suggested by Kass and Wasserman (1995), to approximating the posterior model probabilities is to restrict the choice of priors in such a way that approximations to the posterior model probabilities can be made that have no dependence on the prior distributions and don't involve performing integration. They suggest approximating the posterior model probabilities using the Bayesian Information Criteria, BIC, which was introduced in Section 6.6.2 under its more common use as a measure of model fit. When using the BIC approximation, the posterior probabilities are given by $\hat{P}(M_k|Y) = L_n(\hat{\theta})_k - \frac{r_k}{2} \log n$, where $L_n(\hat{\theta})_k$ is the log likelihood evaluated at the MLE under model k and r_k is the number of parameters in the model. This is an approximation of order $O_P(1)$. For information, the notation, $O_P()$, known as 'big O_P' indicates that the error term series is *bounded in probability*, i.e. for sufficiently large enough n, the sequence of error terms is bounded by some constant, K. The smaller the order of the error term sequence (for example, $1/n, 1/n^2, ...$) the smaller the error terms become for large n, and thus the better the approximation. Further details on the subject of *stochastic order* are available in Bishop, Feinburg, and Holland (1975).

To calculate Bayes factors for generalised linear models, Raftery and Richardson (1996) suggest a Laplace approximation of the form seen in Chapter 5, Section 5.2.

When this is used to approximate the Bayes factor (relative to the null model) the result is

$$\hat{m}_k = L(\hat{\theta}|Y)p(\hat{\theta})2p^{r_k/2}|H|^{1/2} \qquad (6.27)$$

where $\hat{\theta}$ is the mode of the posterior $P(\theta|Y)$ and H is the matrix of second derivatives of the log-posterior evaluated at θ. This approximation has an error of order $O_P(n^{-1})$. In order to use this, the posterior mode $\hat{\theta}$ and the inverse Hessian of the log-posterior evaluated at θ are required, which are not routinely available when using standard statistical software packages.

Raftery and Richardson (1996) suggest a way of using values of the MLE, Deviance and the Fisher information matrix , I_θ which are widely available when fitting generalised linear models (see Chapter 2, Section 2.5 and McCullagh & Nelder, 1989) to estimate Bayes factors for comparing two models, again using the Bayesian Information Criteria. This approximation is of the order, $O_P(n^{-1})$ when the expected Fisher information is used for F_k, but this can be improved to $O_P(n^{-1/2})$ if the observed information is used.

6.9 Bayesian model averaging

In Section 6.6.2, the traditional approach to model 'selection' was considered, that of choosing a single model which is then considered to be correct when drawing inference about the parameters of interest or performing prediction. This ignores the inherent uncertainty associated with the actual modelling process, and in particular the uncertainty which arises through selecting a subset of variables for inclusion, together with that due to distributional assumptions and/or transformations.

Table 6.3: Possible models and posterior model probabilities using combinations of two covariates, with ε_i $(i = 1,2,3,4) \sim$ i.i.d $N(0,\sigma_i^2)$.

Model	Formula	Posterior Probability
M_1	$Y = \beta_0^1 + \varepsilon_1$	$p(M_1|y)$
M_2	$Y = \beta_0^2 + \beta_1^2 X_1 + \varepsilon_2$	$p(M_2|y)$
M_3	$Y = \beta_0^3 + \beta_2^3 X_2 + \varepsilon_3$	$p(M_3|y)$
M_4	$Y = \beta_0^4 + \beta_1^4 X_1 + \beta_2^4 X_2 + \varepsilon_4$	$p(M_4|y)$

This section describes an alternative approach where, rather than choosing a single model, information from several models is combined to obtain averaged estimates and/or predicted values. To illustrate this concept, consider the notation for the regression example introduced in Section 6.6.1 with two potential covariates, X_1 and X_2, and $m = 4$ possible models, which are shown in Table 6.3.

The choice is then to

1. Choose the single model with the highest posterior probability, $p(M_i|Y)$, and use this to draw inference about the effects of the covariates (or to predict), ignoring the model uncertainty in the same way as would be done by picking a single model as the result of some likelihood comparison or stepwise procedure.

2. Examine the estimates from all four of the models separately, which introduces the problem of multiple comparisons and might not be very useful if a single estimate/predicted value is required.

3. Combine the information from the four models.

The last approach is the essence of the Bayesian modelling averaging approach, inference on the parameters of interest and/or prediction is performed using weighted averages of the values from the individual models. The process is carried out within a Bayesian framework; each of the candidate models is assigned a prior probability which is updated using information from the data to obtain posterior model probabilities.

Consideration should be given to the fact that inference about an unknown parameter will be dependent on the structure of the model under examination, M_k, and the set of other variables included in that model. In order to calculate posterior probabilities for the parameters, one should condition not only on the observed data but also on the model in question. This is now incorporated into the model uncertainty framework. Given m candidate models, $M_1, ..., M_m$ ($m = 2^p$), each with prior probability, $p(M_i)$ (the simplest example being where all the models are considered equally likely before any data is observed, where $p(M_k) = 1/m$), then the posterior probability for model, M_k can be found using Bayes' theorem in the following way:

$$p(M_k|y) = \frac{p(y|M_k)p(M_k)}{\sum_{l=1}^{m} p(y|M_l)p(M_l)} \quad\quad (6.28)$$

Contained in each model is the uncertainty associated with its parameters. For ease of illustration, the following description is given in terms of the regression parameters, β. For each of the models, M_k,

$$p(y|M_k) = \int p(y|\beta_k, M_k)p(\beta_k|M_k)d\beta_k \quad\quad (6.29)$$

Therefore, given prior probability distributions for the unknown parameters, $p(\beta|M_k)$, under a particular model, M_k, then, conditioning on the model, gives

$$p(\beta|Y, M_k) = \frac{p(y|\beta, M_k)p(\beta|M_k)}{p(y|M_k)}$$

with $p(y|M_k) = \int p(y|\beta, M_k)p(\beta|M_k)d\beta$.

In order to predict a future observation, $Z = \hat{Y}$, using the model in question, the predictive distribution is given by

$$p(Z|y, M_k) = \int p(Z|\theta, M_k)p(\theta|y, M_k)d\theta \qu\quad (6.30)$$

The Bayes factors comparing two models, M_k and M_l, is then defined as

$$B_{kl} = \frac{p(M_k|y)}{p(M_l|y)} = \frac{\int p(y|\beta_k)p_k(\beta_k)d\beta_k}{\int p(y|\beta_l)p_l(\beta_l)d\beta_l} \tag{6.31}$$

When the posterior model probabilities have been obtained, the weighted posterior distributions of the individual parameters can be calculated, allowing the effect of individual covariates to be assessed. The weighted posterior density, mean and variance of the parameters of interest, β_j can be calculated as follows

$$
\begin{aligned}
p(\beta_j|y) &= \sum_{l=1}^{m} p(\beta_j^l|y,M_l)p(M_l|y) \\
E(\beta_j|y) &= \sum_{l=1}^{m} \overline{\beta}_j^l p(M_l|y) \\
\mathrm{VAR}(\beta_j|y) &= E_{M|y}[\mathrm{VAR}(\beta_j|y,M_l)] + \mathrm{VAR}_{M|y}[E(\beta_j|y,M_l)] \\
&= \sum_{l=1}^{m} (\sigma_j^l)^2 p(M_l|y) + \sum_{l=1}^{m} (\overline{\beta}_j^l - \overline{\beta}_j)^2 p(M_l|y) \\
&= \text{within model uncertainty} + \text{between model uncertainty}
\end{aligned}
$$
$$\tag{6.32}$$
$$\tag{6.33}$$

where $\overline{\beta}_j^l = E(\beta_j|y,M_l)$, is the posterior mean of value of the regression coefficient, β_j, under model M_l, and $\bar{\beta}_j = E(\beta_j|y)$.

When performing prediction of a new observation, $\hat{Y} = E(Z|y,X^*)$, using a particular model, the predictive distribution is given by

$$p(Z|y,M_k) = \int p(Z|\beta^k,X^*,M_k)p(\beta^k|y,M_k)d\beta^k \tag{6.34}$$

When all models are considered, the resulting prediction will be a weighted average of the predicted values from all the possible models

$$p(Z|X^*) = \sum_{l=1}^{m} P_l(Z|M_l,y,X^*)p(M_l|y) \tag{6.35}$$

With expected value

$$
\begin{aligned}
E[Z|y,X^*] &= E_{M|y}[E[Z|y,X^*,M]] \\
&= \sum_{l=1}^{m} E[Z|y,X^*,M_l]p(M_l|y)
\end{aligned}
\tag{6.36}
$$

and variance

$$
\begin{aligned}
V[Z|y,X^*] &= E_{M|y}[\mathrm{VAR}[Z|y,X^*,M]] + \mathrm{VAR}_{M|y}[E[Z|y,X^*,M]] \\
&= \sum_{l=1}^{m} \mathrm{VAR}[Z|y,X^*,M_l]p(M_l|Y) + \\
&\quad \sum_{l=1}^{m} (Z_l - \bar{Z})^2 p(M_l|y)
\end{aligned}
\tag{6.37}
$$

where Z_l is the predicted value using model M_l, and $\bar{Z} = E[Z|y,X^*]$.

6.9.1   Interpretation

It is natural to ask about the interpretation of a coefficient arising from a Bayesian model averaging procedure. For illustration, consider again the case of normal linear regression with just two covariates, X_1 and X_2 as described in Table 6.3. Considering Z, the predicted value given values of X_1 and X_2,

$$E(Z|\beta_0,\beta_1,...,\beta_p,X_1,X_2,y) = \beta_0 + \beta_1 X_1 + \beta_2 X_2 \qquad (6.38)$$

Expanding this expression over all the possible values of β gives

$$\begin{aligned}
E(Z|X,y) &= E_{\beta|y}(E(Z|X,\beta_0,\beta_1,\beta_2,y)) \\
&= E_{\beta|y}(\beta_0 + \beta_1 X_1 + \beta_2 X_2, y) \\
&= E(\beta_0|y) + E(\beta_1|y)X_1 + E(\beta_2|y)X_2 \qquad (6.39)
\end{aligned}$$

Taking expectations over the set of possible models gives

$$\begin{aligned}
E(Z|X) &= E_{M|y}(E(Z|X,M_k,y)) \\
&= E_{M|y}(E_{\beta|y}(E(Z|X,M_k,\beta,y))) \\
&= \sum_{l=1}^{4} E_{\beta|y}(E(Z|X,M_l,\beta,X,y))p(M_l|y) \qquad (6.40)
\end{aligned}$$

where $X = (X_1,X_2)'$. Hence

$$\begin{aligned}
E(Z|X) = \ & E(\beta_0|y,X,M_1)p(M_1|y) \\
+ \ & \{E(\beta_0|y,X,M_2) + E(\beta_1|y,X,M_2)\}\,p(M_2|y) \\
+ \ & \{E(\beta_0|y,X,M_3) + E(\beta_2|y,X,M_3)\}\,p(M_3|y) \qquad (6.41) \\
+ \ & \{E(\beta_0|y,X,M_4) + E(\beta_1|y,X,M_4) + E(\beta_2|y,X,M_4)\}\,p(M_4|y)
\end{aligned}$$

In the above, in model M_2 for example, $E(Z|X,\beta_0,\beta_1) = \beta_0 + \beta_1 X_1$, so that even if the value of X_2 were known, it wouldn't be used in that model. If $\beta^l = (\beta_0^l, \beta_1^l, \beta_2^l)$ is the set of parameter estimates found under model l then

$$\begin{aligned}
E(Z|X_1,X_2,\beta^1,\beta^2,\beta^3,\beta^4,y) = \ & \beta_0^1 p(M_1|y) \\
+ \ & [\beta_0^2 + \beta_1^2 X_1]p(M_2|y) \\
+ \ & [\beta_0^3 + \beta_2^3 X_2]p(M_3|y) \qquad (6.42) \\
+ \ & [\beta_0^4 + \beta_1^4 X_1 + \beta_2^4 X_2]p(M_4|y)
\end{aligned}$$

A unit increase in X_1 will therefore result in a predicted value of Z'

$$\begin{aligned}
E(Z'|X_1+1,X_2,\beta^1,\beta^2,\beta^3,\beta^4,y) = \ & \beta_0^1 p(M_1|y) \\
+ \ & [\beta_0^2 + \beta_1^2 (X_1+1)]p(M_2|y) \\
+ \ & [\beta_0^3 + \beta_2^3 X_2]p(M_3|y) \qquad (6.43) \\
+ \ & [\beta_0^4 + \beta_1^4 (X_1+1) + \beta_2^4 X_2]p(M_4|y)
\end{aligned}$$

The difference is

$$E(Z'|X_1+1,X_2,\beta^1,\beta^2,\beta^3,\beta^4,y) - E(Z|X_1,X_2,\beta^1,\beta^2,\beta^3,\beta^4,y)$$
$$= \beta_1^1 p(M_2|y) - \beta_1^4 p(M_4|y) \qquad (6.44)$$

This weighted average (by the posterior model probabilities) of the effects from each of the models in which X_1 is included, gives a natural interpretation for a regression coefficient obtained using Bayesian model averaging.

Unlike the traditional approach of selecting just one model in which the estimate of a parameter of interest might be highly affected by the choice of which other variables to include/exclude in the model, this estimate allows for the fact that other variables may also be in the model. An estimate of the weight of support for including X_1 in the model can be found by summing the posterior probabilities for the models that include the variable

$$p(X_1 \text{ should be included}) = \sum_{l=1}^{m} P(M_l|y, \text{ model } l \text{ contains } X_1) \qquad (6.45)$$

Only models that contain the particular variable, X_1, and that have non-zero posterior probabilities will contribute to this sum, as is the case for the weighted averages in Equation 6.40.

Example 6.8. *Bayesian model averaging in studies of air pollution and health*

In studies of the effects of air pollution on mortality it is often more appropriate to use Poisson regression, rather than simple linear regression. In this case, the expectation of the predicted value, Z, in the two covariate example is given by

$$E(Z|\beta_0,\beta_1,\beta_2,X_1,X_2,y) = \exp(\beta_0 + \beta_1 X_1 + \beta_2 X_2, y) \qquad (6.46)$$

When this expectation is expanded to condition on the possible models, as in Equation 6.40, it gives

$$
\begin{aligned}
E(Z|X) \;=\; & \sum_{l=1}^{4} E_\beta(E(Z|M_l,\beta,X,y))p(M_l|y) \\
=\; & \exp(\beta_0^1|y)p(M_1|y) \\
+\; & \exp(\beta_0^2 + \beta_1^2 X_1|y)p(M_2|y) \\
+\; & \exp(\beta_0^3 + \beta_2^3 X_2|y)p(M_3|y) \\
+\; & \exp(\beta_0^4 + \beta_1^4 X_1 + \beta_2^4 X_2|y)p(M_4|y) \qquad (6.47)
\end{aligned}
$$

A unit change in X_1 (with X_2 remaining constant) again only results in changes in the predictions from the two models that contain X_1 (2 and 4), the difference between the two predicted values is now represented by

$$
\begin{aligned}
& p(M_2|y)[\exp(\beta_0^2 + \beta_1^2(X_1+1))/\exp(\beta_0^2 + \beta_1^2 X_1)] \\
+\ & p(M_4|y)[\exp(\beta_0^4 + \beta_1^4(X_1+1) + \beta_2^4 X_2)/\exp(\beta_0^4 + \beta_1^4 X_1 + \beta_2^4 X_2)] \\
=\ & p(M_2|y)[\exp(\beta_1^2(X_1+1))/\exp(\beta_1^2 X_1)]) \qquad\qquad (6.48) \\
+\ & p(M_4|y)[\exp(\beta_1^4(X_1+1) + \beta_2^4 X_2)/\exp(\beta_1^4 X_1 + \beta_2^4 X_2)]
\end{aligned}
$$

This nonlinear expression of the regression coefficient for X_1 is not the same as would be obtained by simply calculating a weighted average of the coefficients from each model and taking the exponent, i.e. $\exp(p(M_2|y)\beta_1^1 + p(M_4|y)\beta_1^4)$. Despite this problem, the models can still be used for prediction, as the individual predictions are made under the conditions of the specific models and the resultant predicted values averaged to produce a single final predictive value. For prediction, the fact that the effects of the individual parameters have been lost is not important, as the interest is in the combined effect of all the covariates. In the case of pollution, for example, the ability to interpret the effect of a $25\mu g/m^{-3}$ increase of PM_{10} on the risk of respiratory deaths for all values of PM_{10} would be lost, but it could be found for categories of values, such as 0–25, 25–50, etc. and the ability to predict the additional number of respiratory deaths associated with an increase in PM_{10}, given values for other covariates, such as SO_2, CO, O_3 and temperature is retained.

6.10 Summary

This chapter considers both some of the wider issues related to modelling and the generalisability of results and more technical material on the effect of covariates and model selection. The reader will have gained an understanding of the following topics:

- Why having contrasts in the variables of interest is important in assessing the effects they have on the response variable.
- The biases that may arise in the presence of covariates and how covariates can affect variable selection and model choice.
- Hierarchical models and how that can be used to acknowledge dependence between observations.
- There are issues with using p–values as measures of evidence against a null hypothesis. Basing scientific conclusions on it can lead to non-reproducible results.
- The use of predictions from exposure models including acknowledging the additional uncertainty involved when using predictions as inputs to a health model.
- Methods for performing model selection, including the pros and cons of automatic selection procedures.
- Model selection within the Bayesian setting and how the models themselves can be incorporated into the estimation process using Bayesian Model Averaging.

Exercises

Exercise 6.1. Elliott et al. (1996) performed a study to investigate the possible effects of municipal incinerators on health. Read this paper and consider the following issues:

(i) What was the population under study?

(ii) What factors might you want to know before generalising the results, for example to other countries?

(iii) What was the time period of the study? Are the results still valid today?

(iv) If an incinerator blew up would you be happy applying the risk estimates obtained in this study to the surrounding population?

Exercise 6.2. In modelling the effects of land use covariates on NO_2 concentrations in Europe (Shaddick, Yan, et al., 2013), there were a large number of possible explanatory variables. These variables are listed, together with their mean values, in Table 6.4. The models were fit on a training set with predictions from the models used to predict concentrations at locations in a validation set. The concentrations

Table 6.4: Summary (means) of covariates at locations of NO_2 monitoring sites in Europe; (1) variables constructed at 1 km and (21) at 21 km resolution.

Covariate	Training set	Validation set	EU
Altitude (m)	220	236	410
Distance to sea (m)	202	198	145
Climate factor 1 (1)	0.83	0.87	1.11
Climate factor 2 (1)	−0.42	−0.44	0.15
Climate factor 3 (1)	0.25	0.30	0.18
Climate factor 4 (1)	0.05	0.02	0.05
Climate factor 5 (1)	−0.03	−0.06	0.02
Non-Residential Built Up (21)	3.72	3.47	0.46
Forestry (1)	9.03	10.24	21.47
Agriculture (21)	45.95	46.97	52.08
Major roads (1)	0.65	0.62	0.10
Minor roads (1)	2.42	2.28	0.82
High density residential (1)	11.00	12.14	0.39
Low density residential (1)	38.83	35.9	2.47
Industry (1)	6.12	6.00	0.39
Transport (1)	0.93	0.87	0.04
Sea Ports (1)	0.30	0.44	0.03
Air Ports (1)	0.20	0.03	0.08
Construction (1)	0.56	0.35	0.20
Urban Greenery (1)	2.24	1.72	0.22
Forestry (1)	9.03	10.24	22.24

were known at the validation locations and so a comparison with the predictions could be made.

(i) Are there any substantial differences between the training and validation sets? What effect might such differences have on the assessment of the prediction model built using the training set?

(ii) Often the sample size may not be large enough to split a dataset into a training and validation set in this way. What other methods can you think of to assess the predictive ability of the model?

(iii) How would you choose the best set of variables from those in Table 6.4 to predict NO_2 concentrations?

(iv) Many of the variables seen in Table 6.4 are likely to be highly correlated. How might this affect your choice of which covariates should be included in the model?

(v) The data for this example are included in the online resources. Using this data, perform an appropriate statistical test, or tests, to quantify your answer to (i).

(vi) Perform an automatic stepwise procedure to choose a subset of the variables with which to predict NO_2 concentrations. Comment on whether you are happy with the choice of variables considering your response to question (iv).

Exercise 6.3. Show that in the simplest case where the prior probabilities of two possible models, M_0 and M_1 are equal to 1/2 that posterior odds will equal the Bayes factor.

Exercise 6.4. In the case of the Poisson regression example shown in Example 6.8, consider the case where there are three covariates, X_1, X_2, X_3.

(i) Derive an equivalent expression to that seen in Equation 6.47 for the three covariate case.

(ii) Derive an expression for the relative risk when X_1 is increased by one unit.

(iii) Derive an expression for the relative risk when both X_1 and X_2 are simultaneously increased by one unit.

Chapter 7

Is 'real' data always quite so real?

7.1 Overview

Epidemiological studies require accurate measurements of both health outcomes, potential confounders and estimates of exposures that might drive associations with health (Finazzi, Scott, & Fassò, 2013). These may be measured with varying degrees of error. Considering measurements of air pollution for example, there is a true underlying pollution surface which will form the basis of the exposures experienced by the population at risk. However this surface is not directly observable and instead measurements are taken at locations over space and time. Differences between these exposure measurements and the unknown underlying field are often referred to as *measurement error*. Here the term measurement error is taken to refer to any difference from the underlying true values and what is measured. Traditionally measurement error has been based around the idea of repeated measurements of a value, for example, measuring blood pressure where repeated measurements will contain a component of error, often assumed to be random. In modelling exposures and in spatial epidemiology, error will possibly comprise a number of factors including monitor calibration error and random variation but also variations in the underlying pollution field over time and space which aren't acknowledged in the analyses. This may arise for example when modelling assumptions are too simplistic for the complex surface of the pollution field.

There are also likely to be periods of missing data in the exposure information. Simple methods for handling missing values are commonly used including simply discarding them from the analysis or replacing them by a specific single value, for example the overall mean. By discarding missing values, we may lose useful information and may introduce bias. When replacing missing values by a single value, for example a sample mean of observations or the posterior mean from an exposure model, the intrinsic variability associated with the summary value may be ignored.

Where a health effects analysis uses predictions from an exposure model as substitutes for actual measures of exposures, as with regular measurement error, there is the possibility of bias in the estimation of risks. An additional issue termed the 'change of support' problem by Gelfand et al. (2001) occurs when the exposure and health outcome data are recorded at different levels of aggregation, for example health counts for administrative areas and exposures from monitoring sites at point locations within, or outside, those areas.

It is also common phenomenon that monitors in an environmental monitoring network are often located in a non-random fashion. For example, they are often located to check for adherence to statutory limits and if monitors are positioned close to known pollution sources, such as roads or industrial plants, then the estimated pollution surface is likely to be overestimated. This is known as *preferential sampling*. It is very important that the information coming from networks is accurate and reflects the levels of exposures that may be experienced by the populations at risk. This might be a problem if monitors are placed in locations where pollution might be expected to be high. Formally, preferential sampling occurs when the process that determines the locations of the monitoring sites and the process being modelled are in some ways dependent (Diggle, Menezes, & Su, 2010). For example, in the study of air pollution and health, Guttorp and Sampson (2010) state that the choice of locations for air pollution monitoring sites may because of a number of reasons, include measuring: (i) background levels outside of urban areas; (ii) levels in residential areas and (iii) levels near pollutant sources. Geostatistical methods which assume sampling is non-preferential are often used despite preferential sampling (Diggle et al., 2010). Ignoring preferential sampling may lead to incorrect inferences and biased estimates of pollution concentrations and thus any subsequent estimation of health risks.

In this chapter, we review approaches to classifying missing values and consider measurement error and the effect that it may have on estimates obtained from regression models. We also provide a review of how preferential sampling can affect estimates of exposures that might be used within a health analysis and methods that might be used to adjust for its effects.

7.2 Missing values

In real-life applications, data are often incomplete and there are many reasons why observations may be missing. Here, we follow the notation proposed by Little and Rubin (2014) who classify the process which leads to missing data as either (i) completely random, (ii) random, or (iii) informative. Let Y denote the complete set of data which would be available if there were no missing values. Since this situation is unlikely in practice we split the dataset into $Y = (Y_{obs}, Y_{mis})$ where Y_{obs} are the observed data and Y_{mis} the missing observations.

For the classification of the missing values, we denote R_i as a indicator which takes values 1 or 0 dependent on whether Y_i is a missing value or not. The missing data mechanism can therefore be classified in the following ways:

1. Completely random (MCAR): if R_i is independent of both Y_{obs} and Y_{mis}.
2. Random (MAR): if R_i is independent of Y_{mis}.

3. Informative: if R_i is dependent on Y_{mis}.

If the missing data mechanism can be assumed to be non-informative, ie. MAR or MCAR and if $f(y_{obs}, y_{mis}, r)$ denotes the joint probability density function of (Y_{obs}, Y_{mis}, R) then this can be factorised in

$$f(y_{obs}, y_{mis}, r) = f(y_{obs}, y_{mis})f(r|y_{obs}, y_{mis}) \qquad (7.1)$$

Integrating out the missing values, the joint pdf of the observable random variables, (Y_{obs}, R), can be obtained

$$f(y_{obs}, r) = \int_{y_{mis}} f(y_{obs}, y_{mis})f(r|y_{obs}, y_{mis})dy_{mis} \qquad (7.2)$$

Under the assumption that the missing data mechanism is non-informative we conclude to the following equation

$$
\begin{aligned}
f(y_{obs}, r) &= f(r|y_{obs}) \int f(y_{obs}, y_{mis})dy_{mis} & (7.3) \\
&= f(r|y_{obs})f(y_{obs}) & (7.4)
\end{aligned}
$$

Taking logarithms gives

$$L = log(f(r|y_{obs}) + log(f(y_{obs})) \qquad (7.5)$$

Maximising L here is achieved by maximising the second part of the right hand side, $log(f(y_{obs}))$, meaning that estimation of the model parameters does not depend on the missing data mechanism.

7.2.1 Imputation

Trivial methods of handling missing values include simply discarding them from the analysis or replacing them by a specific single value, for example the overall mean. By discarding missing values we may lose useful information and by replacing a missing value by a single value we may ignore important features of the dataset and ignore intrinsic variability.

A more sophisticated method is multiple imputation (Little & Rubin, 2014). This replaces missing values with a set of plausible values that acknowledge the inherent uncertainty associated with the unobserved value. After the imputation of the missing values, multiple complete-value datasets are produced and a combined analysis of these datasets is performed. This results in valid statistical inferences that properly reflect the uncertainty due to missing values.

Two simple examples of imputation are presented here; the first is based in a regression method that assumes multivariate normality, and a description of how the imputed values are calculated and the complete-values datasets are obtained is given. The second method, set within a Bayesian framework, was proposed by Schafer

(1997). Instead of producing multiple complete-value datasets, the missing values are drawn from a posterior distribution, given the observed data, in each iteration of an MCMC algorithm.

7.2.2 Regression method

The regression method requires that the missing values appear in a monotone pattern. The basic idea is that a regression model is fitted with the variable with missing values as the response, with previous values as covariates. For a variable Y_j, this can be represented with missing values as

$$Y_j = \beta_0 + \beta_1 Y_1 +, \ldots, +\beta_{j-1} Y_{j-1}$$

Since the data has missing values in monotone pattern, $Y_1, \ldots, Y_{j-1}$ do not have missing values. After estimation of the regression parameters $(\hat{\beta}_0, \hat{\beta}_1, \ldots, \hat{\beta}_{j-1})$ with corresponding covariance matrix $\sigma_j^2 V_j$, and after each imputation, new estimates of the parameters are drawn. The imputation is performed by replacing the missing values with

$$\hat{\beta}_0 + \hat{\beta}_1 y_1 +, \ldots, +\hat{\beta}_{j-1} y_{j-1} + z_i \sigma_j$$

where $(y_1, y_2, \ldots, y_{j-1})$ are the covariates and z_i follows a zero-mean normal distribution, $N(0, 1)$.

7.2.3 MCMC method

In Bayesian inference, a prior distribution is assigned to each parameter and after observing the data, information about these parameters is expressed in the form of a posterior distribution. Using imputation to treat the missing values in an MCMC algorithm is akin to treating missing values as unknown parameters. Once we obtain draws from the unknown parameters from their posterior distributions given the observed data, simulation-based estimates can be obtained for the missing values. This method is called Expectation–Maximisation (EM). Basically the EM algorithm is performed in two steps which are as follows:

i. In the Expectation or E-step, the unknown parameters of the model are assumed to have already been estimated. Estimates are drawn from the posterior distributions given Y_{obs}, $P(Y_{mis}|Y_{obs}, \theta^t)$ where t is the current number of iterations in the MCMC algorithm.

ii. In the maximisation or M-step estimation values of the unknown parameters, θ^{t+1} are drawn given the complete-value dataset and are used to perform the following E-step.

Further details of the EM algorithm and its properties can be found in Schafer (1997).

7.3   Measurement error

Measurement error is a general term used to encompass situations where the observed data do not represent the quantity of interest exactly. It can occur in both response variables and covariates. Epidemiological studies require accurate measurements of both health outcomes exposures together with potential confounders. All of these data may be measured with varying degrees of error. Consider measurements of air pollution; there is a true underlying pollution surface that will drive the exposures experienced by the population at risk. However this surface is not directly observable and instead measurements are taken at locations over space and time. Differences between these exposure measurements and the unknown underlying field is often referred to as measurement error. In this sense the term measurement error is taken to refer to any difference from the underlying true values and what is measured.

Traditionally, measurement error has developed around the idea of repeated measurements of a value, for example measuring blood pressure, in which the repeated measurements will contain a component of error that is often assumed to be random. In spatio-temporal epidemiology, error may comprise a number of factors including monitor calibration error and random variation but also variations in the underlying pollution field over time and space which aren't acknowledged in the model. This may arise for example when the modelling assumptions are too simplistic for the complex surface of the pollution field. Another issue that is often contained under the umbrella term of 'measurement error' is the misalignment of locations or times of exposure measurements and health outcomes. This arises because exposure and health data are often drawn from independent sources and not the result of a carefully designed study. In which case a straightforward comparison is not possible without a model to align these elements in the spatial and temporal domains (Gryparis et al., 2009; Peng & Bell, 2010). In such settings, health effects analysis may use predictions from an exposure model as substitutes for actual measures of exposures in the health model.

A brief review of measurement error models is given here while more comprehensive discussions can be found in Fuller (1987); Carroll, Ruppert, and Stefanski (1995). Measurement error models are based on four quantities:

- $Y^{(1)}$ - the response variable;
- Z - the true unobserved exposures;
- $Y^{(2)}$ - the observed exposures which are measurements of Z, potentially incorporating some measure of error;
- X - covariates, which are assumed to be measured exactly.

The joint likelihood of these quantities can be expressed as

$$f(Y^{(1)}, X, Y^{(2)}|Z) = f(Y^{(1)}|X, Y^{(2)}, Z)f(X, Y^{(2)}|Z) \tag{7.6}$$

where the covariates are conditioned on because they are fixed and known. The first element of Equation 7.6 is an exposure-response model, of the form described in Chapter 12, Section 12.5.1 and seen in Chapter 2. It is typically simplified to $f(Y^{(1)}|Z,X)$ by assuming the measurement error is non-differential, i.e. $Y^{(1)}$ and $Y^{(2)}$ are conditionally independent given Z. The second element in Equation 7.6 is a measurement error model, which represents the relationship between the unobserved exposure Z and the measured surrogate $Y^{(2)}$. There are two types of measurement error model; classical and Berkson, which are outlined below.

7.3.1   Classical measurement error

Classical measurement error models rely on the following factorisation

$$f(Z,Y^{(2)}|X) = f(Y^{(2)}|Z,X)f(Z|X)$$

the first element of which is a conditional model for the measured surrogate $Y^{(2)}$ given the true (unobserved) exposure Z. Two common classical measurement error models are (i) additive and (ii) error calibration as follows.

(i) $Y_i^{(2)} \sim N(Z_i, \sigma^2)$ for $i = 1, ..., n$

(ii) $Y_i^{(2)} \sim N(\beta_0 + \beta_z Z_i + \sum_{j=1}^{J} \beta_j X_{ji}, \sigma^2)$ for $i = 1, ..., n$

In the simple additive formulation the observed surrogate is assumed to be correct on average (that is $E[Y_i^{(2)}|Z_i] = Z_i$), while in model (ii) the surrogate is biased. Both models specify an additive relationship between $Y_i^{(2)}$ and Z_i, an alternative being a multiplicative error model $Y_i^{(2)} = Z_i v_i$, where v_i is a zero mean Gaussian error with variance σ^2. The remaining term $f(Z|X)$ can be based on knowledge of that true exposure or represent prior ignorance. In the latter case a common choice is the production of normal distributions each of which has large variance, e.g. $\prod_{i=1}^{n} N(Z_i|\mu, \tau^2)$ where τ^2 is large. In a Bayesian setting $f(Z|X)$ acts as a prior for the unknown exposure Z.

7.3.2   Berkson measurement error

In contrast to the classical case, the Berkson measurement error model relies on the following factorisation:

$$f(Z,Y^{(2)}|X) = f(Z|Y^{(2)},X)f(Y^{(2)}|X)$$

where the first term is a conditional model for the true exposure Z given the measured surrogate $Y^{(2)}$. Again there are two common approaches; (iii) additive and (iv) regression calibration. In common with the classical models $f(Z|Y^{(2)},X)$ is a decomposition of independent distributions for each observation.

(iii) $Z_i \sim N(Y^{(2)}, \sigma^2)$ for $i = 1, ..., n$

(iv) $Z_i \sim N(\beta_0 Y_i^{(2)} + \beta_{Y(2)} Y_i^{(2)} + \sum_{j=1}^{J} \beta_p X_{ji}, \sigma^2)$ for $i = 1, ..., n$

In the simple additive model the true exposure is assumed to be equal to the surrogate on average (that is $E[Z_i|Y_i^{(2)}] = Y_i^{(2)}$), but this is not true for (iv). As with classical models a multiplicative alternative can be used, which is implemented using an additive model on the log scale. In the Berkson model, as $Y^{(2)}$ are known measurements, the distribution $f(Y^{(2)}|X)$ can be ignored. The choice between classical and Berkson models will depend on the structure of the problem as well as the set of available data. Further details can be found in Carroll et al. (1995).

The measurement error models described above can only be used if additional data are available, because the information from $(Y^{(1)}, Y^{(2)}, X)$ is not sufficient to estimate the measurement error process. Examples of such additional data include repeated measurements of $Y^{(2)}$ which in a spatial setting may be measurements at each location over time. Alternatively, external data may be able to inform the process if observed values of Z and $Y^{(2)}$ were available at a subset of locations. The identifiability of a proposed model may also depend on the assumptions made about the measurement error process. There are two generic classes of such assumptions; functional and structural. Functional models are distribution invariant and specify minimal assumptions about the measurement error process. They do not specify a proper likelihood, and estimation is typically based on regression calibration and the SIMEX algorithm (Carroll et al., 1995). In contrast structural models, such as those shown in (i) to (iv), are fully parametric and specify probability distributions for $f(Y^{(2)}|Z,X)$ or $f(Z|Y^{(2)},X)$. The choice between functional and structural models determines the method of estimation and inference that can be used, with structural models enabling likelihood and Bayesian methods to be applied. A brief outline of estimation techniques for measurement error models is given in the next section. For a more comprehensive treatment the reader is referred to Carroll et al. (1995).

7.3.3 Attenuation and bias

Covariate measurement error typically causes non-measurement error models to produce biased estimates of the regression parameters, and except for the simple linear model $Y_i^{(1)} \sim N(\beta_0 + \beta_Z X_i, \sigma_v^2)$, understanding the nature of this bias may not be straightforward. In this simple case replacing Z by $Y^{(2)}$ yields the model $Y_i^{(1)} \sim N(\beta_0 \times +\beta_{Y(2)} Y_i^{(2)}, \sigma_{v2}^2)$, which estimates $\beta_{Y(2)}$ instead of the relationship of interest, β_Z. As Z is unknown β_Z can only be estimated using measurement error methods, and a simple classical model is given by

$$
\begin{aligned}
Y_i^{(1)} &\sim N(\beta_0 + \beta_Z Z_i, \sigma_{v1}^2) \\
Y_i^{(2)} &\sim N(Z_i, \sigma_{v2}^2) \\
Z_i &\sim N(\mu, \sigma_z^2)
\end{aligned}
\tag{7.7}
$$

For this model it can be shown that $\beta_{Y(2)} = \frac{\sigma_z^2}{\sigma_z^2 + \sigma_{v2}^2} \beta_z$ (Fuller, 1987) and so naively replacing Z with $Y^{(2)}$ and ignoring measurement error results in a biased estimate of

β_Z. This bias is known as attenuation, with β_Z being shrunk towards zero by a factor of $\frac{\sigma_z^2}{\sigma_z^2+\sigma_{v2}^2}$. In addition, allowing for measurement error inflates the variance from σ_{v1}^2 using the simple linear model to $\sigma_{v1}^2 + \frac{\beta_Z \sigma_{v2}^2 \sigma_z^2}{\sigma_z^2+\sigma_{v2}^2}$ if Equation 7.8 is used.

In contrast with the Berkson error model

$$Y_i^{(1)} \sim N(\beta_0 + \beta_Z Z_i, \sigma_\varepsilon^2)$$
$$Z_i \sim N(Y^{(2)}, \sigma_u^2) \tag{7.8}$$

there is no attenuation, meaning that $\beta_{Y(2)} = \beta_Z$. However in common with the classical model incorporating measurement error inflates the variance to $Var[Y_i^{(1)}|Y_i^{(2)}] = \sigma_\varepsilon^2 + \beta_Z^2 \sigma_u^2$.

Although the effects of measurement error are well known for the Gaussian linear model, the corresponding effects for more complex linear and non-linear models may be unknown.

7.3.4 Estimation

Here we concentrate on estimation for structural measurement error models. The parameters in the model are collectively denoted by Ω, which can be estimated by maximising the joint likelihood of $(Y^{(1)}, Y^{(2)})$ given by

$$f(Y^{(1)}, Y^{(2)}|X, \Omega) = \int_Z f(Y^{(1)}, Z, Y^{(2)}|X, \Omega)dZ \tag{7.9}$$

As previously described $f(Y^{(1)}, Y^{(1)}|X, \Omega)$ can be factorised according to either classical or Berkson error models.

Classical:

$$f(Y^{(1)}, Z, Y^{(2)}|X, \Omega) = f(Y^{(1)}|Z, X, \omega_1)f(Y^{(2)}|Z, X, \omega_2)f(Z|X, \omega_3) \tag{7.10}$$

or Berkson:

$$f(Y^{(1)}Z, Y^{(2)}|X, \Omega) = f(Y^{(1)}|Z, X, \omega_1)f(Z|Y^{(2)}, X, \omega_2)f(Y^{(2)}|X, \omega_3) \tag{7.11}$$

where $\Omega = (\omega_1, \omega_2, \omega_3)$. However in the Berkson setting $Y^{(2)}$ is assumed to be a known measurement, meaning that Equation 7.9 is replaced by

$$f(Y^{(1)}, Y^{(2)}|X, \Omega) = \int_Z f(Y^{(1)}|Z, X, \omega_1)f(Z|Y^{(2)}, X, \omega_2)dx \tag{7.12}$$

where $Y^{(2)}$ has been conditioned out of the joint likelihood. Likelihood methods estimate Ω by maximising $f(Y^{(1)}, Y^{(2)}|Z, \Omega)$ or $f(Y^{(1)}|X, \Omega)$, where ω_1 is of primary interest because it describes the relationship between the response $Y^{(1)}$ and the true exposure Z. Neither Equation 7.9 nor 7.12 is typically available in closed form. In simple problems where the likelihood can be computed or well approximated analytically Ω can be estimated by iterative numerical methods, for example using the EM algorithm (Schafer, 2001). In more complex cases, Monte Carlo techniques can be used to approximate $f(Y^{(1)}, Y^{(2)}|X, \Omega)$ or $f(Y^{(1)}|X, \Omega)$, an example of which is given by Geyer and Thompson (1992). Alternatively measurement error models can be viewed as missing data problems with Z being a missing covariate and estimation methods from the literature on missing data can be used. Further details of such methods can be found in Little and Rubin (2014).

Bayesian measurement error models comprise one of the likelihoods given by Equations 7.9 or 7.12 and a prior $f(\Omega)$, the latter of which is a product of marginal and conditional distributions. Bayesian inference is based on the posterior distribution of Ω, which for classical and Berkson models is proportional to

$$\text{Classical } f(\Omega|Y^{(1)}, Y^{(2)}, X) \propto f(\Omega) \int_Z f(Y^{(1)}|Z, X, \omega_1) f(Y^{(2)}|Z, X, \omega_2) \times$$
$$\ldots \times f(Z|X, \omega_3) dZ$$

$$\text{Berkson } f(\Omega|Y^{(1)}, Y^{(2)}, X) \propto f(\Omega) \int_Z f(Y^{(1)}|Z, X, \omega_1) f(Z|Y^{(2)}, X, \omega_2) dZ$$

The posterior distribution is typically calculated using MCMC simulation, where Z is treated as additional parameters to be estimated. Both Gibbs and Metropolis–Hastings algorithm have been used, and examples based on conditionally independent models (Richardson & Gilks, 1993), non-linear regression models (Dellaportas & Smith, 1993; Berry, Carroll, & Ruppert, 2002) and Gaussian mixture models (Richardson, Leblond, Jaussent, & Green, 2002).

7.4   Preferential sampling

Preferential sampling is a common phenomenon in environmental studies, as the monitoring locations in a spatial network are often chosen based on a subjective purpose, such as the change of government policies and the intention of monitoring high levels of pollution. For example, if monitors are positioned close to known pollution sources, such as at roadside, near an industrial polluter, or within a city centre, then the estimated pollution surface is likely to be overestimated. Both the number and locations of the pollution monitors will affect the accuracy with which the true exposure surface is estimated. Standard geostatistical methods which assume sampling is non-preferential are often employed despite the presence of a preferential sampling scheme. It is often intrinsically assumed the true exposure surface is based on the random sampling of the complete spatio–temporal pollution field. However, this is extremely unlikely to be the case and the exposure measurements obtained from preferential sampled networks may lead to an inaccurate estimation of exposure to air pollution and consequently to the estimation of relative risks in epidemiological studies.

Recently there have been a small number of papers published on the subject of preferential sampling in an environmental setting, which occurs when the process that determines the locations of the monitoring sites and the process being modelled are in some ways dependent (Diggle et al., 2010; Pati, Reich, & Dunson, 2011; Gelfand, Sahu, & Holland, 2012; Zidek et al., 2014). Diggle et al. (2010) extend the classical geostatistical model in two ways; (i) the monitoring locations are treated as random quantities of a log–Gaussian Cox process rather than being fixed; (ii) the exposures are modelled conditionally on the locations assuming a Gaussian spatial process. Through simulation examples they show that ignoring preferential sampling can lead to misleading inferences, especially with spatial predictions. Pati et al. (2011) adapt this approach within a Bayesian framework and demonstrate its use in a case study of ozone data over the eastern U.S. that shows significant evidence of preferential sampling. Other examples of the application of this approach include Lee, Ferguson, and Scott (2011) who implement it when constructing air quality indicators for a case study set in Greater London.

Gelfand et al. (2012) suggested another approach to deal with the effects of preferential sampling. Again, the locations are treated as a realisation of a random process but they use a deterministic model with informative covariates, such as population density, to indicate the underlying pollution surface. This approach is based on the assumption that if sampling locations are drawn as a reflection of covariate factors, then the covariates should be used in the exposure model to correct the preferential sampling bias. A simulation study shows the spatial predictions of exposures under preferential sampling are substantially biased when compared to those from random sampling. Lee and Shaddick (2010) investigated the influence of preferential sampling to the pollution concentration estimation on spatial prediction using a Bayesian spatio–temporal model, again showing significant biases in spatial predictions.

The majority of research in this area has focused on the predictions of exposure surface in a spatial network. The approach in Diggle et al. (2010) models the spatially continuous unobserved process to be used in the stochastic model of locations, but this process is unknown in practice, so it is difficult to specify its propriety. In addition, only a single realisation of the underlying random field is used to generate the data for the locations in the study region. For the approach proposed by Gelfand, Banerjee, and Gamerman (2005), the difficulty is that it requires the complete information of covariates used in the deterministic model, which is normally unavailable in practice.

An alternative approach to adjusting for preferential sampling from the spatial modelling approaches described above is that of response-biased regression modelling (Scott & Wild, 2001). Zidek et al. (2014) proposed a method to model preferential sampling in environmental networks based on this approach. The idea is based on concepts from survey sampling in which sampling weights define the under- or oversampling of specific demographic groups. Resulting estimates can then be adjusted using the sampling weights to allow for the non-random design. They used the Horwitz–Thomson (HT) estimator to unbias estimates based on preferentially sampled data. In short, the HT estimator weighs each observation against the probability that the particular observation is included in the sample. In the setting considered here however, the sampling weights, which define the process of preferential sampling, are generally not known. The selection probabilities cannot therefore be characterised as they are in survey sampling. The idea of Zidek et al. (2014) was to estimate these probabilities using logistic regression based on concentrations measured in previous years and locations.

Example 7.1. *Long-term monitoring of air pollution in the UK*

Black smoke (BS) has been routinely measured in the UK since the early 1960s as part of the UK Smoke and Sulphur Dioxide network and its predecessor the National Survey. In later years, the network was used to monitor compliance with the relevant EC Directives on sulphur dioxide and suspended particulate matter. The monitoring network, which measures both SO_2 and BS, was established in the early 1960s, and by 1971 included over 1200 sites. As levels of BS and SO_2 pollution have declined, the network has been progressively rationalised and reduced in size and by 2006 when the network ceased to be operational it comprised 65 sites (BS continues to be monitored, but on a much smaller scale (Fuller & Connolly, 2012)).

As the network was reduced in size, there is the possibility of selection bias if there is a tendency for monitoring sites to be kept in the more polluted areas. This may occur for example, if the locations of sites remaining in the network were chosen to assess whether guidelines and policies were being adhered to. This will lead to preferential sampling.

Shaddick and Zidek (2014) considered the change in levels of pollution concentrations in relation to changes in the network from the 1960s until the 2000s; a period for which there was both reduction in the size of the network and in the concentrations being measured. Annual means fell from 237 μgm^{-3} in 1962 to 99 in 1966, 32 in 1976, 19 in 1986, 11 in 1996 and 5 μgm^{-3} in 2006. This period represents a unique period in history containing dramatic declines in the levels of air pollution together with the most marked reduction in the network providing the data on which these reported declines are based. During this period, there were a total of 1466 operational sites of which 35 *consistent* sites were operating throughout the entire period. Figure 7.1 shows the mean concentrations over all sites by year for four groups of sites; (i) the set of 35 consistent sites, (ii) 655 sites which were added (and remained) during the period of study, (iii) 133 sites which were dropped and (iv) 643

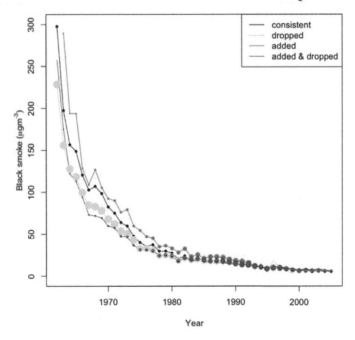

Figure 7.1: Annual means of BS concentrations by year (1962–2006). Dots denote the yearly mean value. The size of the dots is proportional to the number of sites which contribute to the calculation of the mean for each year. Values are given for: (i) sites for which measurements were consistently available throughout the period (black) together with those from sites which were (ii) dropped from (green) and (iii) added to (red) the network within the period of study. Values for (iv) the group of sites which were added and then dropped from the network are given in blue.

sites which were added then dropped. Concentrations are shown for all years for which data were available, with the grouping derived using the main period of analysis. Very similar patterns are seen using different time periods to define the groups, albeit with a smaller number of consistent sites when considering longer periods. There is a clear decline in the level of BS over this period with markedly higher concentrations in the consistent and added sites than those that were dropped, giving the suggestion of preferential sampling.

7.4.1   A method for mitigating the effects of preferential sampling

Zidek et al. (2014) proposes a general framework for dealing with the effects of preferential sampling in environmental monitoring. Strategies for implementation are proposed, leading to a method for improving the accuracy of official statistics used to report trends and inform regulatory policy. An essential feature of the method is its capacity to learn the preferential selection process over time and hence to reduce bias in these statistics, such as annual means and exceedances

As noted in discussion preceding this example, an alternative method can be found from principles underlying survey sampling, in which sampling weights, which are part of the sampling design and may for example over-sample certain age groups, are used to correct the results of the survey (Scott & Wild, 2011). However, in this setting the location of sites in monitoring networks is not often the result of a carefully designed study design but instead site selection is complex involving committees, guidelines and negotiations, and political considerations. In practice then the process that selects the monitoring sites is non-random and generally not known. Thus the selection probabilities cannot be characterised as they are in multi-stage survey sampling, for example. However having a time series of samples of sites from the finite population of N possible sites enables the selection process to be modelled and those probabilities estimated and Zidek et al. (2014) describe a logistic regression approach that may be used when such information is available.

It is assumed that at time t, the sample of S_t among the population of N sites is selected by a PPS (probability proportional to size) sample survey design, $u \in S_t$ being included with probability π_{tu}. That probability is assumed to depend on all responses, both on observed data but also the unmeasured responses at the un-sampled sites over the time period $1 : (t-1)$, the latter of which is treated as latent variables. Thus in terms of the measured and unmeasured responses, Y and vector of binary indicators of selected/rejected sites R, the conditional distribution of the probability of selection is

$$
\begin{aligned}
logit[\pi_{tu}] &= logit[P(R_{tu} = 1 \mid \mathbf{y}^{(1)}{}_{1:(t-1)}, \mathbf{r}_{1:(t-1)})] \\
&= G(\mathbf{y}^{(1)}{}_{1:(t-1)}, \mathbf{r}_{1:(t-1)})
\end{aligned}
\tag{7.13}
$$

for some function $0 \leq G \leq 1$. That function is analogous of the preferential sampling intensity seen in Diggle (2010).

Under the assumption of a superpopulation model (see Zidek, 2014 for details) there will be a predictive probability distribution for the unmeasured responses. Values for these might be obtained using for example geostatistical methods as described in Chapter 9, which under repeated imputation will allow $k = 1, \ldots, K$ replicate datasets. Each replicate enables G to be fitted through logistic regression to get $\hat{G}^k$, $k = 1, \ldots, K$ and in turn $\hat{\pi}_{tu}^k$, $k = 1, \ldots, K$.

From these replicates multiple values of the Horvitz–Thompson estimator can be obtained. In a spatial setting R is defined to be a sampled site indicator so that R_u is

1 or 0 according as site u is selected into the sample or not. Let

$$\pi_u = \pi(y_u, x_u) = P\{R_u = 1 | y_u, x_u\}$$

be the selection probability for site u. The HT approach estimates the first-order parameter β above by solving the estimating equations

$$\sum_u \frac{R_u}{\pi_u} \frac{\partial \log[y_u | x_u, \beta]}{\partial \beta} = 0 \qquad (7.14)$$

assuming $\pi_u > 0, u \in \mathscr{S}$ are known at the sampled sites.

In the simplified case of a strictly decreasing network, at each time point the logistic regression model described in Equation 7.13 is used to predict the inclusion probabilities at each time point, t,

$$logit(\hat{\pi}_{tu}) = \beta_0 + \beta_1 [y_{tu}^{(1)} - \bar{y}_{t\cdot}^{(1)}] \qquad (7.15)$$

for the N binary select–reject indicators for all sites u where $y^{(1)}$ are the observed values at time t which have mean $\bar{y}_{t\cdot}^{(1)}$.

The unconditional site selection probabilities are then calculated, $\hat{\pi}_{tu}^*$. Note that $u \in S_t$ implies $u \in S_{t'}$, $t' \le t$ so under the assumption of no autocorrelation

$$\hat{\pi}_{tu}^* = \Pi_{t=1}^t \hat{\pi}_{tu}$$

Horvitz–Thompson adjusted summaries can then be calculated for each time point. The estimate of the annual mean will be

$$\hat{\mu}_t = \sum_{u \in S_t} \frac{y_{tu}^{(1)}}{N \hat{\pi}_{tu}^*}$$

Example 7.2. *Adjusting annual means and exceedances of black smoke*

As seen in Example 7.2, there is evidence of preferential sampling in the long-term network of black smoke in the UK. Here we present adjustments for two characteristics associated with the responses which are of interest. The first is the set of annual averages as this information could be published to show the effect of regulatory policy over time. The left hand panel of Figure 7.2 shows the estimated geometric annual mean levels over time (blue line) together with the Horvitz–Thompson adjusted ones (red line). It clearly shows the adjustment reduces the estimates of the average levels. Since the standard unit for calculating relative risks of particulates in health effects analysis is $10\mu gm^{-3}$ the difference seems important, being more than one of these standard units over much of the period.

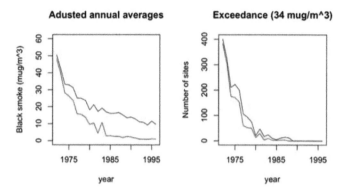

Figure 7.2: Changes in the levels of black smoke within the UK from 1970-1996 and the effects of adjusting estimates of annual indices. The left hand panel shows the annual means over all sites (blue line) together with adjusted values (red line). The right hand panel shows the number of sites exceeding the EU guide value of 34 μgm^{-3} (blue line) together with the corrected values (red line).

The second characteristic may potentially be of even greater operational importance; the number of sites in non-attainment, that do not comply with air quality standards in any year. This number is a surrogate for the cost of mitigation for putting the BS concentrations into compliance. The right hand panel of Figure 7.2 shows the number of sites each year that exceed the 1980 EU guide value of 34 μgm^{-3} (EC, 1980). The blue line is the number of exceeding sites based on the recorded data with the red line the numbers after adjustment for the preferential sampling. The unadjusted numbers are the fraction in the monitoring network out-of-compliance multiplied by the finite population total of $N = 624$. Their adjusted counterparts are found by using the Horvitz–Thompson weights in the summation used to calculate that fraction. For example, in 1974, the crude estimate gives 211 of the 624 sites out of compliance with the 34 μgm^{-3} criterion, while its adjusted counterpart is just 189. This is a substantial difference, especially considering the large costs that can be involved in achieved compliance.

7.5 Summary

This chapter considers some of the issues that will arise when dealing with 'real data'. Data will commonly have missing values and may be measured with error. This error might be random or may be due to systematic patterns in how it was collected. The reader will have gained an understanding of the following topics:

- Classification of missing values into missing at random or not at random.
- Methods for imputing missing values.

- Various measurement models including classical and Berkson.
- The attenuation of regression coefficients under measurement error.
- Preferential sampling, where the process that determines the locations of monitoring sites and the process being modelled are in some ways dependent.
- How preferential sampling can bias the measurements that arise from environmental monitoring networks.

Exercises

Exercise 7.1. The `Amelia` R package provides several methods for estimating missing values using multiple imputation.

(i) Investigate the use of multiple imputation by analysing the `freetrade` dataset supplied with the package. Produce ten simulated datasets with missing values imputed.

(ii) Perform an analysis of the variation between the simulated datasets and show your results graphically.

(iii) Show how you might use these multiple datasets in a regression analysis where `tariff` is the response variable. How will you ensure that the variability between the datasets is reflected in the standard errors associated with the regression coefficients?

Exercise 7.2. Show that the attenuation factor for the measurement error model shown in Equation (7.8) is $\beta_{Y(2)} = \frac{\sigma_z^2}{\sigma_z^2 + \sigma_{v2}^2} \beta_z$

Exercise 7.3. (i) Simulate a set of true exposures, Z based model shown in Equation (7.8) using a fixed variance, σ_Z^2.

(ii) For a value of $\beta_Z = 2$ generate a set of $Y^{(1)}$'s (using a fixed σ_{v1}^2) and perform a linear regression. Note your estimate of β_Z.

(iii) Generate a set of data measured with error, $Y^{(2)}$, for a fixed σ_{v2}^2 and perform a linear regression of $Y^{(1)}$ on $Y^{(1)}$.

(iv) Compare the two values of β you get from the two models. Does the difference between them correspond to the attenuation factor you would obtain if you used the equation in part (i)?

(v) Show the effect of the attenuation graphically.

Exercise 7.4. Repeat the simulation exercise in Exercise 7.3 using different values for the variances, σ_{v1}^2 and σ_{v2}^2. What do you conclude about the relative importance of these two variances?

Exercise 7.5. Repeat the simulation exercise in Exercise 7.3 but now generate the set of responses, $Y^{(1)}$, using a Poisson distribution. Note, the linear part of the model will now be on the log scale and so care should be taken in regards to the magnitude of the values that will be exponentiated. What do you conclude about the effects of measurement error in this case? For further reading on the subject of measurement error in non-linear models see Carroll et al. (1995)

Exercise 7.6. The data and code to perform the preferential sampling approach given in Diggle et al. (2010) is available on the R–INLA website. Use this to perform an analysis of the data from Galicia shown in the paper, with and without adjustment for preferential sampling.

Chapter 8

Spatial patterns in disease

8.1 Overview

Disease mapping has a long history in epidemiology starting with John Snow's map of cholera cases in London in 1854 (Hempel, 2014). The aims of disease mapping range from simple spatial description of health data and hypothesis generation to the estimation of risks over space. In the latter, allowance should be made for differing sized populations which may provide varying levels of uncertainty in the estimation of risks. Area-based mapping has also been used for the assessment of inequalities and the allocation of health care resources.

In terms of the areas used for disease mapping, there will be a trade-off in relation to the geographical scale that used: rates calculated using larger geographical areas will be more stable but summaries of risk, such as relative risks, may be affected by the aggregation of large numbers of individuals. If there is substantial variation between risks within a particular area this information will be lost. An effect of a high risk in a subregion will be diluted under aggregation. Detecting elevated risks in such subregions will not be possible unless data are available at a lower level of aggregation.

For small areas, and in particular rare diseases, estimates of risk may be dominated by sampling variability. This issue has led to methods being developed to produce 'smoothed' estimates of risk using hierarchical/random effects models. These use data from the individual areas together with global measures of risk estimated using data from all the areas to provide more reliable estimates in each of the constituent areas.

In disease mapping we assume that the spatial data is in the lattice domain $L = \mathscr{L} = \{1, \ldots, N_L\}$ as described in Chapter 1, Section 1.6. We assume that the spatial data is in the lattice domain $L = \mathscr{L} = \{1, \ldots, N_L\}$. An important body of theory underlies these kinds of spatial process and this is described in this chapter. In particular we will see the importance of the Hammersley–Clifford theorem in the underlying theory of the models that are commonly used for disease mapping, such as conditional autoregressive models.

8.1.1 Smoothing models

We assume that disease counts, Y_l, in an area l are Poisson distributed with the rate being a combination of the overall relative risk, μ, multiplied by the expected number of health outcomes, E_l (as seen in Chapter 2, Section 2.8) and the risk in that particular area, θ_l. For simplicity, here we assume there are no covariates. The model for the health counts in each is given by

$$Y_l|\theta_l,\beta \sim_{ind} \text{Poisson}(\mu E_l \theta_l) \tag{8.1}$$

for all $l \in L$ where *ind* means 'independently distributed' and μ is the overall relative risk that reflects differences between the reference rates and the rates in the study region as a whole.

At the second stage, the random effects θ_l are assigned a distribution which will reflect the deviations of the relative risks from the overall mean, μ. A common choice of distribution is the Gamma distribution which is conjugate to the Poisson as seen in Chapter 5, Example 5.2. They are modelled by

$$\theta_l|a \sim_{lld} \text{Gam}(a,a) \tag{8.2}$$

Here, the relative risks for each individual area have a gamma distribution with mean 1, and variance $1/a$.

In this case, the marginal distribution of $Y_l|\mu,a$ is negative binomial. This is obtained by integrating out the random effects θ_l; $p(Y|\mu,\theta) = \int_\theta p(Y|\mu,\theta,a)d\theta$.

Marginally, the mean and variance are given, respectively, by

$$\begin{aligned} E[Y_l|\mu,a] &= E_l\mu \\ \text{Var}(Y_l|\mu,a) &= E[Y_l|\mu,a](1+E[Y_l|\mu,a]/a) \end{aligned} \tag{8.3}$$

Therefore the variance increases as a quadratic function of the mean, and the scale parameter α can accommodate different levels of 'overdispersion' (see Chapter 2, Section 2.8.2).

8.1.2 Empirical Bayes smoothing

In empirical Bayes the prior distribution is estimated from the data (Carlin & Louis, 2000). This is in contrast to fully Bayesian analyses in which the prior distribution is chosen before any data are observed. If we have estimates $\widehat{\mu}$, $\widehat{a}$ then the posterior distribution would be

$$\theta_l|\mathbf{y},\widehat{\mu},\widehat{a} \sim \text{Ga}(\widehat{a}+y_l,\widehat{a}+E_l\widehat{\mu})$$

The relative risk that would be applied to the expected numbers of health outcomes,

E_l, is given by $RR_l = \mu \theta_l$, and has mean

$$
\begin{aligned}
\widehat{RR_l} &= \hat{\mu} \times E[\theta_l | \text{by}, \hat{\mu}, \hat{a}] = \hat{\mu} \left(\frac{\hat{a} + Y_l}{\hat{a} + \hat{\mu} E_l} \right) \\
&= E[RR_l] \times (1 - w_l) + SMR_l \times w_l
\end{aligned}
\tag{8.4}
$$

This is a weighted combination of the prior estimate $E[RR_l] = \mu$, which will be the global RR over the entire set of areas, $l \in L$, and the SMR (Y_l / E_l) in area l.

The *weight*,

$$
w_l = \frac{E_l \hat{\mu}}{\hat{a} + E_l \hat{\mu}}
$$

given to the observed SMR using the data increases as E_l, which represents the size of the population, increases. Therefore, for areas with large populations the estimate is dominated by the data in that area rather than the overall RR. If a is large then there will be less variability in the random effects than with small a and so there will be more shrinkage towards the overall RR.

It can be seen that the estimates will be less variable than the original SMRs. However, this does mean that a very high SMR, which might be important in detecting potential risk factors, may be shrunk if it is based on a small population and so may be overlooked.

Example 8.1. *Empirical Bayes and Bayes smoothing of COPD mortality for 2010*

Here we consider hospitalisation for a respiratory condition, chronic obstructive pulmonary disease (COPD), in England for 2010. There are $N_l = 324$ local authority administrative areas each with an observed, Y_l and expected, E_l, number of cases, $l = 1, ..., 324$. As described in Section 2.4 the expected numbers were calculated using indirect standardisation by applying the age–sex specific rates for the whole of England to the age–sex population profile of each of the areas. In order to perform empirical Bayes smoothing we will use the eBayes function in the SpatialEpi package.

```
# requires SpatialEpi package
library(SpatialEpi)
RRs = eBayes(Y,E)
plot(RRs$SMR, RRs$RR, xlim=c(0,2),ylim=c(0,2), xlab="SMRs",
    ylab="RRs")
abline(a=0,b=1, col="red", lwd=3)
```

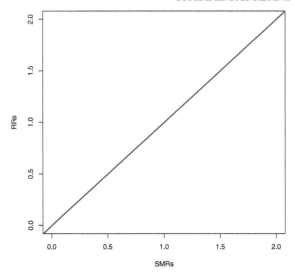

Figure 8.1: The effect of smoothing SMRs using an empirical Bayes approach. The red line has a slope of one and helps show the shrinkage effect with estimates being brought towards the overall mean relative risk.

The result can be seen in Figure 8.1 in which the smoothing can be seen with lower and higher SMRs being brought close to the overall average, μ which in this case is $\exp(-0.0309) = 0.9696$. Note that in this example, the areas are relatively large and would be expected to have substantial populations, so the effect of the smoothing is limited. In this example, α was estimated to be 14.6.

For a fully Bayesian, rather than empirical Bayes analysis, we can use WinBUGS. The code is as follows:

```
model
{
  for (i in 1 : N) {
      Y[i]    ~ dpois(mu[i])
      mu[i] <- E[i]*exp(beta0)*theta[i]
      RR[i] <- exp(beta0)*theta[i]
      theta[i] ~ dgamma(alpha,alpha)
  }
# Priors
  alpha ~ dgamma(1,1)
  beta0 ~ dflat()
# standard deviation of non--spatial
  sigma.theta <- sqrt(1/alpha)
    overall.risk <- exp(beta0)
}
```

8.2   The Markov random field (MRF)

We now introduce some basic theory that underlies the processes used in disease mapping. The starting point is the seminal paper of Besag (1974) which introduced what has proven to be an very important idea for modelling areal data, the Markov random field. For a fixed time point a lattice process, Z_l, over domain $L = \{1, \ldots, N_L\}$, $Z_l, l \in L$ has a distribution determined by its conditional distribution given its values at the neighbouring points. Thus if for example Z_l were observed with error, yielding Y_l then the conditional distribution could be used for inference about Z_l by borrowing strength from data collected at the neighbouring lattice points.

In his celebrated paper, Besag builds on earlier work, including one of the most famous unpublished papers in the history of statistical science, the 1971 paper of Hammersley and Clifford. In personal communications with the second author, Peter Clifford said he was a postdoctoral summer visitor at the University of California, Berkeley when the by then distinguished Hammersley delivered a series of lectures, one of which contained a false claim that sparked the collaboration that led to the famous result. As we understand it, the authors were not satisfied with the progress they made that summer and did not feel the result worthy of submission for publication!

As with Hammersley and Clifford, Besag makes an important assumption of positivity throughout his paper. More precisely, after ordering the lattice points in an arbitrary order (which won't matter as far as the results are concerned) let $\mathbf{Z} = (Z_1, \ldots, Z_{N_L})$, and f denote its joint PDF, i.e. $f_{\mathbf{Z}}(\mathbf{z}) = f(\mathbf{z})$ while $f(z_l)$ denotes that of Z_l for lattice point l. Then positivity means that $f(z_l) > 0$ for all l implies $f(z_1, \ldots, z_{N_L}) > 0$ over the range of Z that is $\mathscr{Z} = \{\mathbf{z} : f(z_1, \ldots, z_{N_L}) > 0\}$.

Under the positivity assumption, Besag presents the surprising result now referred to as Brook's lemma (Brook, 1964), that for any two points $\mathbf{z}, \mathbf{z}^0 \in \mathscr{Z}$, we have

$$f(\mathbf{z}) = f(\mathbf{z}^0) \Pi_{j=1}^{N_L} \frac{f(z_j \mid z_1, \ldots, z_{j-1}, z_{j+1}^0 \cdots, z_{N_L}^0)}{f(z_j^0 \mid z_1, \ldots, z_{j-1}, z_{j+1}^0 \cdots, z_{N_L}^0)} \qquad (8.5)$$

As observed in Banerjee et al. (2015), this means that the joint distribution is determined by the full set of conditional densities for $\mathbf{Z}$, with each coordinate conditional on all the others. Knowing these conditionals, fixing $\mathbf{z}^0$, then determining the second factor in Equation (8.5) as a function $\mathbf{z}$ and finally integrating the right hand side over $\mathbf{z}$ would determine $f(\mathbf{z}^0)$. That is because the latter does not depend on $\mathbf{z}$ while the left hand side of the equation would become 1. That discovery is important for Gibbs sampling as it means sampling from those conditionals in a sequential manner would yield samples from the joint distribution, which is at the heart of Gibbs sampling.

Besag recognises some important implications of Brook's lemma in modelling lattice random fields. He states that the lemma highlights a fundamental difficulty:

"... concerning the specification of a system through its conditional probability structure. ...the labelling of individual sites in the system being arbitrary implies that many factorizations of $P(\mathbf{x})/P(\mathbf{y})$ are possible. All ... must be equivalent and this ... implies the existence of severe restrictions on the available functional forms...to achieve a mathematically consistent joint probability structure."

In other words, the form of the distribution must not depend on the way in which we happened to label the lattice points from 1 to N_L.

The Brook's lemma would not be very useful for modelling the joint distribution of a random field when N_L is large as specifying all these conditionals would not be easy. More preferable would be a simplified model where the conditional distribution of a site depends on just a small subset of them. Besag approaches this using a neighbourhood structure. Using the notation in Banerjee, Gelfand, and Carlin (2004), the set $\partial_l \subset L$ is a neighbourhood of l if it satisfies

$$[Z_l \mid \{Z_{l'} : l' \in L\}] = [Z_l \mid \{Z_{l'} : l' \in \partial_l\}] \tag{8.6}$$

where in general the bracket notion $[U]$, $[U \mid V = v]$ and $[U \mid A]$ for arbitrary random variables U and V and an event A means respectively the distribution of U, the conditional distribution of U given $V = v$ and the conditional distribution of U given that the event A occurs.

It would greatly simplify things if the joint distribution were determined by conditional distributions given just the responses for the neighbourhoods. For example, if the neighbourhood of each response were the null set, meaning the responses at each lattice point were completely independent of all the other responses. If this were the cases, then the joint distribution would be completely specified just by its marginals.

The problem of finding a set of such neighbourhoods proves challenging. Yet in practice that neighbourhood structure would need to be specified by the modeller for all $l \in L$, as in Equation 8.6, along with the conditional distributions. In a regular lattice for example, these may be the nearest four points to the north, south, east and west of l in the lattice. Then again there are cases where this not be appropriate, for example widely separated lattice points on high peaks on opposite sides of a valley may have very similar temperatures (see Example 13.14 for further discussion on this point).

Even at the theoretical level finding admissible neighbourhood structures is difficult. Initially, based on Besag's quotation above we anticipate a strong restriction that these neighbours must be invariant in some sense under permutations of the site labels. However, given such a set of invariant neighbourhoods, what kind of joint distributions would be consistent with them? Does such a joint distribution even exist?

There is certainly no reason to be optimistic, after all specifying just local dependence properties of the process is far from specifying a *bona fide* joint distribution over the whole lattice domain. That is the central issue Besag addressed in his paper and his representation of joint lattice process probabilities is a central result.

That result is given first in the discrete case, albeit by a somewhat heuristic argument. A rigorous, constructive proof of that representation is given in a general theorem of Hosseini, Le, and Zidek (2011) where like Besag they have to exclude 0 from the list of possible values.

Theorem 1. Suppose, $h : \prod_{i=1}^{p} M_l \to R$, M_l being finite with $|M_l| = c_l$ and $0 \in M_l$, $\forall l$, $1 \leq l \leq p$. Let $M_l^* = M_l - \{0\}$. Then there exists a unique family of functions

$$\{G_{i_1,\cdots,i_k} : M_{i_1}^* \times M_{i_2}^* \times \cdots \times M_{i_k}^* \to R, \ 1 \leq k \leq r, \ 1 \leq i_1 < i_2 < \cdots < i_k \leq p\}$$

such that

$$\begin{aligned}
h(z_1,\cdots,z_p) = \ & h(0,\cdots,0) + \sum_{i=1}^{p} z_l G_l(z_l) + \cdots + \\
& \sum_{1 \leq i_1 < i_2 < \ldots < i_k \leq r} (z_{i_1} \cdots z_{i_k}) G_{i_1,\cdots,i_k}(z_{i_1},\cdots,z_{i_k}) \\
& + \cdots + (z_1 z_2 \cdots z_{N_L}) G_{12\cdots p}(z_1,\cdots,z_{N_L})
\end{aligned} \qquad (8.7)$$

In an extension of this theorem, Hosseini et al. (2011) replace 0 by an arbitrary point to get a more general representation. They apply their result to derive a class of Markov chain models in a temporal context. There, since time is ordered and invariance under permutations would not make sense, the issues around invariant neighbourhood structures do not arise.

Leaving the general result above and returning to the case of lattice processes, we note that Besag applies his result to the function

$$h = \log \{f(\mathbf{z})/f(\mathbf{0})\} \qquad (8.8)$$

assuming $f(\mathbf{0}) > 0$ and embraces the need expressed in his quotation above by incorporating in his expansion, the idea of a 'clique'. This is a set, C, of location indices with the property that every one of its points is in the neighbourhood of every other point in C thus ensuring its invariance under a permutation of their site labels. In the context of his application, the functions G in the expansion above are non-null only if the subscripts correspond to the members of a clique, for otherwise the required invariance will be lost.

We now turn to the Hammersley–Clifford theorem and the work of Geman and Geman (1984), which gives its inverse. Banerjee et al. (2015) give a very clear account of that work. We start with the set of all cliques of size $l = 1,\ldots,N_L$ in the expansion shown in Equation 8.7. Any clique $c_j \subset L$ of size j corresponds to a vector $\mathbf{z}_{c_j}$ of values. By inverting Besag's function h in the form given by Geman and Geman (and presented by Banerjee et al., 2015) we get the density function of the so-called Gibbs distribution named after Josiah Willard Gibbs (1803–1903) from his original work in statistical mechanics:

$$f(\mathbf{z}) = \exp\left\{ \tau \left(\sum_{c_1 \subset L} H_1(\mathbf{z}_{c_1}) + \sum_{c_2 \subset L} H_2(\mathbf{z}_{c_2}) + \ldots + c_{N_L} H_{N_L}(\mathbf{z}) \right) \right\} \qquad (8.9)$$

The H's, which are required to be invariant under permutations of their arguments, are called potentials, τ the 'temperature', and the expression in brackets in Equation 8.9, the energy function. The Hammersley–Clifford theorem says a valid MRF model must yield a Gibbs distribution while Geman's paper proves the inverse. An update to this theory can be found in Kaiser and Cressie (2000).

Example 8.2. *The Ising model*

One well known Gibbs distribution comes from the Ising model where the process is binary and each $z \in \{-1, 1\}$. The distribution models pairwise interactions between the sites through

$$f(\mathbf{z}) = \exp(\tau \sum_{(l,l') \in L_2} z_l z_{l'})$$

where L_2 defines pairs of neighbouring sites, which are neighbours of each other. In other words, $c_2 = \{l, l'\}$, $\mathbf{z}_{c_2} = (z_l, z_{l'})$ and $H_2(\mathbf{z}_{c_2}) = z_l z_{l'}$. When the temperature is high, the probability tends to higher when the values at different sites are alike.

An assessment of the MRF approach

The MRF approach would be considered in situations where processes of interest manifest themselves through observable responses that generate the data. It is assumed that such data are available for all indices $\{l\}$ and it provides a natural way of building up a Bayesian hierarchical model in such situations.

However, identifying appropriate neighbourhoods may be difficult. (See Section 8.4.1 for discussion of this issue). Unlike classical geostatistical methods, which allow extrapolation as well as smoothing, the MRF will only be of value when there are neighbours. Finally, the change of support problem can make this approach difficult to use, in other words Z may be measured at point-referenced points s while Y is for lattice points l.

8.3 The conditional autoregressive (CAR) model

Since space, unlike time, is not ordered we cannot use the classes of models such as the AR(1) that depend on that ordering. However, Besag proposes a spatial autoregressive model that resembles its temporal counterpart that is known as the conditional autoregressive (CAR) model.

We begin with the case of a Gaussian random field . The CAR model is built on the lattice domain L. Assume the model for the observables and process are the same, in other words no measurement error. So we have an observable response of interest Y_l at the point l, i.e. the process and measurement models coincide.

To get to the autoregressive part of the model let $\mathbf{Y}_{-l}$ denote the set observables after removing Y_l and let ∂_l denote the neighbours of l. Then the CAR model is defined as follows for all l:

$$[Y_l \mid Y_{-l} = y_{-l}] \quad = \quad N\left(\mu_l, \sigma_l^2\right) \tag{8.10}$$

where $\mu_l = E(Y_l|Y_{-l} = y_{-l}) = \sum_{l' \in \partial_l} c_{ll'} Y_{l'}$, $\sigma_l^2 = Var(Y_l|Y_{-l} = y_{-l})$. It now follows that

$$f(\mathbf{y}) \propto \exp\left[-\frac{1}{2}\mathbf{y}^T L^{-1}(I - C)\mathbf{y}\right] \tag{8.11}$$

where $D = diag\{\sigma_1^2, \ldots, \sigma_{N_L}^2\}$, $C = (c_{ll'})$ with $c_{ll} = 0$, and $c_{ll'}\sigma_{l'}^2 = c_{l'l}\sigma_l^2$, the last condition being to ensure that $L^{-1}(I - C)$ is a positive definite symmetric matrix (Gelfand et al., 2005). The result can be extended to include covariates as follows for all l

$$Y_l \quad \sim \quad N\left(\mu_l, \sigma_l^2\right) \tag{8.12}$$

where $\mu_l = \mathbf{x}_l^T \beta + \sum_{l'} c_{ll'}(y_l - \mathbf{x}_{l'}^T \beta)$ and σ_l^2 is defined as before to be the residual covariance matrix, $c_{ll'} > 0$, $l \neq l'$ and $c_{ll} = 0$ (De Oliveira, 2012).

Equation 8.10 defines both the marginal response models for each site response, as well as conditional distribution for each such response given its neighbours. There is the question of whether these equations define a joint distribution for all the responses indexed by the elements of L, that is of the vector $(Y_1, \ldots, Y_{N_L})$. The answer is that they do under reasonable conditions for the Gaussian random field (Besag, 1974).

In the CAR model, the advantage of the neighbourhood structure has the disadvantage that the coefficients $c_{ll'}$ now need to be specified. An obvious and intuitively appealing solution to this problem is based on the idea the conditional mean in Equation 8.10 should be a weighted average of means in the neighbouring points. For this purpose we need the idea of adjacency weights $w_{ll'}$ that are non-negative only if l and l' are adjoining points while $w_{ll} = 0$. We now take $c_{ll'} = w_{ll'}/w_{l\cdot}, l \neq l'$ and $\sigma_l^2 = \tau/w_{l\cdot}$, the $\cdot$ notation meaning summation over that subscript. This way we can take account of the degree to which i is seen as like l when it comes to borrowing strength. We now find that $D^{-1}(I - C) = \tau^{-2}(C_w - W)$ in Equation 8.11 where $C_w = diag\{w_{1\cdot}, \ldots, w_{N_L\cdot}\}$. We also find this distribution is singular. That is because the sum of the columns of $(C_w - W)$ is a vector of zeros (Banerjee et al., 2015),

$$\mathbf{y}^T D^{-1}(I - C)\mathbf{y} = \sum_{l \neq l'} w_{ll'}(y_l - y_l')^2 \tag{8.13}$$

This singularity makes this joint distribution unsuitable since its integrable would be infinite even though all its conditional distributions are proper. Nevertheless it is used in practice due to the latter property and it is called an intrinsically autoregressive (IAR) process.

One way around this difficulty is to introduce an additional parameter and replace W with ϕW in this case with a suitable ϕ so that the covariance matrix has full rank, that is the columns of the $(C_w - \phi W)$ no longer sum to one (De Oliveira, 2012).

Other versions of the CAR model have been developed, notably a multivariate extension called the MCAR model (Gelfand et al., 2005).

8.3.1   The intrinsic conditional autoregressive (ICAR) model

A common approach is to assign the spatial random effects an intrinsic conditional autoregressive (ICAR) prior.

Under this specification it is assumed that

$$Y_l | Y_{l'}, l' \in \partial_l \sim N\left(\overline{Y}_l, \frac{\omega_u^2}{m_l}\right)$$

where ∂l is the set of neighbours of lattice point l, m_l is the number of neighbours, and $\overline{Y}_l$ is the mean of the spatial random effects of these neighbours. The parameter ω_u^2 is a conditional variance and its magnitude determines the amount of spatial variation. The variance parameters σ_v^2 and ω_u^2 are on different scales, σ_v is on the log odds scale while ω_u is on the log–odds scale, *conditional* on $U_{l'}', l' \in \partial_l$; hence they are not comparable (in contrast to the joint model in which σ_u is on the same scale as σ_v). Notice that if ω_u^2 is 'small' then although the residual is strongly dependent on the neighbouring value the overall contribution to the residual relative risk is small. This is a little counterintuitive but stems from spatial models having two aspects, strength of dependence and total amount of spatial dependence, and in the ICAR model there is only a single parameter which controls both aspects.

8.3.2   *The simultaneous autoregressive (SAR) model*

The simultaneous autoregressive (SAR) model, a very natural choice, resembles the CAR (Whittle, 1954) but is specified through a regression model resembling the AR(1) model. Thus it is defined this way when we ignore measurement error:

$$Y_l - \mu_l = \sum_{l'} c_{ll'}(Y_{l'} - \mu_{l'}) + v_l \qquad (8.14)$$

where $v_l \sim indN(0, \sigma_l^2)$ and $c_{ll} = 0$. Thus Y is regressed on all the remaining responses. In vector–matrix form:

$$\mathbf{Y} - \mu = \mathbf{C}(\mathbf{Y} - \mu) + \mathbf{v} \qquad (8.15)$$

where $\mathbf{C}$ is chosen to be invertible. It follows that

$$(\mathbf{I} - \mathbf{C})(\mathbf{Y} - \mu) = \mathbf{v}$$

from which it follows that

$$\mathbf{Y} = \mu + \mathbf{v}^*$$

where $\mathbf{v}^* \sim N_{N_L}(0, (\mathbf{I} - \mathbf{C})^{-1}\Sigma(\mathbf{I} - \mathbf{C}^T)^{-1})$ with $\Sigma = diag\{\sigma_1^2, \ldots, \sigma_{N_L}^2\}$. This model captures spatial independence through the mean structure—a moving average of the $\{v_l\}$.

Banerjee et al. (2015) note that the SAR is commonly used to model regression residuals where $\mu = \mathbf{X}\beta$ incorporates through the design matrix $\mathbf{X}$, the covariates that might explain the spatial variation in the process. Then we have for the residuals $\mathbf{U} = \mathbf{Y} - \mathbf{X}\beta$ the equation $\mathbf{U} = \mathbf{C}\mathbf{U} + v$ or equivalently $\mathbf{Y} = \mathbf{C}\mathbf{Y} + (\mathbf{I} - \mathbf{C}^T)\mathbf{X}\beta + \mathbf{v}$. In this form we see $\mathbf{Y}$ as a weighted mixture of $\mathbf{Y}$, recalling that the diagonal elements of $\mathbf{C}$ i.e. $c_{ll} = 0$, and a regression model for the spatial mean are based on covariates. Thus we are borrowing strength from the neighbours as well as the covariates.

8.4   Spatial models for disease mapping

In general we might expect residual relative risks in areas that are 'close' to or more similar, than in areas that are further apart. Ideally we would exploit this information and provide more reliable estimates of relative risk in each area. This is analogous to the use of a covariate; if we knew that a covariate, X, had an effect on risk then we might expect areas with similar X values to have similar relative risks. The idea here is that spatial location is acting as a surrogate for unobserved covariates that will induce a spatial pattern.

8.4.1   *Poisson–lognormal models*

The Poisson–gamma model is analytically tractable but does not easily allow the incorporation of spatial random effects. Spatial random effects can be incorporated in a Poisson log–normal model in a straightforward and interpretable fashion. Whereas in the Poisson–Gamma model we have $\theta \sim \text{Ga}(\alpha, \alpha)$, here we have $\theta = e^{V_l} \sim \text{LogNormal}(0, \sigma^2)$. A Poisson–lognormal non-spatial random effect model is given by

$$Y_l | \beta, V_l \sim_{ind} \text{Poisson}(E_l \mu_l e^{V_l}) \quad V_l \sim_{lld} N(0, \sigma_v^2) \qquad (8.16)$$

where V_l are area-specific random effects that capture the unexplained (log) relative risk of disease in area $l \in L, L = l_1, \ldots, l_{N_L}$.

However, the model shown in Equation 8.16 does not give a marginal distribution of known form. The marginal variance is of the same quadratic form as that seen in Equation 8.3 and Chapter 2, Section 2.8.2. It order to implement this model it is easier to consider a fully Bayesian analysis rather than an empirical Bayes approach. Therefore we need to specify prior distributions.

We need to specify priors for the regression coefficients, β, and the variance of the random effects σ_v^2. For a rare disease, a log–linear link is a natural choice for the link function (between the response and the linear predictor):

$$\log \mu(\mathbf{x}_l, \beta) = \beta_0 + \sum_{j=1}^{J} \beta_l' x_{lj}$$

where x_{ij} is the value of the j-th covariate in area l. For regression parameters $\beta = (\beta_0, \beta_1, ..., \beta_J)$, vague, e.g. $N(0, 1000)$ or improper priors $p(\beta) \propto 1$ are often used. However, in some circumstances that latter choice may lead to an improper posterior. If there are a large number of covariates, or high dependence amongst them, more informative priors may be required.

Extending the standard Poisson log–linear model to include random effects, U_l with spatial structure in addition to the purely random ones, V_l, gives,

$$Y_l | \beta, \gamma, U_l, V_l \sim_{ind} \text{Poisson}(E_l \mu_l e^{U_l + V_l}) \qquad (8.17)$$

where, in addition to the effects of (local) covariates, the mean term may also capture large-scale spatial trend.

Here we consider *conditional* models that are amenable to area level (lattice) data as discussed in Section 8.3. As discussed there, in this approach we specify the distribution of each U_l as if we knew the values of the spatial random effects $U_{l'}$ in 'neighbouring areas'. In order to do this, we need to be able to specify the 'neighbours' of each area. A simple approach is to define areas l' and l as neighbours if they share a *common boundary*. In Chapter 9 we describe *joint* models for spatial effects using point level spatially referenced data.

Example 8.3. *Fitting a conditional spatial model in WinBUGS*

In this example, we see how to fit the Poisson log–normal model seen in Section 8.4 with spatial random effects coming from the ICAR model described in Section 8.3.1.

The ICAR model can be specified via the function car.normal:

$$U[1:N] \sim \text{car.normal}(\text{adj}[], \text{weights}[], \text{num}[], \text{tau})$$

where:

- adj[] : A vector listing the ID numbers of the adjacent areas for each area.
- weights[] : A vector the same length as adj[] giving unnormalized weights associated with each pair of areas.
- num[] : A vector of length N (the total number of areas) giving the number of neighbours n_l for each area.
- The car.normal distribution is parameterized to include a sum-to-zero constraint on the random effects. A separate intercept term must be used in the model and this must be assigned an improper uniform prior using the dflat() distribution (see full code below).

We now use the Poisson log–normal model with the data for respiratory admissions seen in Example 8.4.

The WinBUGS code is as follows:

```
model {
   for (i in 1 : N) {
       Y[i]    ~ dpois(mu[i])

       log(mu[i]) <- log(E[i]) + beta0 + V[i] + U[i]
       RR[i] <- exp(beta0  + V[i] + U[i])
       V[i]  ~ dnorm(0,tau.V)
   }
# ICAR prior for the spatial random effects
   U[1:N] ~ car.normal(adj[], weights[], num[], tau.U)

   for(k in 1:sumNumNeigh) {
       weights[k] <- 1
   }
# Other priors
     alpha0   ~ dflat()
     alpha1 ~ dnorm(0.0, 0.001)
     tau    ~ dgamma(0.5, 0.0005)
     sigma <- sqrt(1 / tau)
}
```

Example 8.4. *Fitting a conditional spatial model in R*

Here we consider fitting the Poisson log–normal with spatial effects to the data for respiratory admissions seen in Example 8.4 using the R package

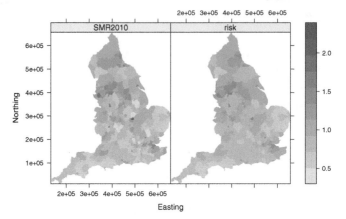

Figure 8.2: Maps of the spatial distribution of risks of hospital admission for a respiratory condition, chronic obstructive pulmonary disease (COPD), in the UK for 2001. The left hand map shows the smoothed risks which were estimated using a Bayesian model. Darker shades indicate higher rates of hospitalisation allowing for the underlying age–sex profile of the population within the area.

CARBayes. Again, the spatial effects come from the ICAR model described in Section 8.3.1 and this requires a set of observed and expected values together with an adjacency matrix. In this example, we created the required adjacency matrix from a shapefile using the spdep and shapefiles packages (although the maptools package could also be used). CARBayes performs MCMC simulation in a similar way to using WinBUGS but has the distinct advantage that the simulations are performed within R. WinBUGS is of course a much more general platform for fitting a wide variety of complex models using MCMC as described in Section 5.5.

```
# requires CARBAyes, spdep and shapefiles libraries
library(CARBAyes)
library(spdep)
library(shapefiles)

# read in the shape file
shp <- read.shp(shp.name="england local authority.shp")
# read in the details of the areas
dbf <- read.dbf(dbf.name="england local authority.dbf")

# calculate the SMRs and combine with the spatial
    information
SMR <- Y/E
SMRspatial <- combine.data.shapefile(SMR, shp, dbf)

# Create the neighbourhood matrix
W.nb <- poly2nb(SMRspatial, row.names = rownames(SMR))
W.list <- nb2listw(W.nb, style="B")
```

```
W.mat <- nb2mat(W.nb, style="B")

# Fit a  CAR smoothing model
formula <- observed~offset(log(expected))

model <- iarCAR.re(formula=formula, family="poisson", W=W.
   mat,
      burnin=20000, n.sample=100000, thin=10)
risk <- model$fitted.values[ ,1] / expected
```

The CARBayes package allows a number of CAR models to be used including non-intrinsic models. For further details see Lee (2013).

Again, Figure 8.2 is used to show the resulting map of the estimates of relative risks from applying the intrinsic CAR model as the results are the same as those using WinBUGS.

Example 8.5. *Fitting a conditional spatial model using R–INLA*

In this final example of implementing the Poisson log–normal with spatial effects, we use R–INLA. Again, the spatial effects come from the ICAR model described in Section 8.3.1. This form of spatial data, e.g. areas, is already in the form that can be used by R–INLA. Chapter 9, Section 9.13 considers the case when spatial data takes the form of points rather than areas and details are given of the methods that need to be employed to transform point referenced data into a form that INLA can use.

The intrinsic, and non-intrinsic, CAR models are two of a set of defined latent models in R–INLA. As such the syntax is very simple with the random effects being defined following the form f(ID, model="besag", graph="UK.adj") where UK.adj is an adjacency matrix that is suitable for use with R–INLA. The code to perform the model is as follows and follows on from the component of the code seen in Example 8.4 which sets up the data and adjacency matrix.

```
# requires INLA
library(INLA)

### Create the neighbourhood matrix
W.nb <- poly2nb(SMRspatial, row.names = rownames(SMR))
W.list <- nb2listw(W.nb, style="B")
W.mat <- nb2mat(W.nb, style="B")

#Convert the adjacency matrix
 into a file in the INLA format
nb2INLA("UK.adj", W.nb)

#Create areas IDs to match the values in UK.adj
data=as.data.frame(cbind(Y, E))
data$ID<-1:324
```

```
# run the INLA model
m1<-inla(Y~f(ID, model="besag", graph="UK.adj"),
    family="poisson", E=E,data = data,
    control.predictor=list(compute=TRUE))
```

As mentioned in the previous two examples, Figure 8.2. shows the resulting map of the estimates of relative risks from applying the intrinsic CAR
model.

8.5 Summary

This chapter introduces disease mapping and contains the theory for spatial lattice
processes and models for performing smoothing of risks over space. The reader will
have gained an understanding of the following topics:

- Disease mapping, where we have seen how to improve estimates of risk by borrowing strength from adjacent regions which can reduce the instability inherent in
  risk estimates (SMRs) based on small expected numbers.

- How smoothing can be performed using either the empirical Bayes or fully
  Bayesian approaches.

- Computational methods for handling areal data.

- Besag's seminal contributions to the field of spatial statistics including the very
  important concept of a Markov random field.

- Approaches to modelling areal data including conditional autoregressive models.

- How Bayesian spatial models for lattice data can be fit using WinBUGS, R and
  R–INLA.

Exercises

Exercise 8.1. Find a Gibbs distribution when all cliques are of size 1. Repeat when
all cliques are of size 2 that tends to put high probability on sites in cliques of size 2
where the two values tend to be different.

Exercise 8.2. Show that a random walk process of order 1 can be expressed in terms
of an intrinsic CAR model, i.e. if $p(\theta_t|\theta_{t-1}) \sim N(\theta_{t-1},\sigma_w^2)$ then

$$p(\theta_t|\theta_{-t},\sigma_w^2) \sim \begin{cases} N(\theta_{t+1},\sigma_w^2) & \text{for } t=1 \\ N\left(\frac{\theta_{t-1}+\theta_{t+1}}{2},\frac{\sigma_w^2}{2}\right) & \text{for } t=2,...,T-1 \\ N(\theta_{t-1},\sigma_w^2) & \text{for } t=T \end{cases}$$

where θ_{-t} represents the vector of θ's with θ_t removed. Pay particular attention to
any assumptions that need to be made when $t=1$ and $t=T$.

Exercise 8.3. Using Brook's lemma, prove the assertion in and Equation 8.11.

Exercise 8.4. Prove the assertion in and Equation 8.13.

Exercise 8.5. The data used to produce the map for hospital admissions for COPD in 2001 are included in the online resources. Using that data, reproduce the map of smoothed risks shown in Figure 8.2.

(i) Perform the analysis using WinBUGS, the CARBayes package in R and R–INLA. Compare the results from the three approaches.

(ii) Investigate the sensitivity of the results to the choice of priors for the variance terms. For example, in WinBUGS change

```
alpha1 ~ dnorm(0.0, 0.001)
tau    ~ dgamma(0.5, 0.0005)
```

(iii) Make the same changes using CARBayes and R–INLA.

Exercise 8.6. The online resources also have data for COPD admissions for other years.

(i) Choose two other years of data and reproduce your analysis from Exercise 8.6. Is there any evidence of any changes in the patterns of risk over time?

(ii) After covering the material in Chapter 11, construct a spatio–temporal model for this data and fit it in WinBUGS and/or R–INLA.

(iii) Do the results from your model in part (ii) agree with your initial findings from considering the years separately?

Chapter 9

From points to fields: modelling environmental hazards over space

9.1   Overview

The study of exposures to environmental hazards and their potential effects on health outcomes begins with understanding the underlying structure of, and variation in, the hazard over space and time. Changes in exposures to the hazard will be used to estimate the associated health effects together with associated measures of uncertainty. An important aspect of this is to provide the contrasts in exposures over space and time that will drive the health outcome. In addition, being able to model the underlying structure of the exposure field will enable predictions to be made at locations where data are not available due to the absence of monitoring. Often there will be locations and periods of time in which exposure data will not be available. This may be due to a fault in monitoring equipment or may be due to design.

In many epidemiological studies the locations and times of exposure measurements and health assessments do not match, in part, because the health and exposure data will have arisen from different data sources. The ability to predict levels of exposures at all locations for which health data is available can therefore maximise the use of the available health data. In this chapter we concentrate on spatial models for exposures and in the next two chapters we will encounter exposure models that produce similar contrasts over time (Chapter 10) and both space and time (Chapter 11).

This chapter focuses on processes for point referenced spatial data point referenced data, i.e. exact spatial locations rather than area-level data on a lattice as seen in Chapter 8. Methods for characterising underlying spatial processes and for modelling exposures over space are described.

9.2   A brief history of spatial modelling

Techniques for modelling random spatial fields began in earnest in the 1950's when the foundation of the subject of geostatistics was laid by Krige and Matheron (Cressie, 1990). Its origins lay in the mining industry and the need to predict ore deposits beneath the surface of the earth. The approach was to use a few core samples taken at a few selected locations to map ore deposits by predicting deposits at locations which were unsampled. The method has come to be known as kriging, after Krige, a South African mining engineer. His method relied on estimating spatial correlation between observed concentrations at the sampling sites (Krige, 1951).

The concept of spatial variation goes back a long way, at least to Kolmogorov (1941) although it was Matheron (1963) in the 1960s who first published a detailed account of geostatistics. He called the optimal linear unbiased prediction of responses at unsampled locations 'Kriging' to honour Krige's contributions. Kriging expanded to include the environmental sciences with Eynon and Switzer (1983) amongst the first statisticians to see its potential in that area. Spatial statistics has since become an extremely important topic within statistical science (Cressie, 1993).

9.3   Exploring spatial data

A number of exploratory methods have been developed for the analysis of spatial data and in this section we introduce a selection of them through a series of examples. A number of R packages have been designed specifically to display and model spatial data including gstat and geoR. Where such packages are used, they are indicated within the given R code. The first stage of a spatial analysis is to investigate the distribution of the exposure of interest, for example concentrations of lead, in order to assess whether the assumptions necessary for applying subsequent methods are tenable.

Example 9.1. *Spatial patterns in lead concentrations in the soil of the Meuse River flood plain*

The Meuse River is one of the largest in Europe and a great deal of research has been performed in relation to potential environmental hazards within its flood plain (Ashagrie et al., 2006). A comprehensive survey of concentrations of a variety of elements in the river was collected at 155 sampling sites in 1990. We now consider how the measurements at these locations can be visualised using R.

Figure 9.1 shows the locations of the 155 sampling locations and indicates the lead concentrations measured at those sites. The plot suggests the possibility of spatial patterns in the concentrations. Figure 9.2 shows a set of four plots of the Meuse valley lead concentrations using geoR. The four plots show: (i) the locations of the sampling sites (as seen in Figure 9.1); the concentrations in relation to (ii) the *x* and (iii) *y* coordinates and (iv) a histogram of the concentrations indicating the distribution of concentrations together with an estimate of the density. The latter of these shows that the distribution is very skewed to the right. The code to produce both of these plots is included in the online resources.

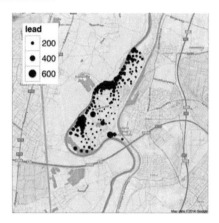

Figure 9.1: Bubble plot showing the size of lead concentrations measured in samples taken at 155 locations in the Meuse River flood plain.

9.3.1 *Transformations and units of measurement*

Based on the distribution of the concentrations shown in Figure 9.2 (bottom right panel), before embarking on the analysis we will log–transform the lead concentration due to the strong right skew in the data.

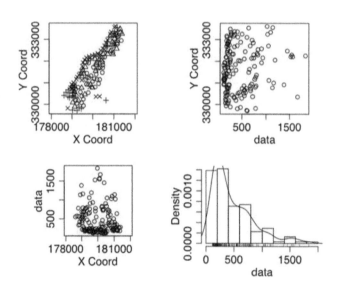

Figure 9.2: Summaries of the concentrations of lead measured at 155 sampling sites within the Meuse River plain. The top left panel shows the location of the sampling sites, the off-diagonal panels show scatterplots of concentrations against x and y coordinates and the fourth panel shows the distribution of lead concentration levels. The plots were produced by the R package geoR.

Before performing such a transformation, we need to consider the effects on the units of measurement. The logarithm is defined as the reciprocal operation of exponentiation, that is

$$x = \log_b(y) \text{ if } y = x^b$$

where b, x, and y are real numbers, b being the base of the logarithm. This definition precludes the association of any physical dimension to any of the three variables, b, x or y. Therefore the data, y, needs to be 'normalised' before transforming it. Failure to do so can lead to results that can be difficult, or even impossible, to interpret. For example, consider a weight of 10.2 kg. Applied directly, its log would be $\log(10.2\text{kg})$ which could be expressed as $\log(2) + \log(5.1\text{kg})$ which has a less obvious interpretation! Similarly, if the mean, μ and standard deviation, σ, of a log–Normally distributed variable, $Y \sim LN(\mu, \sigma^2)$, had units of measurement then the expectation $E(Y) = E(e^{\log(Y)}) = e^{(\mu + \frac{1}{2}\sigma^2)}$ would have units which were a mixture of two different scales.

In order to normalise the data, we can simply divide each observation by $m = 1$ mg/kg, the units for the soil concentrations. Were these units to be changed, both denominator and numerator would then have to be changed but the result is that $\log(\text{lead}/m) = \log(\text{lead})$, i.e. it is invariant under changes of the unit of measurement.

Example 9.2. *Examining the log concentrations of lead in the Meuse River flood plain*

In Figure 9.3 we see a histogram and a qq-plot for the log values of the concentrations of lead. Even after taking logs the data still doesn't follow a Gaussian distribution exactly, but the assumption (of normality) doesn't seem entirely unreasonable.

Investigating further the possible relationship between concentrations and spatial location, Figure 9.4. shows the result of using the coplot function

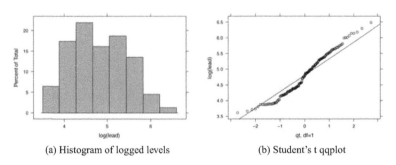

(a) Histogram of logged levels   (b) Student's t qqplot

Figure 9.3: Assessment of the assumption of normality for logged concentrations of lead in the Meuse River plain. Plot (a) shows a histogram of the logged concentrations and plot (b) a qq-plot using a t-distribution with 10 degrees of freedom.

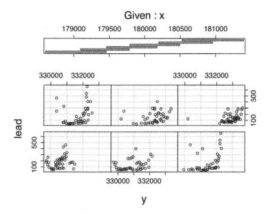

Figure 9.4: Trends in the level of lead concentrations in the soil of the Meuse River flood plain in relation to location. The top panel shows the distribution of x coordinates for a set of selected y coordinates. The corresponding scatterplots of lead concentrations against y coordinates is given for a set of six fixed x coordinates. The plots were produced by the R package `gstat` and show distinct spatial patterns in lead concentrations.

in the `gstat` package. This fixes the x coordinate at six levels and, for each, shows how the concentrations change as y increases. The code for this plot is included in the online resources. The top panel shows the six x coordinates on which we are conditioning and for each of these we see the scatterplots of lead concentrations against y coordinates. We see intriguing patterns in these scatterplots, suggesting a need for a spatial mean function when modelling the lead concentrations, i.e. one that includes x and y coordinates.

9.4   Modelling spatial data

When modelling spatial data, we distinguish between the underlying spatial process and the process in which measurements are made. A spatial random field is a stochastic process over a region, $Z_s : s \in \mathscr{S}$. This underlying process is not directly measurable, but realisations of it can be obtained by taking measurements, possibly with error, at a set of N_S known locations, $S \in \mathscr{S}$ where the points in S are labelled $s_1, ..., s_{N_S}$ and commonly $s_l = (a_l, b_l)$. One way of representing a random field is as a combination of an overall spatial trend together with a process that has spatial structure. In this case we have,

$$
\begin{aligned}
Y_s &= Z_s + v_s \\
Z_s &= \mu_s + m_s \\
\mu_s &= \sum_{j=1}^{J} \beta_j f_j(X_s)
\end{aligned}
\tag{9.1}
$$

where v is measurement error. The first component of the second line, μ_s, is the mean of the underlying process. The latter is modelled as a function of the location, information about which is contained in X_s, with associated coefficients, β. The second term in the middle line, m_s, is a process with spatial structure.

The observable data, Y_s, at the first level of the model are considered conditionally independent given the value of the underlying process, Z_s. In this case, spatial structure is incorporated through the spatial component of the underlying process, m_s, which will have spatial structure in its covariance, $Cov(m_s, m_{s'})$. In a fully Bayesian analysis, a third level of the model would assign prior distributions to the hyperparameters from the first two levels. An alternative approach is to estimate them from the data (Diggle, Tawn, & Moyeed, 1998; Le & Zidek, 2006).

In a purely spatial analysis, repeated observations at a specific location over time are treated as independent realizations of the underlying process. The mean and covariance function are then estimated from the data and used to describe the spatial trend and association and for prediction at unsampled locations.

9.5   Spatial trend

A spatial trend refers to the case where the mean term (in Equation 9.1) is allowed to vary over space. The mean, μ_s, may be modelled directly as a function of s (Diggle & Ribeiro, 2007) which is often done by using a polynomial regression model with the coordinates of s used as explanatory variables. When replicates are available over time these can be used, after incorporating appropriate temporal structure in the model, to improve the modelling of the spatial trend (Guttorp, Meiring, & Sampson, 1994; Sahu & Mardia, 2005; Sahu, Gelfand, & Holland, 2006, 2007; Giannitrapani, Bowman, Scott, & Smith, 2007; Fanshawe et al., 2008; Bowman, Giannitrapani, & Scott, 2009; Paciorek, Yanosky, Puett, Laden, & Suh, 2009). Varying coefficient models have also been used in order to allow the mean to vary over space by assuming that the model parameters vary over the study region (Gelfand, Kim, Sirmans, & Banerjee, 2003).

When values of potential explanatory variables are available these may be used to model the spatial trend. This method offers an appealing approach that allows the mean trend to vary over the study region and to be modelled as a function of explanatory variables, the effect of which can be scientifically interpreted. When the same set of explanatory variables is also available at locations for which predictions might be required, then we can use values of the explanatory variable at the new location(s) to produce the mean at that location. One drawback of this method is that it relies highly on data availability. It is often the case that such explanatory variables are only available at the locations where measurements are made so prediction will have to rely solely on spatial prediction methods as described in the following sections.

9.6   Spatial prediction

Typically fields of environmental hazards are sampled at a sparse set of spatial sites over an area of interest. For example, the US Environmental Protection Agency is charged with setting air quality standards that must be met. Ozone is strongly associated with adverse health outcomes and therefore must be regulated according to the US Clean Air Act of 1970 with concentrations monitored, in part, to ensure adherence with the act. Figure 9.5 shows an example of an ozone monitoring network, showing the locations of ozone monitoring sites within New York State. Note the relatively small number of monitors for such a large area. In many areas, especially rural ones, monitoring may be sparse despite the need for accurate information related to exposure levels given the potentially negative impacts of air pollution on human health and welfare.

Example 9.3. *Mapping the locations of ozone monitoring sites in New York State*

In this example we show how the locations of monitoring sites can be superimposed onto a background map. The R code to produce the map shown in Figure 9.5 is given together with commented details of how the background map is defined in Google maps and then downloaded into R.

```
> library(sp)
> library(ggmap)

## Load the metadata giving the site coordinates
> ny_data <- read.csv("NY.metadata.txt", sep="")

## Now copy ny_data into ny_data_sp and convert data
  to "sp" format
> ny_data_sp <- ny_data
> coordinates(ny_data_sp) <- ~Longitude+Latitude

  ## assign a reference system to ny_data_sp
> proj4string(ny_data_sp) <-
          CRS("+proj=longlat +ellps=WGS84")

### We next specify a bounding box - a 2 x 2 matrix of
### corners of the geographic area. Then specify the
### range of locations within the box.
###  Note: location must be in left-bottom-right-top
### bounding box format
> latLongBox = bbox(ny_data_sp)
> location = c(latLongBox[1, 1]-0.2, latLongBox[2, 1]-0,
          latLongBox[1, 2]+0.2, latLongBox[2, 2]+0.2)

######## Now create the map with location dots
> NYmap <- get_map(location = location, source = "google",
  color="bw", maptype="roadmap")
> NYmap <- ggmap(NYmap)
> NYmap <- NYmap + geom_point(data=ny_data,
  aes(x=Longitude, y=Latitude), size=4)
```

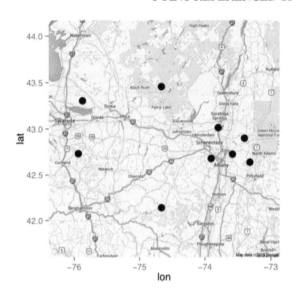

Figure 9.5: Locations of ozone (O_3) monitoring sites in New York State.

Here the ggmap library, which is a geographical mapping tool for the ggplot2 package, is used. It allows maps such as the one shown to be obtained from a variety of sources and in this example the map is obtained from Google maps.

This sparsity of such sites has led to the need to interpolate or extrapolate measured values using spatial prediction. For pollutants such as ozone, a photooxidant produced by atmospheric chemistry, pollution fields are relatively flat meaning that measurements at any one site are highly correlated with those at other nearby sites meaning that interpolating such fields between sites is feasible.

A simple approach for spatial prediction would be to calculate a weighted average of all points within a certain neighbourhood of a chosen point, s_0. The weights may be chosen to reflect the distance between the point in question and a set of monitored locations, S containing points labelled $s_1, ..., s_{N_S}$. There are many possible ways of weighting the samples, all leading to different estimators. For example, the inverse of the squared distance between the points may be used, $\sum_{i=1}^{N_S} (||s_i - s_0||)^{-2}$.

Kriging is one of the most popular methods used for spatial prediction. It is a method of interpolation in which interpolated values are modelled by a Gaussian process. Under suitable assumptions, it gives the best linear unbiased prediction of the intermediate values (Cressie, 1985). The class of kriging models includes *simple kriging, ordinary kriging, universal kriging, indicator kriging, probability kriging, disjunctive kriging* and *cokriging* amongst others (Cressie, 1993). Further details on kriging can be found in Section 9.10.

9.7 Stationary and isotropic spatial processes

A stationary random field is a stochastic process over a region, $Z_s, s \in \mathscr{S}$ where s is a location in Euclidean space, $\mathscr{R}^d$. Commonly in spatial analysis, d will be equal to two and represent x, y coordinates, for example longitude and latitude, UTM or some other coordinate system.

The joint cumulative distribution function of a realisation of the spatial process $\mathbf{Z}_s = (Z_{s_1}, \ldots, Z_{N_S})$ for any integer N_S is given by

$$F_{1,\ldots,N_S}(\mathbf{z}) = F_{1,\ldots,N_S}(z_1, \ldots, z_{N_S}) \equiv P\{Z_1 \leq z_1, \ldots, Z_{N_S} \leq z_{N_S}\},$$

for all $\mathbf{z} \in R^{N_S}$.

Definition 1. Strict stationarity.

A spatial process, Z, is strictly stationarity if

$$F_{1,\ldots,N_S}(\mathbf{z}) = F_{1+h,\ldots,N_S+h}(\mathbf{z})$$

for any vector h and arbitrary N_S when $\mathscr{S} \subset \mathscr{R}^d$ is a continuum.

Definition 2. Second-order stationarity.

A weaker assumption is one of *weak* or *second-order* stationarity. Here, the mean is constant over space and the covariance depends only on the distance between locations and not their actual locations in space. This, on its own, would not imply strict stationarity.

A spatial process, Z, is second-order stationary if for all $s \in \mathscr{S}$ and arbitrary h,

$$\begin{aligned} \mu_s &= E[Z_s] \equiv \mu \text{ and} \\ Cov(s,s') &= C(s+h, s'+h) \equiv C(h) \end{aligned}$$

Note that when we have $h = 0$, we have $Cov(Z_s, Z_{s'}) = C(s,s) = C(0) = Var[Z_s]$. The covariance kernel $C(h)$ here has a number of important properties. First, observe that for any pair of points s and s', $Cov(s,s') = C(||s - s'||) = C(h_{ss'})$ so that all inter-site correlations can be obtained from the kernel. The kernel must be positive definite meaning that if $\Sigma = C(h_{ss'})$ is the $N_S \times N_S$ dimensional covariance matrix for $(Z_1, \ldots, Z_{N_S})$ then it must be positive definite. This means that for any vector a that

$$\sum_i \sum_j a_i a_j C(h_{ij}) > 0 \tag{9.2}$$

Definition 3. Isotropic stationary processes.

An *isotropic* stationary process is one where the covariance kernel depends only on the distance between two points s and s' irrespective of the direction of one from the other. Any process that is not isotropic is called *anisotropic*.

A process, Z, is an isotropic second-order stationary if it is second-order stationary and in addition

$$C(h) \quad = \quad C(|h|)$$

where $|h|$ denotes the length of the vector h.

Example 9.4. *Gaussian random fields (GRFs)*

A special case of a stationary process is the *Gaussian random field (GRF)* in which the realisations come from the multivariate normal distribution with mean zero, $E(Z_s) = 0$, and covariance, $Cov(Z_s, Z_{s'}) = \sigma_s^2 \rho(||s - s'||)$. It has strong stationarity, where the joint distribution between locations is constant over the entire region, which implies second-order stationarity. Using Gaussian assumptions makes the theory for estimation and prediction considerably more straightforward than might otherwise be the case.

Definition 4. A relaxed form of stationarity is *intrinsic stationarity* which is based on the difference in the process between locations. For a choice of two locations, the difference in means will be zero and the difference in variances is defined through the *semi-variogram*.
A process Z is intrinsically stationary if

$$\frac{1}{2} Var(Z_s - Z_{s'}) = \gamma(||s - s'||) = \gamma(h) \tag{9.3}$$

where $h = ||s - s'||$ and $E[Z_s - Z'_s] = 0$.

If two locations are close together their difference would typically be small, and hence so would the variance of their differences. As the locations get farther apart, their differences get larger and usually the variance of the difference will increase. This is similar to the definition of second-order stationarity although here stationarity is defined in terms of the variance of the difference between the process at different locations and not the covariance.

Here $\gamma(h)$ is the semi-variogram which plays a major role in modelling spatial processes.

9.8   Variograms

The covariance function and the semi-variogram are both functions that summarise the strength of association as a function of distance and, in the case of anisotropy, direction. When dealing with a purely spatial process where there are no independent realisations, patterns in correlation and variances from different parts of the overall region of study are used as if they were replications of the underlying process.

Under the assumption of stationarity, a common covariance function for all parts of the regions can then be estimated.

The semi-variogram will be zero at a distance of zero as the value at a single spot is constant and has no variance. It may then rise and reach a plateau, indicating that past a certain distance, the correlation between two units is zero. This plateau will occur when the semi-variogram reaches the variance of Z. Assuming second-order stationarity, the relationship between the covariance and the semi-variogram is as follows:

$$
\begin{aligned}
\gamma(d) &= \frac{1}{2}Var(Z_s - Z_{s'}) \\
&= \frac{1}{2}\{Var(Z_s) + Var(Z_{s'}) - 2Cov(Z_s, Z_{s'})\}
\end{aligned}
\tag{9.4}
$$

If $Var(Z_s) = Var(Z_{s'}) = \sigma_s^2$ and $C(Z_s, Z_{s'}) = C(h)$ for all s, s', then

$$
\begin{aligned}
\gamma(h) &= \frac{1}{2}(2\sigma_s^2 - 2C(h)) \\
&= \sigma_s^2 - C(h)
\end{aligned}
$$

Therefore, at $h = 0$, the semi-variogram is $\gamma(0) = \sigma_Z^2 - C(0) = \sigma_Z^2 - \sigma_Z^2 = 0$, where $C(0) = Cov(Z_s, Z_s) = Var(Z_s)$ when d is large, i.e. when $C(d) = 0, \gamma(d) = \sigma_s^2 - C(d) = \sigma_s^2$. Valid variograms must be *conditional negative definite* (Cressie, 1993), namely $\sum_{i=1}^{m}\sum_{j=1}^{m} a_i a_j 2\gamma(Z_s, Z_{s'}) \leq 0$, for any $\sum_{i=1}^{m} a_i = 0$. Given an intrinsically stationary process, they must satisfy the condition,

$$
2\gamma(d)/d^2 \to 0, \quad \text{as} \quad d \to \infty
\tag{9.5}
$$

In practice, the semi-variogram is often preferred to the covariance function because of the relaxed rules of stationarity and also because it only uses pairs of locations d units apart, and doesn't involve the overall mean, so if there is a shift in the mean (trend) that is not explicitly modelled, the semi-variogram is likely to be less affected than the covariance function.

The variogram is a realisation of the covariance function and under second–order stationarity the functions are related as follows,

$$
\begin{aligned}
2\gamma(h) &= Var(Z_{s+h} - Z_s) \\
&= Var(Z_{s+h}) + Var(Z_s) - 2Cov(Z_{s+h} - Z_s) \\
&= C(0) + C(0) - 2C(h) \\
&= 2[C(0) - C(h)]
\end{aligned}
\tag{9.6}
$$

Thus $\gamma(h) = C(0) - C(h)$ and from this relationship it can be seen that from the covariance C we can easily determine the semi-variogram γ. In general obtaining C from γ is not as straightforward. If $C(h) \to 0$ as $|h| \to \infty$ then the covariance goes to zero as distance goes to infinity. If we take the limit on both sides of Equation 9.6 then we get $\lim_{h\to\infty} \gamma(h) = C(O)$. However, the limit may not exist, an example of this being in the case of a linear variogram. In such cases, additional assumptions about the spatial process, such as ergodicity, will be required (Cressie, 1993; Banerjee et al., 2015).

9.8.1   The nugget

The semi-variogram of two observations taken at the same place is zero by definition since $Var(Z_s - Z_{s'}) = 0$. However there may be 'microscale variations' and measurement error, so that two measurements taken at the same location are not exactly the same. This *nugget effect*, or nugget variance, is a discontinuity in the semi-variogram at the origin, the difference between zero and at a lag distance just greater than zero.

9.8.2   Variogram models

Although it is possible to create empirical estimates of the semi-variogram, these are often less than ideal for modelling spatial data as there may only be a few points at particular lag distances and so estimates of the covariance may not be stable. In addition, the matrix of sample covariances, $\hat{C}$, may not be positive definite and thus may not be a valid covariance matrix $(Var(q'\hat{C}q) \geq 0)$. Often it will be better to pool the data at different lags and fit a smooth curve through the resulting estimate of the semi-variogram using an appropriate model. Such semi-variogram models do not always produce well defined covariance matrices for the underlying processes as often the exact variance function is not specified.

Figure 9.6 shows variograms for NO_2 concentrations throughout Europe in 2001 and for the residuals after fitting a set of land-use covariates (Shaddick, Yan, et al., 2013). It illustrates three important features of a variogram:

(i)   The *nugget* - when $\gamma(0) = 0$ by definition, $\gamma(0^+) \equiv lim_{h\to 0^+} \gamma(h) = \tau^2$, this quantity is the nugget, the variation of the difference of observations at same sites.

(ii) The *sill* - the asymptotic value of the semi-variogram $lim_{h\to\infty}\gamma(h) = \tau^2 + \sigma_s^2$. The the difference between the sill and τ^2, and called the *partial sill*.

(iii) The *range* - the distance at which the plateau starts and after which there is little association between correlation and distance.

Here we provide a brief discussion of a subset of possible variogram models. Details of a number of other models for this purpose can be found in Cressie (1993) and Le and Zidek (2006). The linear semi-variogram has the simplest functional form, but the sill and range are both infinite. The spherical variogram yields a stationary process and the corresponding covariance function is easily computed. However it fails to correspond to a valid spatial covariance matrix when the dimension is $d \geq 4$.

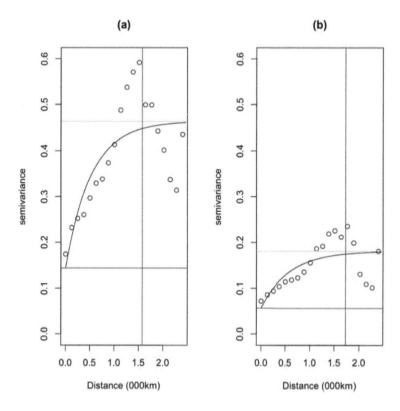

Figure 9.6: Variograms for (a) log values of Nitrogen Dioxide (NO_2) measured at monitoring sites throughout Europe in 2001; (b) residuals after fitting a model with land-use covariates. Lines show fitted exponential curve (black), nugget (blue), partial sill (green) and effective range at which correlation falls to 0.05 (red). Note the apparent lack of intrinsic stationarity suggested by the points on the right hand side of the plots where the variogram appears to decline as the distance between locations increases.

A common class of models used for variogram models is the Matern class (Matern, 1986).

Matern class of models

$$\gamma(h|\theta) = \begin{cases} 0 & h = 0 \\ \theta_1 \frac{1}{2^{\theta_2 - 1}\Gamma(\theta_2)} \left(\frac{2d\sqrt{\theta_2}}{\theta_1}\right)^{\theta_2} K_{\theta_2}\left(\frac{2d\sqrt{\theta_2}}{\theta_1}\right) & h > 0 \end{cases} \qquad (9.7)$$

where $\theta_1 > 0$ is a scalable parameter controlling the range of the spatial correlation, and $\theta_2 > 0$ is the smoothness parameter. K_{θ_2} is the modified Bessel function of order θ_2. The exponential model is a special case of this, with $\theta_2 = 1/2$.
Exponential model

$$\gamma(d|\theta) = \left\{ \begin{array}{ll} 0 & d = 0 \\ \theta_1 \exp(-\theta_2 d) & d > 0 \end{array} \right.$$

The limiting case of the Matern class of models, when $\theta_2 \to \infty$, is the Gaussian model.

9.9   Fitting variogram models

The simplest method of fitting a variogram model is by least squares, although this assumes linearity and that the observations are uncorrelated with each other and have the same variance. That is unlikely to be the case as estimates for different distances may be based on different numbers of pairs and will be correlated with each other. Simple weighted least squares can be used under certain stationarity and sampling conditions (Cressie, 1993), or the observations can be transformed to ensure that they are not correlated, i.e. $X^* = (V')^{-1/2}X$, where $V = Var(Z)$, which results in performing generalised least squares regression.

It is often not easy to judge which variogram model is the most appropriate for a set of data. The 'goodness' of fit (based on nugget, sill and range) can be judged visually or by using criterion such as least squares. More formal approaches to this problem are detailed in Cressie (1993); Cressie and Hawkins (1980); Omre (1984).

Matheron (1963) proposed a simple nonparametric estimate of the semi-variogram:

$$\hat{\gamma}(t) = \frac{1}{2N(d)} \sum_{(s,s')\in N(d)} [Z_s - Z_{s'}]^2 \tag{9.8}$$

where $N(d)$ is the set of pairs of points such that $||s - s'|| = h$, and $|N(d)|$ is the number of pairs in this set. Notice that the difference between the sites will all be different unless the observations fall on a regular grid and therefore a modification is used,

$$N(d_k) = (s, s') : |s - s'| \in I_k, \text{for}, k = 1, ..., K$$

where $I_1 = (0, d_1)$, $I_2 = (d_1, d_2)$, and so on, up to $I_k = (d_{k-1}, d_K)$. See Banerjee et al. (2015) for further details. For each interval, it has been suggested at least 30 pairs of data should be included (Journel & Huijbregts, 1978).

It is noted that when fitting a variogram model a good characterisation of spatial trend, i.e. spatial mean function, is vital. Otherwise the semi-variogram may include the square of a large bias (spatial mean estimation error) and hence give a misleading characterisation of inter-site correlation.

Example 9.5. *Spatial modelling of temperature*

This example is based on temperature, which may reach extreme values, both high and low, and have a great effect on human health. Li et al. (2012) review the literature on this topic and report an excess death toll of 70,000 people due to high temperatures in Europe in 2003 and an increased number of hospital admissions for five different causes due to extreme temperatures in Madison, Wisconsin.

This emerging topic in environmental health risk analysis motivated this example on spatial mapping which is based on daily maximum temperatures measured in California on Apr 1, 2012 at eighteen locations. Figure 9.7 shows an exponential variogram with a sill of 12000 (degrees squared) and a range of 5 miles.

The small number of sites makes any conclusion drawn from this variogram quite speculative. The apparent lack of fit may reflect the failure to correctly model the underlying spatial mean function or it may suggest non-stationarity. The range and/or sill may differ in the north–south and east–west directions, a common feature of geostatistical fields that exhibit anisotropy (Webster & Oliver, 2007). In such cases the usual Euclidean measure of distance between sites may not be appropriate. We return to this topic in

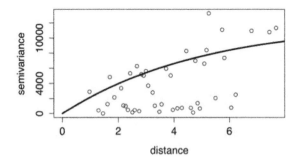

Figure 9.7: A semi-variogram based on daily maximum temperatures observed on April 1, 2012 at 18 locations in California. The solid black line represents a fitted exponential variogram. This fits one group of site pairs fairly well, however there is a second group of sites with a smaller sill but similar range suggesting the need for further analysis.

Chapter 14, Section 9.14. This example shows the difficulty of modeling the temperature field and why it is the subject of current research (Kleiber, Katz, Rajagopalan, et al., 2013).

9.10 Kriging

Kriging has been used extensively as a method for spatial prediction. As originally conceived it was a distribution free method for developing predictive distributions. In particular, if we assume the process is measured without error, i.e. $Y = Z$, then it produces the best linear unbiased predictor, $\hat{Z}_0 = \sum_{s=1}^{N_S} \alpha_s z_s$ of Z_0, i.e. the one that minimises $E[(Z_0 - \hat{Z}_0)^2]$ under certain conditions (Cressie, 1993). No distributional assumptions are explicitly made so this is a robust method in that sense.

The choice of a linear predictor is largely made for simplicity, but this may not necessarily be a sensible assumption unless Z has a multivariate normal distribution. In general therefore, modelling a geostatistical analysis should begin with an exploration of the distribution of Z, based on available data and a decision on how best to proceed based on that analysis. If the assumption of a Gaussian random field is not justified another approach will be required. A number of these are described in Section 9.11.

If the use of linear prediction is deemed appropriate then the two-step approach suggested by Webster and Oliver (2007) can be used.

Step 1. Model the spatial variation: Determine if the assumption of a stationary random field is valid and if it is then model its correlation structure.

Step 2. Develop the spatial predictor: The classical technique is kriging, which incorporates spatial correlation and yield the best linear unbiased predictor (BLUP) for unsampled sites.

Although on the face of it Step 1 is based on variances and correlations, these second-order properties are not really separable from the first-order properties. The spatial trend must therefore also be characterised as part of this step. To clarify this point, suppose the spatial trend given by $E(Z_s) = \mu_s$, $s \in \mathcal{S}$ is mis-specified as μ_s^*. In this case, the assumed variance would be $E(Z_s - \mu_s^*)^2 = Var(Z_s) + b_s^2$ where b denotes the bias $b_s = (\mu_s - \mu_s^*)$. A badly misspecified trend surface can greatly distort the apparent variance (and covarianceed) structure of the random field and hence the accuracy of the BLUP.

As described in Section 9.6 spatial prediction can be performed by using a weighted average of measurements from all the points within a certain range. In ordinary kriging, the estimator incorporates correlations among the Z_s into the weights for predicting Z_0. The weights are based on the covariances among points in the sample and the covariances between sample points and the point to be predicted. To compute kriging estimates the covariances among all points, $\tilde{C}_{ss'} = Cov(Z_s, Z_{s'})$, and between each of the observed points and the point to be predicted, $C_{s0} = Cov(Z_s, Z_0)$, need to be estimated using a covariance function or semi-variogram. The kriging estimator is the optimal estimator in that it minimizes the prediction variance or mean squared prediction error, $E(Z_s - \hat{Z}_s)^2$. Ordinary kriging gives optimal predictions under the assumption that the mean is constant (but unknown), with the weights obtained by solving the system of *kriging equations*,

$$
w = \begin{bmatrix} w_1 \\ w_2 \\ \vdots \\ w_{N_s} \\ \lambda \end{bmatrix} = \begin{bmatrix} C_{11} & \cdots & C_{1p} & 1 \\ \vdots & \cdots & \vdots & \vdots \\ C_{p1} & \cdots & C_{pp} & 1 \\ 1 & \cdots & 0 & 0 \end{bmatrix}^{-1} \times \begin{bmatrix} C_{11} \\ \vdots \\ C_{p1} \\ 1 \end{bmatrix}
\tag{9.9}
$$

where λ is a Lagrange multiplier that imposes the constraint $\sum w_s = 1$. Only intrinsic stationarity is required to obtain the ordinary kriging estimator, and so the results can thus be derived using semi-variograms instead of covariances.

Using kriging, the prediction variances $Var[Z_s|z]$ are generally small at locations close to the sampling locations because the location close to observation sites obtains greater influence through the weights w_s.

Example 9.6. *Spatial prediction of temperatures in California*

We return to the topic of modelling daily temperature as introduced in Example 9.5. Here we assess whether measurements from the eighteen temperature monitors in that example could accurately predict temperatures over the rest of the state. Note that this example is meant to be illustrative; a fully fledged analysis would look at all sites in the state and not just this subset.

We now describe how to map the maximum daily temperatures. First we load the appropriate R libraries: geoR, sp, utils and gstat. Next we load the data, changing the names of the latitude and longitude in the data file to x and y respectively and augment it with the spatial locations of the sites.

```
CAmetadata<- read.table("metadataCA.txt",header=TRUE)
CATemp<-read.csv("MaxCaliforniaTemp.csv",header=TRUE)
CATemp_20120401<-CATemp[CATemp$Date==20120401,2:19]
CATemp_20120401<-t(CATemp_20120401)

### Change names of lat, long
names(CAmetadata)[names(CAmetadata)=="Long"]<-"x"
names(CAmetadata)[names(CAmetadata)=="Lat"]<-"y"

### Augment data file with coordinates
CATemp_20120401 <-cbind(CAmetadata,CATemp_20120401)
names(CATemp_20120401)[names(CATemp_20120401)=="92"]<-"Temp"
coordinates(CATemp_20120401) <- ~x +y
```

We are now ready for spatial prediction that starts with fitting the variogram.

```
### Now construct the variogram
CAgstat.vario <- variogram(Temp~1,CATemp_20120401, print.SSE
    =TRUE)
plot(CAgstat.vario)

### Get the fit
CATemp.variofit <- fit.variogram(CAgstat.vario, vgm(5000,"
    Exp",2.9,0))

### Examine the fit
summary(CATemp.variofit)
```

Then a grid is created on which the predictions will be made:

```
### Create the grid on which to make the spatial predictions
CAPred.loci<-expand.grid(seq(-125,-115,.1),seq(32,42,.1))
names(CAPred.loci)[names(CAPred.loci)=="Var1"]<-"x"
names(CAPred.loci)[names(CAPred.loci)=="Var2"]<-"y"

coordinates(CAPred.loci) = ~ x + y
gridded(CAPred.loci) = TRUE
```

The final step is to produce the spatial predictions:

```
mod <- vgm(5000, "Exp", 2.9, .04)
# ordinary kriging:
x <- krige(Temp ~1, CATemp_20120401, CAPred.loci, model =
    mod)
spplot(x["var1.pred"], main = "ordinary kriging predictions"
    )
spplot(x["var1.var"],  main = "ordinary kriging variance")
```

The resulting images are shown in Figure 9.8 and were produced using the spplot function from the sp package. This function allows labels to be attached to regions with greater ease than with standard R plotting functions (Murrell, 2005). The predictions and their variances were produced using the gstat package in R.

The result of mapping the California maximum daily temperature field can be seen in Figure 9.8 and shows areas with relatively high (and low) daily maximum (and minimum) temperatures. In the right hand panel we see the prediction variances that are smaller at points near the monitoring sites as might be expected. In Chapter 13 we will see how the locations of sites being added to a monitoring network can be determined in order to minimise prediction variances.

9.11   Extensions of simple kriging

We now describe some of the enhancements of the basic kriging principle that have been made for spatial prediction.

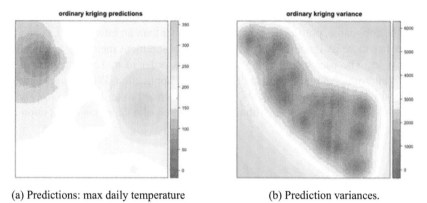

(a) Predictions: max daily temperature   (b) Prediction variances.

Figure 9.8: Predictions of the maximum daily temperature field in California on April 1, 2012 using kriging based on observed temperatures at eighteen monitoring sites. Panel (a) shows predicted temperatures using kriging and panel (b) shows the prediction variances. Here we see lower variances near the temperature monitoring sites where there is more information than areas further away from the monitoring locations.

9.11.1 Co-kriging

If one or more covariates are measured along with the response variable, and if the cross correlation functions are known or can be estimated, then the covariates can be used to improve prediction, even when they are not measured at the same locations as the response variable. This is especially useful when there are few observations of the variable to be predicted, but many observations of some other variable which is highly correlated with the first. The co-kriging prediction is then a weighted average of nearby values of all the variables.

A function for multivariate or co-kriging is included in the gstat package for the case where a multiplicity of responses are present at each spatial site. This topic has not received nearly the attention it deserves in the spatial statistical literature, since even when interest lies in only one of the random responses, strength can be borrowed from the other responses in making spatial predictions.

9.11.2 Trans–Gaussian kriging

Transformed-Gaussian kriging is based on the assumption that if the random field is not Gaussian then it can be transformed to a Gaussian one by a suitable transformation function. It applies the kriging method on Box–Cox transformed data where y_s is transformed into $(y_s^\lambda - 1)/\lambda$ or $\log(y_s)$ when $\lambda = 0$, the result of letting $\lambda \to 0$.

At the root of the problem is the requirement $y_s > 0$ for the Box–Cox transformation to be applicable, putting it on a ratio scale rather than an interval one. A positive scale σ has to be assigned and the size of the measurement measured against it by taking y_s/σ. The user has to choose the scale in order to make the results of the analysis interpretable in the context of the specific application. With this understanding a positive field of measurements can be transformed to an interval scale by the Box–Cox transformation or logarithm for application of a Gaussian distribution and kriging.

9.11.3 Non-linear kriging

Long ago geostatisticians recognised the need to spatially predict the presence or absence of material and developed a method called indicator or probability kriging. And non-linear kriging or disjunctive kriging was introduced where the predictor was $\hat{Z}_0 = \sum_{s=1}^{N_s} f_s(Y_s)$ where the fs are selected to minimise $E[\hat{Z}_0 - Z_0]^2$. The resulting solutions are more difficult to find than those for trans-Gaussian kriging, but less restrictive (Cressie, 1993; Wackernagel, 2003).

9.11.4 Model-based kriging

A major advance in geostatistics came with model-based kriging (Diggle & Ribeiro, 2007). If the random field is assumed to be Gaussian, then *linear-kriging* can be performed. Conditional on Z, the data, Y are mutually independent Gaussian random variables with expectations $\mu_s + Z_s$ and common variance, σ_y^2. The kriging predictor, $\hat{Z}_0$, is then a linear function of Z (Diggle et al., 1998). If the predictor $\hat{Z}_0$ is not linear in Z, as would be the case if the outcome variable were in the form of counts or proportions, then $E(Y_s|Z_s) = \mu_s = g(\sum_{j=1}^{J} \beta_j f_s(X_s))$, for some known function, f and explanatory variables, X_s at s, with associated coefficients, β. In such cases, a *generalised linear predictor* can be calculated (Diggle et al., 1998).

Example 9.7. *Model-based kriging with binary responses*

Suppose Y_s, $s \in \mathscr{S}$ is a measurable binary response representing presence or absence of an outcome, e.g. disease, at a location s and Z_s is a realisation latent Gaussian spatial process (at s). A predictive model for unmeasured binary responses may be constructed by assuming that for any two sites s, s', these responses are conditionally independent as follows:

$$[Y_s = y_s, Y_{s'} = y_{s'} \mid Z_s = z_s, Z_{s'} = z_{s'}, \theta]$$

$$= [Y_s = y_s \mid Z_s = z_s, \theta][Y_{s'} = y_{s'} \mid Z_{s'} = z_{s'}, \theta]$$

where θ denotes the vector of model parameters. More specifically we may assume that given $Z_s = z_s$,

$$\log \frac{p_s}{1 - p_s} = \beta z_s$$

where $p = P[Y_s = y_s \mid Z_s = z_s, \beta]$. Embedding this model in a hierarchical Bayesian model with a parameter model for β provides a very flexible model for the binary field.

9.11.5 Bayesian kriging

Bayesian kriging is a strong competitor to classical methods and one that has seen a great deal of development over the past few decades, including work from Kitanidis (1986) to that of Kazianka and Pilz (2012). It is the focus of much current research.

Here, a hierarchical model as shown in Equation 9.1 is assumed with prior probability distributions being assigned to the unknown parameters in $\mu_s = \sum_{j=1}^{p} \beta_j f_j(X_s)$. Bayesian kriging then uses this Bayesian hierarchical model to produce predictions (Cressie, 1993).

The posterior distributions for the underlying level at point not included in the sampled locations is

$$
\begin{aligned}
p(Z_0|Y) \propto p(Z_0, Y) &= \int \dots \int p(Z_0, Y, Z,) dZ \\
&= \int \dots \int \left\{ \prod_{s=1}^{N_S} p(Z_s|Y_s) \right\} p(Z_0|Z) p(Z) dZ \quad (9.10)
\end{aligned}
$$

This form can be further expanded to incorporate the conditioning on the parameters within the model, i.e. $p(Z|\beta, \theta)$, where β are the coefficients in the mean term and θ those in the variogram/covariance function. Samples from the posterior can be obtained by MCMC simulation , or in the case where $p(Y|Z)$ and mean $p(Z)$ are Gaussian, the posterior will be Gaussian. Within a Bayesian framework, Berger, De Oliveira, and Sansó (2001) provide an interesting discussion of the choice of priors in spatial models and in particular distinguish between non-informative priors that lead to proper posteriors and those that do not. Further details of how samples from the posterior distributions of the predictions can be found using MCMC can be found in Section 9.12 and of how INLA can be used to obtain predictions in Section 9.13.

Example 9.8. *Modelling the spatial effects of SO_2 using Bayesian kriging*

Holland, De Oliveira, Cox, and Smith (2000) use Bayesian kriging in a hierarchical framework to model the spatial effects of SO_2 in the Eastern U.S.

At the first level, the logs of weekly measurements of SO_2 are smoothed as functions of the period and the temperature using a GAM (see Chapter 2, Section 2.6). The model was used to estimate the true (unobserved and unknown) value of a Gaussian random field. In a non-Bayesian analysis, the analytical estimates of the variances are likely to be underestimated as the covariance parameters are assumed to be known, when in fact they are estimated. This problem is not easily remedied within a MLE framework, but by adopting a Bayesian approach, posterior distributions can be computed with allowance for the extra uncertainty. Holland et al. (2000) used MCMC methods to draw samples from the posterior distribution, although it was not a fully Bayesian analysis, as the estimates obtained from the GAM in the first level are treated as given, albeit with measurement error.

9.12   A hierarchical model for spatially varying exposures

In this section we give details of a hierarchical model described by Shaddick and Wakefield (2002) at the heart of which is a Bayesian kriging component. The aim is to consolidate many of the ideas and concepts introduced within the chapter and to provide details of possible methods for implemention.

As described in Section 9.4 there are three stages to the model: (i) the observation, or data, model; (ii) the process model that in this case describes the form of the underlying spatial process and (iii) assigning prior distributions to the unknown parameters.

Stage 1 - Observation model

At the first stage of the model, the observed data is related to an underlying spatial process, which is not observable but may be measured, possibly with error.

$$y_s = z_s + u_s$$

where y_s denotes the observed level of the exposure at spatial location s for $s = 1,..,N_S$ where N_S is the number of monitoring sites. In this model u_s represents measurement error which is assumed to be independent and identically distributed, $N(0, \sigma_u^2)$ or equivalently $\mathbf{u} \sim MVN_S(0_{N_S}, \Sigma_u)$ where Σ_u is a diagonal matrix.

Stage 2 - Process model

In this stage the underlying levels of the exposure are modelled as a stationary spatial process.

$$z_s = \mu_s + m_s$$

where μ_s is a spatial mean term, $\mu_s = \sum_{j=1}^{J} \beta_j X_{js}$ and m_s is a spatial process. It is assumed that the spatial random effects $\mathbf{m} = (m_1, ..., m_{N_S})'$ at locations s arise from the multivariate normal distribution

$$\mathbf{m} \sim MVN_S(0_{N_S}, \sigma_m^2 \Sigma_m)$$

where 0_{N_S} is an N_S x 1 vector of 0s, σ_m^2 is the between location variance and Σ_m is the N_S x N_S correlation matrix, in which element (s, s') represents the correlation between sites s and s'. The covariance between the sites s and s' is assumed to be a function of the distance, $h = ||s - s'||$, between them. Often a member of the Matern class is used, the most common choice being the exponential covariance function

$$f(h, \phi) = \exp(-\phi \times h)$$

where $\phi > 0$ describes the strength of the correlation.

Stage 3 - Hyperpriors

The prior distribution for the vector of regression coefficients, β is assumed to be a $N(\mu_\beta, \sigma_\beta^2)$. For the precision parameters a Gamma prior is selected, for instance $\sigma_u^{-2} \sim Gam(a_u, b_u)$ and $\sigma_m^{-2} \sim Gam(a_m, b_m)$. A uniform prior is used for the parameter of the strength of the correlation between the sites, ϕ, with limits based on beliefs about likely possible values.

9.12.1   Implementation

In the following sections we will consider methods for implementing this model using both MCMC and INLA.

The posterior distribution is analytically intractable but samples may be generated in a straightforward fashion using Markov Chain Monte Carlo (Smith & Roberts, 1993). If the variances σ_v^2 and σ_w^2 were known then the Kalman filter could be applied (Meinhold & Singpurwalla, 1983; Fahrmeir & Tutz, 1994) for efficient estimation.

Gibbs sampling is described in Chapter 5 as an efficient way to obtain samples from complex posterior distributions. Gibbs sampling requires the full conditional distribution of each parameter that is described in this section. The majority of them are in closed form and so sampling is relatively straightforward, however a notable exception is the parameter, ϕ, that controls the relationship between correlation and distance. If the prior distribution of ϕ is a uniform distribution, the full conditional cannot be found in a closed form but a Metropolis–Hasting step can be used. Other approaches are available and have been shown to offer successful alternatives to using a continuous uniform prior which often results in problems with convergence and can be computationally demanding (Finley et al., 2007; Diggle & Ribeiro, 2007).

Four such approaches are:

(i) **Continuous** - drawing independent samples from a continuous uniform distribution as proposed values.

(ii) **Discrete** - drawing independent samples from a discrete uniform distribution as proposed values. The discrete method is straightforward resulting in a much faster algorithm because many of the required calculations, such as the calculations of the matrix inversions, can be done off-line, i.e. not at each iteration of the simulation. Despite the possible loss of information when approximating a continuous distribution by a discrete one, it is also likely to give better mixing.

(iii) **Adaptive Metropolis within Gibbs** - taking as a proposed value the next step in a random walk process with a certain variability that can be tuning in order to have a suitable acceptance rate. Details of this method can be found in Roberts and Rosenthal (2009).

(iv) **Marginalised** - marginalise out the spatial effects and also use steps from a random walk process to propose values for the posterior distribution of ϕ (Finley et al., 2007). The basic idea is that we use a simpler model

$$y_s = X_s \beta_s + v_s$$

with the spatial effects component marginalised out. Now $y. \sim MVN(X\beta., \sigma_m^2 \Sigma_m + I_S \sigma_u^2)$ where I_{N_S} is the identity matrix of size $N_S \times N_S$. The idea is that the matrix $S = \sigma_m^2 \Sigma_m + I_{N_S} \sigma_u^2$ is more stable than $\sigma_m^2 \Sigma_m$ and is expected that the posterior distribution of ϕ will converge. This does mean that more of the full conditionals of the parameters can not be found in closed form and here a Metropolis-Hasting step is used for the variables $(\sigma_u^2, \sigma_m^2, \phi)$.

9.12.2   Prediction at unmeasured locations

Based on the posterior estimates of the site effects, m_s and the variance-covariance matrix $\sigma_m^2 \Sigma_m$, it is possible to estimate the spatial random effects, and thus exposure levels, at locations where there is no monitoring site. For a site at a new location, $m_{S+1}, (m_1, ..., m_{N_S}, m_{N_S+1})$ follows a multivariate normal distribution with zero mean and $(N_S + 1) \times (N_S + 1)$ variance-covariance matrix.

The conditional distribution of $m_{N_S+1}|\mathbf{m}$ is normal with mean and variance given by

$$E[m_{S+1}|\mathbf{m}] = \sigma_m^{-2} \Omega' \Sigma_m^{-1} m,$$

$$var(m_{S+1}|\mathbf{m}) = \sigma_m^2 (1 - \Omega' \Sigma_m^{-1} \Omega)$$

For exploratory purposes, the posterior medians may be substituted into these expressions, although this will ignore the inherent uncertainty in the estimates. The uncertainty in the estimation of the parameters should be fed through to that associated with the predictions. Prediction at arbitrary locations can be carried out using either single site joint/simultaneous prediction. The difference between them is that the single site prediction yields marginal prediction intervals, i.e. ignoring correlation between prediction locations, whereas joint prediction yields simultaneous prediction intervals for the set of target locations. The predicted means should be the same under joint or single site prediction but the intervals for joint prediction will tend to be narrower than the marginal prediction intervals. The disadvantage of joint prediction is that it is very slow as the computational time is of order P^3, where P is the number of prediction sites.

Example 9.9. *Fitting an exponential spatial model to PM$_{10}$ concentrations in London using WinBUGS*

The methods presented for spatial estimation and prediction can be implemented using MCMC and here we discuss how this can be performed using WinBUGS. Specifically, we show how the model with an exponential correlation structure as described in Section 9.12 can be implemented. We consider the case of collection of N_S monitoring sites measuring an environmental hazard and use the structure shown in Section 9.12 to represent the link between the measured data and the underlying spatial field.

In order to fit the exponential model in WinBUGS we use the spatial.exp function. In the arguments for this function, ϕ and κ in the correlation function, $\exp(-(\phi h)^\kappa)$ are labelled $\phi = \phi_1$ and $\kappa = \phi_2$, where h denotes the distance between points s and s'. As an example, we apply this to data on PM$_{10}$ concentrations for eight locations within London as seen in Shaddick and Wakefield (2002).

```
model {

for (site in 1:NS) {
y[site] ~ dnorm(mean.site[site],tau.v[site])
mean.site[site] <- beta0 + m.adj[site]

m[1:S] ~ spatial.exp(mu[], xcoords[],ycoords[],
tau.m,phi1,phi2)

# Set up the priors for the spatial model
# here we fix the second parameter to be one
phi2 <- 1
phi1 ~ dunif(0.005,0.115)
tau.m ~ dgamma(1,0.01)
sigma.m <- 1/sqrt(tau.m)

# Set up the site specific observation precisions
for (site in 1:NS) {
tau.v[site] ~ dgamma(1,0.001)
sigma.v[site] <-1/sqrt(tau.v[site])
}

# prior for the intercept term

beta0 ~ dnorm(0,1000)
```

Where there is an intercept term in the model, as there is here, the spatial random effects will need to be constrained otherwise the model will be non-identifiable. Here, the site effects are constrained to sum to zero. Note that the standard deviation is then calculated using $N_S - 1$ degrees of freedom.

```
for (site in 1:NS) {
mu[site]<-0
m.adj[site] <- m[site]-mean(m[1:NS])
sigma.m.adj <- sqrt(pow(sigma.m,2.0)*NS/(NS-1))

} # end of model
```

In this case, the posterior median values of ϕ indicates that correlation falls to 0.3 after 20 km, i.e. correlation $= \exp(-0.05675 * 20)$. The difference in the spatial effects for each site can also be seen, ranging from $+0.1341$ (on the log scale) above the mean for the site at Bloomsbury to 0.1210 below the mean for the monitoring site at Brent.

Example 9.10. *Spatial prediction of NO_2 in Europe*

In conjunction with fitting the spatial.exp model to a set of observed data, joint and single site predictions can be performed using the spatial.pred and spatial.unipred functions respectively. In this example, we apply the functions to data on NO_2 concentrations in Europe as seen in Shaddick, Yan, et al. (2013). The WinBUGS code to obtain these predictive distributions is as follows:

```
## Joint prediction:
        y.pred[1:NP] ~ spatial.pred(mu.T[], xcoord.pred[],
        ycoord.pred[], y[])

## Single site prediction:
        for(j in 1:NP) {
                y.pred[j] ~ spatial.unipred(mu.T[j], xcoord.
                    pred[j],
                ycoord.pred[j], y[])
                }
```

where N_P is the number of prediction locations, mu.T[] is vector of length N_P (or scalar for single site version) specifying the mean for each prediction location that should be specified in the same way as the mean for the observed data as seen in Section 9.5. The x and y coordinates of the location of each prediction point are contained in x.pred[] and y.pred[] and finally y is the vector of observations to which the spatial.exp model has been fitted.

The result is a set of samples from the posterior distributions at each of the prediction locations. Using these samples, it is straightforward to obtain not just a single measure of the prediction at each point, e.g. the median, but also functions of the predictions, e.g. probability of exceeding a threshold.

In addition, the samples from the posterior distributions give us a way of obtaining measures of the associated uncertainty, both for the distribution of the predictions but also distributions of functions based on them.

In the example of NO_2 in Europe, predictions were made at points on a grid defined over the entire study region. Due to computational considerations when using the large number of locations that were considered in this example, single site prediction was used with calculations being performed off-line or out-of-simulation using posterior medians of the required parameters. Whilst computationally efficient, to perform such a large number of predictions within MCMC would be computationally prohibitive, it does ignore the inherent uncertainty in the parameter estimates. The uncertainty associated with the predictions will therefore be underestimated. Predictions at these points can then be used to produce maps of concentrations such as that seen in Figure 9.9. In this case, the results from WinBUGS were imported (via R) to ESRI ArcGIS.

An important motivation for producing information about air quality is to support national governments in environmental policy making. The point estimates from the posterior distributions shown in Figure 9.9 give a summary of air quality for each 1×1 km^2 grid cell within the EU, with the most heavily polluted regions being in the north of Italy, the Be-Ne-Lux region, and the western part of Germany with other large cities being evident. At each of these locations there is an underlying (posterior) probability distribution which incorporates information about the uncertainty of these estimates.

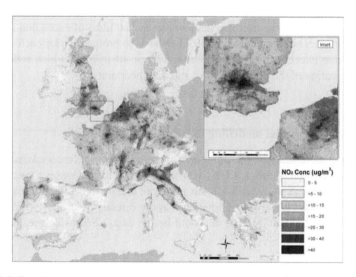

Figure 9.9: Predictions of nitrogen dioxide (NO_2) concentrations in Europe for 2001. Predictions are from a Bayesian spatial model and are estimates of posterior medians for each 1 km $\times$ 1 km geographical cell in Europe.

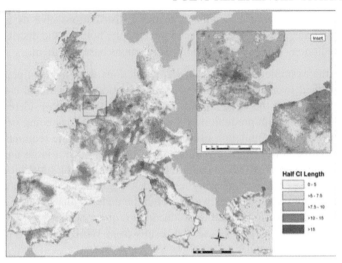

Figure 9.10: Uncertainty associated with predictions of nitrogen dioxide (NO$_2$) over Europe. Here the uncertainty associated with the prediction at each point is represented by half the length of the estimated 95% credible interval.

There are a number of ways of presenting this information and Figure 9.10 shows one of these; half of the length of the 95% credible intervals (Denby, Costa, Monteiro, Dudek, & Erik, 2007). As might be expected, higher uncertainty is observed in areas with higher concentrations. The distributions for each cell can also be used to examine the probabilities of exceeding any particular concentration and Figure 9.11 shows the probability for each cell that the value exceeds the WHO/EU guideline of 40 μgm^{-3}, which reflects the combination of the predicted level and the uncertainty.

9.13   INLA and spatial modelling in a continuous domain

The methods presented in this chapter are for use with point-referenced data and particularly cases where there is a Gaussian field (GF) with responses measured with error. A GF itself has no natural Markov structure and so INLA, as originally developed and described in Chapter 5, does not apply directly as it does with Gaussian Markov random fields (GMRFs) as would be the case when using areal data as in Chapter 8.

It is possible to use a bridge between a GF and a GMRF to which INLA can be applied. This is done using the stochastic partial differential equation (SPDE) approach presented by Lindgren et al. (2011). This starts with a GF over a continuous domain of arbitrary dimension and from it induces a GMRF. Key elements of this

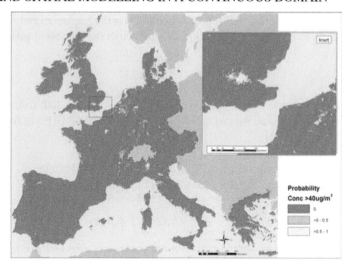

Figure 9.11: Probability that the level of nitrogen dioxide (NO_2) exceeds 40 μgm^{-3} based on samples from the posterior distributions of predictions at each 1 km $\times$ 1 km geographical cell in Europe.

method are the use of GFs, that are characterised by their second order properties, and the restriction that the field must have a Matern covariance structure. Lindgren et al. (2011) show that such a field can be expressed as the solution of an SPDE that can be approximated using a finite element method whose elements are triangles over the field's domain. The induced Gaussian random weights attached to its vertices then determine the joint distribution of the induced GMRF representation of the original GF. The precision matrix for this GMRF is approximated by a sparse precision matrix, $\mathbf{Q}$, that represents the covariance Σ of the GF well, i.e with $\mathbf{Q}^{-1}$ close to Σ, in order to achieve computational simplicity.

The result is a GF model for the process but with an associated GMRF that can be used (by INLA) for performing the computations that would be computationally prohibitive using the GF directly. The resulting algorithm is implemented in R–INLA (www.r-inla.org), an extensive R library of programs which accesses the core INLA computational engine.

9.13.1 Implementing the SPDE approach

We now describe the SPDE–GRMF approximation following Lindgren et al. (2011). INLA assumes the GF $Z_s, s \in \mathscr{S}$ has a Matern spatial covariance as given by (9.7) that is the solution of the SPDE

$$(\kappa^2 - \Delta)^{\alpha/2} Z_s = v_s, \alpha = v + d/2, \kappa > 0, v > 0 \qquad (9.11)$$

where $(\kappa^2 - \Delta)^{\alpha/2}$ is a pseudo–difference operator, Δ is the Laplacian and v is spatial white noise with unit variance. The marginal variance of the process is given by

$$\sigma^2 = \frac{\Gamma(v)}{\Gamma(v+d/2)(4\pi)^{d/2}\kappa^{2v}} \tag{9.12}$$

Representing the process in this way is key to the developments that follow; it provides the bridge over which we can cross from the GF to the GMRF via an approximate solution to the SPDE.

An infinite dimensional solution, Z_s, of the SPDE over its domain, $\mathscr{S}$, is characterised by the requirement that for all members of an appropriate class of test functions, ϕ,

$$\int \phi_{js}(\kappa^2 - \Delta)^{\alpha/2} Z_s dZ = \int \phi_{js} v_s ds \tag{9.13}$$

However in practice, only approximate solutions are available and Lindgren et al. (2011) use the conventional finite element approach, which uses a Delauney triangulation (DT) over $\mathscr{S}$. Initially the triangles are formed with vertices at the points of the sparse network where observations are available with additional triangles added until $\mathscr{S}$ is covered, leading to an irregular array of locations (vertices).

Example 9.11. *Creating a mesh: black smoke monitoring locations in the UK*

Figure 9.12 shows the mesh that was constructed using Delauney triangulation for the locations of black smoke monitors in the UK. The R code for producing the mesh is as follows.

```
    mesh = inla.mesh.create(locations[,1:2],
extend=list( offset=-0.1), cutoff=1,
# Refined triangulation,
# minimal angles >=26 degrees,
# interior maximal edge lengths 0.08,
# exterior maximal edge lengths 0.2:
refine=(list(min.angle=26,
max.edge.data = 100,
max.edge.extra=200))
)
```

where `locations` is a matrix with two columns, containing the x and y coordinates of the monitoring sites. With the plot created as follows:

```
ukmap <- readShapeLines("uk_BNG.shp")
plot(mesh, col="gray", main="")
lines(ukmap)
points(locations, col="red", pch=20,bg="red")
```

Here the file `uk_BNG.shp` is the shapefile for the UK using the British National Grid projection that provides the outline of the UK coastline and is included in the online resources. In this case, the distance between the points

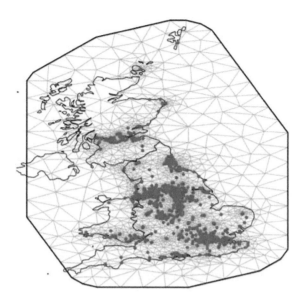

Figure 9.12: Triangulation for the black smoke network in the UK. The red dots show the locations of black smoke monitoring sites.

is expressed in metres and so the distances used in creating the mesh to ensure the plots will overlay should also be in metres.

In this case, there are 3799 edges and the mesh was constructed using triangles that have minimum angles of 26 and a maximum edge length of 100km. There are 1466 monitoring locations being considered over the period of study and these are highlighted in red. This lattice underlies the GRMF and gives a finite element representation of the solution of that shown in Equation 9.11,

$$Z_s = \sum_{k=1}^{n} \psi_{ks} w_k \qquad (9.14)$$

where n is the number of vertices of the DT, $\{w_k\}$ are Gaussian weights and ψ_{ks} are piecewise linear in each triangle (1 at vertex k and 0 at all other vertices). The ψ_{ks} then need to be linked to the class of test functions and in order to obtain an approximate solution to the SPDE.

In practice, the implementation in R–INLA takes this approximation one step further by requiring n test functions in order to obtain a finite dimensional approximation to the SPDE. Specific details can be found in Lindgren et al. (2011), but briefly, $\phi_k = (\kappa^2 - \Delta)^{\alpha/2} \psi_k$ is used with $\alpha = 1$. Substituting these test functions into Equation 9.13 along with the approximation shown in Equation 9.14 gives a set of n equations which may be solved. These equations characterize the elements of that approximation, including a sparse precision matrix for the GMRF distributed over the vertices of the irregular lattice and defined by the random Gaussian weights $\{w_k\}$.

Example 9.12. *Fitting an SPDE model using R–INLA: black smoke monitoring locations in the UK*

Using the mesh set up in Example 9.11 we now create the INLA SPDE object that will be used as the model in the same form as seen in Chapter 5. The following R code creates the object for the UK black smoke monitoring data using the `inla.spde2.matern()` command, which in its simplest form would be `spde = inla.spde2.matern(mesh, alpha=2)` for the case where $\alpha = 2$ from Equation 9.11.

```
# Field std.dev. for theta=0
  sigma0 = 1
# find the range of the location data
  size = min(c(diff(range(mesh$loc[,1])),
              diff(range(mesh$loc[,2]))))
# A fifth of the approximate domain width.
  range0 = size/5
  kappa0 = sqrt(8)/range0
  tau0 = 1/(sqrt(4*pi)*kappa0*sigma0)
  spde = inla.spde2.matern(mesh,
    B.tau=cbind(log(tau0), -1, +1),
    B.kappa=cbind(log(kappa0), 0, -1),
    theta.prior.mean=c(0,0),
constr=TRUE)
```

The value of `kappa0` represents the prior belief of the distance at which the correlation, ρ, is expected to fall to 0.1, and is equal to $\rho = \frac{\sqrt{8\nu}}{\kappa}$ (Lindgren et al., 2011), where here $\nu = 1$. The value for the spatial standard deviation, `tau0` is the square root of the spatial variance, σ^2, given in Equation 9.12; $\frac{1}{\sqrt{4\pi\kappa\tau}}$.

The model is then fit by defining a formula and then running the `inla` command using the SPDE object in the random effects term, `f(spde)`.

```
formula = logbs ~ 1+ urban.rural +f(site, model=spde)

model = inla(formula, family="gaussian", data = BSdata,
control.predictor = list(compute=TRUE),
control.compute = list(dic = TRUE, config=TRUE))
```

where `urban.rural` is an indicator variable which represents whether the monitoring site is in a rural (0) or urban area (1). Here the optional argument in the `control.predictor` means that predictions for any missing values, including predictions at unmonitored locations, will be stored and the first argument in control.compute means that the information required to compute the DIC will be retained. The second argument allows information to be stored to enable joint samples from the posterior to be obtained, the result of which means that predictions together with associated measures of uncertainty are 'automatically' available for all data points, including cases where data was and was not originally available.

The fixed effects of the model can be obtained as follows (results rounded to 3 dp):

```
model$fixed.effects
```

```
                mean      sd  0.025quant  0.5quant  0.975quant
(Intercept)    0.042  0.707      -1.346     0.042       1.430
ur            -0.066  0.013      -0.092    -0.066      -0.040

    mode      kld
   0.042      0
  -0.066      0
```

Here we can see that concentrations are lower in rural areas than urban ones.

We are interested in the posterior marginals of the latent field,

$$\pi(x_i|y) = \int \pi(x_i|\theta,y)\pi(\theta|y)d\theta \qquad (9.15)$$

$$\pi(\theta_j|y) = \int \pi(\theta|y)\pi(\theta_{-j})d\theta \qquad (9.16)$$

where $i = 1,...,N_S + P$ where N_S is the number of monitoring locations and P the number of predictions to be made and $j = 1,...J$ is the number of parameters in the model. With regards to spatial prediction, the SPDE–INLA algorithm provides the posterior conditional distribution of the random effects terms at all the vertices of the triangulation. Given these, there is a mapping to the response variable which allows samples of predictions to be obtained (Cameletti, Lindgren, Simpson, & Rue, 2011).

In order to produce a map displaying the spatial predictions of the model we first need to define a lattice projection starting from the mesh object back to the grid on which the data lies and will be plotted. This is performed using the inla.mesh.projector command. Then the posterior mean and standard deviations can be extracted and then projected from the latent field space to the grid, using inla.mesh.project and then plotted as a map. The example of the result can be seen in Figure 9.13. There are a number of ways of converting the output from SPDE/R–INLA to a form that can produce maps and the package geostatsp provides many routines for doing this.

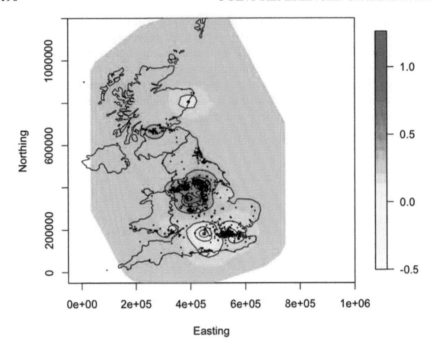

Figure 9.13: Map of predicted values of black smoke in the UK. Values are medians of posterior predicted distributions on the logarithmic scale from an SPDE–INLA model.

9.14   Non-stationary random fields

A number of approaches for dealing with non-stationary spatial random fields have been developed. The case of spatio–temporal processes is considered in Chapter 11. There one has replicates over time which, after adjusting for correlation over time, allows more scope for dealing with non-stationarity. A purely spatial process is more challenging since with only one replicate of the spatial process, estimating second-order properties such as spatial covariances is difficult. In this section we review some of the approaches that have been developed in this situation.

9.14.1   Geometric and zonal anisotropy

Directional variogram plots are constructed by placing an origin in the spatial domain of interest $\mathscr{S}$ from which radial lines emanate at a set of angles. For any one of those lines, the empirical variogram computed for sites within the wedge between the lines is the *directional variogram*.

One approach to dealing with geometric anisotropy is to assume the plot of the ranges on the radial direction lines form an ellipse with the diameters of the major and minor axes determined respectively by the largest A and smallest B ranges (Webster & Oliver, 2007). Suppose as well that the major axis makes an angle φ with the horizontal axes. Rotate the horizontal axes so that it now lies along the major axes. Then inflate the diameter of the minor axes by the ratio $R = A/B$ to get a sphere in the new coordinate system.

In summary the distance between geographic sites, $s = (s_1, s_2)$ and $s' = (s'_1, s'_2)$ is defined

$$\|s - s'\| = \sqrt{(s - s')^T (s - s')} = \sqrt{\sum_k (s_k - s'_k)^2} \sqrt{\sum_k (s_k - s'_k)^2}$$

which is replaced by

$$\sqrt{(s - s')^T \zeta (s - s')}$$

in the new measure of distance where ζ is a positive definite 2×2 matrix.

Zonal anisotropy, where both the sill and the range depend on direction, proves more challenging and various methods have been suggested. The simplest is to select a small, fixed, number of zones or strata and model the field separately in each zone.

Example 9.13. *Plotting directional variograms for temperatures in California*

Recall that the estimated variogram for maximum daily temperatures for Apr 1, 2012 at 18 sites in California shown in Example 9.5 appears to exhibit anisotropic features. In such cases, a directional variogram as shown in Figure 9.14 may be plotted in order to explore anisotropy.

The plot shown in Figure 9.14 was produced using the `variog4` function in the geoR packaged to make the plot as follows:

```
### Load data
> CAmetadata<- read.table("metadataCA.txt",header=TRUE)
> CATemp<-read.csv("MaxCaliforniaTemp.csv",header=TRUE)
> dimnames(CATemp)[[2]][1]="Date"
> CATemp20120401<-subset(CATemp, Date==20120401)

### Augment data file with coordinates, note date is omitted
    as is the
### value for Bakersfield for which there is no meta-data
>   CAmetadata=cbind(CAmetadata,  t(CATemp20120401[c(2:14,
    16:19)]))

### Create geoR data geofile
>   library(geoR)
```

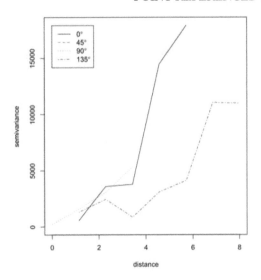

Figure 9.14: Direction variogram for temperatures in California. This plot revisits that seen in Example 9.6. Here we see a directional variogram for pairs of sites lying within a bin around four geographical direction lines for the eighteen California sites whose maximum daily temperatures were recorded on Apr 1, 2012.

```
CATemp.geo<-as.geodata(CAmetadata,coords.col = 3:4,data.col
   =7)

### Compute and plot the directional variogram
> CA.vario4<-variog4(CATemp.geo, uvec = seq(0,8,l=8))
> plot(CA.vario4)
```

The two variograms for the 135 and 45 degree directions seem to correspond to the two groups of sites seen originally in the aggregate estimate of Example 9.5. However any conclusion would be speculative given the small number of sites in the dataset.

9.14.2   Moving window kriging

Moving window kriging (Haas, 1990) is perhaps the simplest approach to dealing with non-stationary data. Here we divide the region into a number of subregions, each of which has a sufficient number of points to estimate a (stationary) variogram. The overall spatial process and its covariance structure can then be represented by a weighted average of the locally stationary processes (Cressie, 1986). Ordinary kriging assumes that the mean is constant over the entire region of study, but in *universal kriging* a non-constant mean is represented as a function of the location, a polynomial function of finite order, k. The assumption that $E(Z_s - Z_{s'}) = 0$ no longer holds and the method assumes that the variogram and the order k of the polynomial are known. Unfortunately k is never known and has to be guessed, not estimated, from the data and γ has to be estimated from the residuals. Other approaches include *median polish kriging* and *intrinsic random functions of order k* (Cressie, 1986). The former involves modelling the mean term using the medians of rows and columns of a spatial grid in such a way that the resulting residual terms are stationary and thus can be used with traditional geostatistical techniques. The choice of the spatial grid is critical to this approach, and attempts are often made to define it in a meaningful way with the precise formulation being determined by the particular application. The latter has more general model assumptions than the kriging methods, and involves estimating k and the *generalised covariance function* from k^{th} order differences (Cressie, 1993), however its use in practice is limited as it is not generally possible to estimate the covariance function and the order k of the differencing again has to be guessed, rather than estimated from the data.

9.14.3 Convolution approach

We now give a brief description of the convolution approach to modelling non-stationary fields. Calder and Cressie (2007) review this approach starting with the basic spatial process model

$$Z_s = \mu_s + m_s + v_s, \ s \in \mathscr{S}$$

where μ_s is the spatial mean and v_s is a random noise term that represents the nugget. The convolution formulation derives from the assumption

$$m_s = \mu_s + \int_{R^d} k(s,u)W_{du}, s \in \mathscr{S} \tag{9.17}$$

where k is a kernel function and W, a general zero mean spatial process with independent increments. The spatial covariance function, which for sites s and s', is

$$C(s,s') = \int_{R^d} k(s,u)k(s',u)du$$

must be positive definite. This model can be discretised in for discrete domains S and it can produce a variety of different non-stationary models, two of which we now describe.

A number of approaches rely on two-dimensional, non-stationary, Gaussian processes and allow spatial dependence to vary with location (Higdon, 1998; Higdon, Swall, & Kern, 1999; Higdon, 2002). In this case, W in Equation 9.17 will be a white noise process giving,

$$\int_C W_s ds \sim N(0, \tau^2 |C|))$$

for every C where $|C|$ is the area of subregion C and $\tau > 0$ is a constant. Moreover $k(s,u) = k(s-u)$ has the form of the Gaussian probability density function

$$k(s) = \frac{1}{2\pi} exp\left\{-\frac{1}{2}s^T s\right\}$$

As a result m_s would have the usual isotropic Gaussian correlation function,

$$\rho(h) = exp\{-h^T h\} \tag{9.18}$$

The above representation can be generalised by using a smoothing kernel, denoted by k_s, which depends on spatial location s. Higdon et al. (1999) chose a bivariate Gaussian kernel for their application:

$$k_s(s) = \frac{1}{2\pi} |\Sigma_s|^{-1/2} exp\left\{-\frac{1}{2}s^T \Sigma_s^{-1} s\right\}$$

where Σ_s is a function of location s. The one standard deviation ellipsoids of concentration for this kernel become ellipsoids whose major and minor axes vary in direction and length from one location to another along with the degree of smoothing induced by the kernel over regions of the spatial domain (Paciorek & Schervish, 2006).

Fuentes (2002a, 2002b) proposes another approach where now m in Equation 9.17 represents a weighted average of locally isotropic stationary processes that are uncorrelated with each other. The geographical region is divided into well-defined subregions, each of which has a locally isotropic stationary process.

More precisely,

$$m_s = \sum_{i=1}^{k} V_{is} w_{is}$$

for $i = 1, \cdots, k$, i.e., where $Cov(V_{is}, V_{js}) = 0$ for $i \neq 0$. The geographical region is divided into k well-defined subregions $A_1, \ldots, A_k$ and (V_{is}) is a local isotropic stationary process in a subregion A_i. The weights, w_{is} come from a positive kernel function centred at the centroid of A_i.

The covariance between any two locations s_1 and s_2 in the geographical region can be written as

$$\begin{aligned}
Cov(m_s, m_{s'}) &= \sum_{i=1}^{k} w_{is} w_{is'} cov(V_{Is}, V_{Is'}) \\
&= \sum_{i=1}^{k} w_{is} w_{is'} C_{\theta_i}(|h|)
\end{aligned}$$

depends only on the distance $|h|$ due to the isotropic stationarity. Since θ_i can change from subregion to subregion, $Cov(m_{s,s'})$ can depend not just on that distance but also on s and s'. Thus the process m_s is non-stationary.

9.15 Summary

This chapter contains the basic theory for spatial processes and a number of approaches to modelling point-referenced spatial data. The reader will have gained an understanding of the following topics:

- Visualisation techniques needed for both exploring and analysing spatial data and communicating its features through the use of maps.
- Exploring the underlying structure of spatial data and methods for characterising dependence over space.
- Second-order theory for spatial processes including the covariance. The variogram for measuring spatial associations.
- Stationarity and isotropy.
- Methods for spatial prediction, using both classical methods (kriging) as well as modern methods (Bayesian kriging).
- Non-stationarity fields.

Exercises

Exercise 9.1. This exercise relates to units of measurement.

(i) The US ozone standard set in about 1997 was set at 0.08 (ppm, parts per million). Why did it not contain another decimal place?

(ii) The 1997 standard was changed to the more restrictive upper limit of 0.075 (ppm). Why not set this to be 75 (ppb - parts per billion) instead to give it a simpler form?

(iii) Ground level ozone particulate concentrations $PM_{2.5}$ are measured in units of μgm^{-3} (micrograms per metre cubed). These distributions of these measurements are highly right skewed and that has led some statistical analysts to transform these measurements as $y = \log PM_{2.5}$. What units of magnitude should be attached to y? Why?

(iv) For levels of particulates, y, risk assessments are commonly made with $\Delta y = 10$ μgm^{-3} while the log transform ($y' = \log(y)$) is often employed in statistical analysis to get $\alpha \exp(\beta y')$ for the concentration response function (CRF). Would the units of measurement used for y matter when looking at the effect of such a change in the CRF? In other words, what would happen if the US EPA were to change their measurements from $10\ \mu gm^{-3}$ to $10,000$ milligrams m^{-3}?

Exercise 9.2. This exercise is about measuring distance on the earth's surface, as one must do in developing variogram models when large geographical domains are involved. The central problem stems from the fact that the lines of longitude are not parallel, unlike the lines of latitude.

(i) Approximately how many kilometres is a degree of latitude on the Earth's surface?

(ii) Given a circle of radius r centred at the origin, and two points on that circle, P_1 and P_2, find a formula for the distance between them along the arc of that circle between them.

(iii) Repeat the first calculation, but this time for two points on the Earth's surface whose coordinates are given in latitude and longitude.

(iv) Use your formula to calculate the distance in kilometres between Whitehorse and Toronto (both in Canada), using the latitudes and longitudes provided by Google maps. How does this compare to the naive result you would get if you assumed a degree of longitude was about the same as a degree of latitude in size in kilometres?

Exercise 9.3. Produce a map as shown in Figure 9.5 for another state or area.

Exercise 9.4. (i) What adverse effects on human populations does cadmium have in sufficiently high concentrations?

(ii) Using the R gstat library, perform a trend and spatial distribution analysis for the cadmium concentrations in the Meuse River flood plain.

Exercise 9.5. (i) For any $N_S \times 1$ random vector $\mathbf{W}$, prove that its covariance matrix $Cov(\mathbf{W})$ must be positive definite.

(ii) Suppose Z is second order stationary process and $\mathbf{Z}$, a $N_S \times N_S$ process vector over points s_i, $i = 1, \ldots, N_S$. Show that Equation 9.2 implies that $Cov(\mathbf{Z})$ must be positive definite.

(iii) Show that $C(h) = \sigma(\exp - \| h \|)$, $h \in \mathscr{R}^2$ is positive definite. *Hint:* Find two zero mean random vectors $\mathbf{U}$ $\mathbf{V}$ of dimension 2 such that $C(\mathbf{U}, \mathbf{V}) = C(h)$.

Exercise 9.6. (i) Repeat the ozone examples shown in Examples 9.5 and 9.13 for Jul 1, 2012 and compare the results to the ones from April 1, 2012 as given in the text. Compute directional variograms using variog4 in geoR and explain the results.

(ii) Redo the directional variogram plots for using gstat.

Exercise 9.7. Give a theoretical example of a spatial field that is intrinsically stationary but not second order stationary.

Exercise 9.8. Using the R package, geoRglm, carry out a Bayesian analysis of the lead data seen in Example 9.1. Use ggmap to display your predictions on a Google map.

Exercise 9.9. Determine explicitly the form of the matrix ζ in Section 9.14 which is required to correct for geometric anisotropy.

Exercise 9.10. Prove the assertion in Equation 9.18.

Exercise 9.11. Using WinBUGS

(i) Compare the results from the model using the joint spatial model (using spatial.exp) with one that assumes that the site effects are conditionally independent.

(ii) Compare the results using spatial.pred and spatial.unipred (for a given set of locations).

Exercise 9.12. (i) Compare the results from the model using the joint spatial model (using spatial.exp) to the London PM_{10} data with one that assumes that the site effects are conditionally independent.

(ii) Compare the results using `spatial.pred` and `spatial.unipred` (for a given set of locations) paying particular attention to the width of the credible intervals. What might be the reason for any differences?

Exercise 9.13. Using R–INLA

(i) Fit a spatial model to the European NO_2 concentrations and produce a map of the spatial effects from the model.

(ii) Use the results from this model to produce a map of the probabilities that concentrations exceed $40\mu gm^{-3}$.

(iii) Land–use regression uses information on factors that might affect levels of pollution at the locations of monitoring sites. By choosing a suitable selection of factors, fit a model that estimates the effects of those factors and fits a spatial model to the residuals.

(iv) Under what circumstances could this model be used for predicting concentrations at unmonitored locations?

Exercise 9.14. RESEARCH PROBLEM: Determine the source of the anisotropy seen in Example 9.5. A useful approach would be to download data for many more sites, fit a spatial mean function $\mu(s)$ and then build spatial dependence into the residual process.

Chapter 10

Why time also matters

10.1 Overview

Modelling time series data and the temporal processes that generate them is of paramount importance in environmental epidemiology where we find several areas of application. These include modelling underlying patterns in exposures, for example to air pollution, as well as temporal patterns in health outcomes. Predictions may be made both within the time frame of the given data and in the future, the latter of which it is known as *forecasting*. An example of forecasting in environmental processes is where urban areas produce 24 hour ahead forecasts of air pollution levels (Dou, Le, & Zidek, 2012). This chapter contains a background to the study of temporal processes, which replaces space, the subject of Chapter 9, as the domain of interest whilst drawing on many of the concepts from the previous chapter.

This chapter begins by giving a general perspective on the role of time series in environmental epidemiology. Then we turn to exploratory and mainly classical methods for handling temporal data, notably on how to separate low and high frequency components of a process that evolves over time and also on characterising dependence in the series. Finally we show how Bayesian methods can be used for incorporating temporal dependence into health effect analyses.

10.2 Time series epidemiology

One of the areas in which time series methods are most extensively used in environmental epidemiology is in assessing the short-term effects of changes in air pollution on health outcomes. In this case, the data commonly takes the form of daily measurements of one, or more, pollutants that are related to daily counts of mortality or morbidity. The outcomes are likely to exhibit temporal correlation, i.e. there is likely to be a high degree of dependence between values of the process, Z_t, especially within short periods of time. This is not necessarily because the outcome, e.g. the number of daily deaths, is causally related to the number the day before, as might be the case with a contagious disease, but because the underlying risk factors, such as temperature and pollution tend to be highly correlated day-to-day. The values of Z_t are then not mutually independent and this must be taken into account in any analysis that involves them. Classical time series composition and analysis is primarily interested in modelling the behaviour of the response variable, rather than its relationship with a set of explanatory variables. However, the methods can be extremely valuable in understanding the nature of any temporal dependence which might manifest itself for example in model residuals and thus in constructing suitable models. There are many comprehensive texts on the subject of time series and forecasting and only a brief review is presented here. For a more complete treatment of the subject see Harvey (1993), Diggle (1991), Hamilton (1994), Chatfield (2013) and Chatfield (2000).

10.2.1   Confounders

In addition to the possible effects of the exposure of interest, counts of mortality or morbidity will depend on a set of other risk factors, i.e. confounders, and if the influence of these factors is not adequately accounted for then the estimated exposure–mortality association may be biased. Confounding may induce long-term trends, seasonal variation, over-dispersion and short-term temporal correlation. Confounders may include meteorological conditions such as temperature, humidity, wind speed, and rainfall. In addition to measured confounders there may also be unmeasured factors. When modelling, these are often represented by proxy variables such as functions of calendar time and variables that indicate the day of the week.

10.2.2   Known risk factors

The health problems that result from exposure may be felt immediately, that is on the same day (Moolgavkar, 2000), after a lag of one or two days (Peters et al., 2000), or from continued exposure over preceding weeks or from long-term exposure over several decades (Elliott et al., 2007). The choice between different lags is a long-standing research problem without consensus with regards to which should be used. Numerous approaches have been suggested including selecting the lag that is associated with the most significant effect (Lumley & Sheppard, 2000) or the one that minimises an objective criteria (such as the AIC). Alternatively, results for multiple lags have been presented, for example by Burnett et al. (1994).

In studies of the short-term effects of environmental hazards, such as air pollution, it is commonly the case that risk factors will be related to meteorological covariates. These might include temperature (Mar, Norris, Koenig, & Larson, 2000), humidity Lee et al. (2000), precipitation (Spix et al., 1993), and pressure (Vedal, Brauer, White, & Petkau, 2003). Temperature can play a central role as it drives part of the seasonal variation typically present in health data where higher counts of morbidity and mortality are observed during cold periods. Although meteorological data are routinely available including them in an epidemiological analysis requires a number of decisions including which lag should be used and the exact nature of the shape of their relationship with health, i.e. can it be presented by a linear relationship or is something more complex required?

10.2.3 Unknown risk factors

Unknown risk factors may result in long-term trends and seasonal variation in time series studies and large-scale spatial trends, for example north to south gradients, in spatial studies. As they cannot be added to regression models in the same way as known factors allowing for the influence of unknown risk factors is less straightforward than for known, measured, factors. In early temporal studies, Schwartz, Slater, Larson, Pierson, and Koenig (1993) and Spix et al. (1993) modelled seasonal variation with pairs of sine and cosine terms at different frequencies, and long-term trends with parametric functions such as cubic polynomials of calendar time. Other early approaches modelled these factors with indicator variables (Verhoeff, Hoek, Schwartz, & van Wijnen, 1996) which, as with parametric functions, may be overly restrictive and lack the necessary flexibility to model excessive variation in mortality. For example the sinusoidal terms force the peak in mortality to occur at the same time each year, while the monthly indicator variables do not allow for within month variation. More recently, these unmeasured risk factors have been represented using smooth functions of calendar time, which can be more flexible than fixed parametric alternatives. Such functions have been implemented using parametric and nonparametric methods, including regression splines (Daniels, Dominici, Zeger, & Samet, 2004) and smoothing splines (Dominici, Samet, & Zeger, 2000b).

Categorical, or indicator, variables are often used as proxies for factors that may be confounded with the relationship of interest. These may include variables for 'day of the week' (Kelsall, Zeger, & Samet, 1999), for times of influenza epidemics (Peters et al., 2000) and public holidays (Schwartz, 2001). In spatial studies, where the relative risk (Chapter 2, Section 2.3.1) is driven by differences in health counts between different areas, confounding variables might for example represent the effects of socio-economic deprivation which has been shown to be a strong predictor of both health (Kleinschmidt et al., 1995) and air pollution (Elliott et al., 2007).

10.3 Time series modelling

In modelling time series data, as with spatial data seen in Chapter 9, we distinguish between the underlying temporal and measurement processes. Although time is a continuous measure over $\mathcal{T}$, data will only be collected at N_T discrete points in time, $T \in \mathcal{T}$ where these points are labelled $T = \{t_0, t_1, \ldots, t_{N_T}\}$. Note that time, unlike space, does have a natural ordering.

The underlying process is not directly measurable, but realisations of it can be obtained by taking measurements, possibly with error at times in T. One way of expressing the random field is as a combination of the overall trend together with a spatial structure, for example

$$
\begin{aligned}
Y_t &= Z_t + v_t \\
Z_t &= \mu_t + \gamma_t \\
\mu_t &= \sum_{j=1}^{J} \beta_j f_j(X_t)
\end{aligned}
\tag{10.1}
$$

where ε_t is measurement error, μ_t is the mean of the underlying process, modelled as a function of covariate information which might include time itself with associated coefficients β and γ_t is a process with temporal structure.

For clarity of exposition, in the following introduction to classical time series methodology, we assume that measurement error is not present, i.e. $Y_t = Z_t$ in Equation 10.1. We return to the hierarchical modelling approach in Sections 10.7 and 10.16.

Classical time series modelling aims to decompose the variation in the series into:

- Trend - long-term movements in the mean.
- Seasonality - annual cyclical fluctuations.
- Cycles - other cyclical variations, at different frequencies which can be greater than or less than a year.
- Residuals - other random or systematic fluctuations.

Formally, the trend, μ_t, of a time series is defined as the expectation of the set of random variables Z_t, i.e. $\mu_t = E(Z_t)$. It is the long-term change in the underlying process and indicates the general pattern of rise or fall of the outcome variable. A simple linear trend, $\mu_t = \alpha + \beta t$ is often referred to as a *deterministic* or *global* trend. Alternatively, a *local* trend can be modelled that evolves through time, allowing the parameters to be dependent on time, $\mu_t = \alpha_t + \beta_t t$. A local trend can also take the form of a recursive equation, such as in state space models (see Section 10.7), e.g. $\mu_t = \mu_{t-1} + \beta_t t$.

After modelling the trend, seasonality and cyclic components, there may still be autocorrelation in the residual term, due to short-term dependencies in the data. These dependencies over time can be explicitly modelled and details of models for this purpose are given in Example 10.9.

Example 10.1. *Ground level ozone concentrations*

Ozone is a colourless gas produced through a combination of photo-chemistry, sunlight, high temperatures, oxides of nitrogen, NO_x, emitted by motor vehicles. Levels are especially high during morning and evening commute hours in urban areas. It is one of the criteria pollutants regulated by

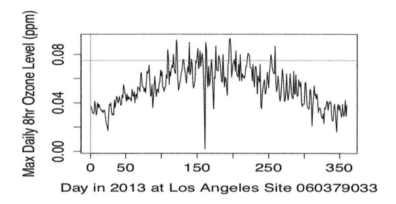

Figure 10.1: Daily concentrations of ozone (ppm) at the EPA monitoring site 060379033 in Los Angeles. The horizontal line is plotted at the current for 8 hour regulatory standard of 0.075 (ppm).

the US Clean Air Act (1970) because of a substantial body of literature that shows a strong association with human morbidity, notably respiratory diseases such as asthma, emphysema, and chronic obstructive pulmonary disorder (COPD) (EPA, 2006). Periods of high ozone concentrations can lead to acute asthma attacks leading to increased numbers of deaths, hospital admissions, and school absences. High levels of ozone concentrations can also lead to reduced lung capacity.

Figure 10.1 shows daily concentrations of ozone measured at sites located in the geographical region of Los Angeles, California. Clear seasonal patterns can be observed due to the higher temperatures in summer.

Figure 10.2 depicts the same series at the hourly level, but restricted to the so-called 'ozone season', which is taken to be May 1–Sep 30. Here a clear 24-hour daily cycle can be seen along with a period of missing data. The latter is likely due to the monitoring systems being checked each night by the injection of a calibrated sample of air. If the instrument reports the ozone incorrectly, an alert is sounded and that instrument is taken off line until it is repaired. This can lead to gaps in the series as seen around hour 1100.

This example shows the existence of regular, systematic, or low frequency, patterns together with irregular components. The regular components include trends over time and periodic components, e.g. 24-hour cycles and day of the week effects while the irregular components are randomly distributed around the regular patterns.

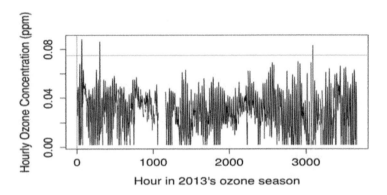

Figure 10.2: Hourly concentrations of ozone at the Los Angeles ozone (ppm) monitoring site 840060370113 during the summer ozone season May 1–Sep 30, 2013. The horizontal line is plotted at the current 8 hour regulatory standard of 0.075 (ppm).

10.3.1   Low-pass filtering

In some cases like that in Example 10.1, there is a well established scientific foundation on which to build a model for the regular components and in such cases analysis reduces to estimating the model parameters (Huerta, Sansó, & Stroud, 2004) by standard statistical methods. In other cases, exploratory analysis will be needed to identify these components. This will begin with estimating the trend by smoothing the data. There are a number of methods for doing this including moving averages (Section 2.6) fitting locally linear models and Shumway's 19 day symmetrically weighted moving average, $\sum_{-9}^{9} \psi_j y_{i-j}/\bar{y}$ (where the weights, $\psi_0, ..., \psi_9$, are symmetric and sum to one). These methods are called low-pass filters since they do not allow the irregular variation to 'pass' through them.

Example 10.2. *Low-pass filtering of carbon monoxide levels using the moving average*

Carbon monoxide (CO) is commonly called the 'silent killer' and when inhaled in large quantities it causes death by asphyxiation with minimal symptoms that include tiredness and dizziness. As it is inhaled, it passes through the lung's gas exchange membrane into the blood stream where it binds tightly to haemoglobin to form carboxyhaemoglobin (COHb) in the red blood cells. Oxygen is thus blocked from binding to the haemoglobin distribution throughout the body. Concern about the effects of CO has made it one of the criteria pollutants regulated by the US Clean Air Act. It is therefore subject to air quality regulations which require that the eight-hour moving daily

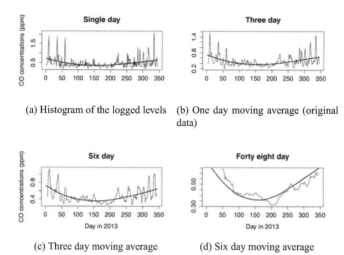

(a) Histogram of the logged levels   (b) One day moving average (original data)

(c) Three day moving average   (d) Six day moving average

Figure 10.3: Moving averages of the daily maximum eight-hour moving average of carbon monoxide (CO) concentrations (ppm) at a site in New York during 2013. The panels show: (a) the one (original data); (b) three; (c) six and (d) forty-eight day moving averages. To help visualise the underlying trend in the process, a smooth loess curve has been added. Regular components of the time series are seen in the results for six and forty-eight day results.

average level of CO concentrations must remain below 9 ppm. Figure 10.3 shows daily average CO levels for an EPA monitoring site in New York State (Site ID:340130003) during 2013. In addition to showing the original data, which is essentially a one-day moving average, different levels of smoothing are shown (three, six and forty-eight day moving averages). In each case, a locally linear smoother known as loess is added to the plots to help identify longer term trends.

The smoothed six-day version shows a pattern of cyclical episodes of CO concentrations, each of about twenty-four days duration. Using a forty-eight day moving average smooths out the bumps to reveal a longer cycle of about six months in duration, one that dips to its minimum in the summer months. with higher levels of CO in the winter months.

Example 10.3. *Moving average smoothers in epidemiological studies of air pollution*

A moving average approach was used by Mazumdar, Schimmel, and Higgins (1982), utilising a 15-day moving average when analysing daily counts

of death in winters in London, UK in the 1960s. This took into account long
wavelength patterns in the data, but rather than using density estimation, the
weights dropped from 1 to 0 between the seventh and eighth day which created distortions in the smoothed data. Moving averages were also used by
Kinney and Ozkaynak (1991) who investigated the relationships between
daily death counts (all cause, respiratory and cardiovascular) and levels of
five pollutants (O_x, SO_2, NO_2, CO and KM, a measure of optical reflectance)
together with temperature variables in Los Angeles County during the period
1970–79. They explored the temporal structure of any associations between
mortality and the pollutants (and temperature) by pre-whitening the data and
then examining the cross-correlations. The pre-whitening consisted of transforming both the input and output series using a filter designed to remove all
the autocorrelation in the individual variables, in this case an AR(2) function
and 365 day differencing was applied to each variable, leaving residuals that
had little autocorrelation. They then used multiple regression to assess the effect of the short-term changes in the pollutants on the different outcomes, after
subtracting weighted 19-day moving averages from each variable, including
the outcome. In both these examples, the data were assumed to be Gaussian,
which may have been plausible in large cities such as London with mean daily
deaths of almost 300 (Schwartz & Marcus, 1990) and Los Angeles with mean
daily deaths of 152, but most studies of air pollution and health involve far
fewer daily counts and whilst filtered Gaussian data is still Gaussian, this is
not generally the case for other distributions, such as the Poisson.

Given the somewhat arbitrary choice of smoothing functions such as loess, to remove long-term time trends or the number of degrees of freedom when fitting spline
functions, it is important that sensitivity analysis is performed to assess the possible effects. This is particularly true where decisions have to be made as to which
data should be included in the analysis. Schwartz (1994c), for example, presents results from several models using different smoothers. Different sets of outliers are
excluded, e.g. very hot or very humid days and different temperature variables, e.g.
mean daily or nightly minima. Although it is good practice to present the results from
a number of different possible models, only a small selection can be presented and if
they are not nested, or have any common structure, then it is difficult to make sensible comparisons between the different models, in addition to the theoretical issues of
testing between nonparametric models of this type.

10.4   Modelling the irregular components

After modelling the regular features of a temporal process, i.e. trend, seasonality
and cycles, we need to consider the random or irregular components. This has been
a central focus of classical temporal process modelling. Such modelling was typically based on a single long record of measurements, for example levels of ash in the
atmosphere following a volcanic eruption. This contrasts with longitudinal data analysis in which there are replicates of short times series of measurements arising from

different subjects, e.g. patients. These two modelling paradigms have evolved in very different ways.

10.4.1   Stationary processes

The concept of stationarity in time series analysis is similar to that in spatial analysis as seen in Chapter 9. A time series is said to be stationary if the distributional structure of Y_t is unaffected by a shift in time. *Strict stationarity* means that for any choice of h, and times $t_1 < \ldots < t_k$, the joint probability distribution of $(t_1, \ldots, t_k)$ is identical to that of $(t_1 + h, \ldots, t_k + h)$. Therefore, if a process is stationary, observations from any time period can be used to make inference about the overall underlying structure.

A weaker assumption is one of *weak* or *second order* stationarity where the mean, $E(Y_t) = \mu_t$ is constant for all t, and the autocovariance function denoted for any two times t, t' by $\rho_{tt'}$ is defined by $\rho_{tt'} = Cov(Y_t, Y_{t'}) = E(Y_{t'} - \mu_{t'}, Y_t - \mu_t)$ depends only on elapsed time between t and t' and not their actual location, e.g. $\rho_{1,5} = \rho_{11,15}$. This on its own would not imply strict stationarity.

10.4.2   Models for irregular components

A number of models have been developed for the irregular components of temporal components and here we present a few examples of both stationary and non-stationary processes. In these examples, $Var(Y_t) = \sigma_t^2$ and for simplicity we restrict to a discrete time domain of $T = t_1, \ldots, t_{N_T}$ for our exposition.

Example 10.4. *White noise*

$$Y_t = W_t \sim (\mu_W, \sigma_W^2) \tag{10.2}$$

where the W_t are independent and identically distributed, usually with $\mu_W = 0$.

Example 10.5. *A random walk*

$$Y_t = Y_{t-1} + W_t \tag{10.3}$$

where the W_t come from a white noise process. Thus $\mu_t = t\mu_W$ and $\sigma_t^2 = t\sigma_W^2$. Hence Y cannot be stationary. Examples of the use of such a model include modelling the logarithm of the ratio of a stock's closing price for tomorrow over that for today. Commonly it is used in process modelling for convenience since few parameters are needed to describe it (Huerta et al., 2004).

Example 10.6. *Autoregressive processes*

These process models capture local dependence in time through a Markov like model. They do not capture long memory processes where dependence can persist over days or even centuries as described in Example 10.12.

The simplest autoregressive process is the AR(1). It is Markovian in nature: given Y_{t-1} and Y_{t+1}, Y_t is independent of all other responses, past and future. It is defined at time t as

$$Y_t = \alpha_1 Y_{t-1} + W_t$$

where $\alpha_1 = \rho_{t-1,t}$ for all t, i.e. it is a stationary process. The w_t are a set of realisations of a white noise process.

This model extends to the multivariate autoregressive process denoted by MAR(1):

$$\mathbf{Y}_t = \alpha_1 \mathbf{Y}_{t-1} + \mathbf{W}_t.$$

Example 10.7. *Autoregressive processes in environmental epidemiology*

The autoregressive model is commonly seen in environmental epidemiology. For example the random hourly PM_{10} concentrations over Vancouver, Canada were found to have an AR(3) form (Li, Le, Sun, & Zidek, 1999). Kinney and Ozkaynak (1991) explore the temporal structure of any associations between mortality and the pollutants (and temperature) by pre-whitening the data and then examining the cross-correlations. The pre-whitening consisted of transforming both the input and output series using a filter designed to remove all the autocorrelation in the individual variables, in this case an AR(2) function and 365 day differencing was applied to each variable, leaving residuals that had little autocorrelation. They then used multiple regression to assess the effect of the short-term changes in the pollutants on the different outcomes, after subtracting weighted 19-day moving averages from each variable, including the outcome. In both these examples, the data were assumed to be Gaussian, which may have been plausible in large cities such as London with mean daily deaths of almost 300 (Schwartz & Marcus, 1990) and Los Angeles with mean daily deaths counts of 152. However most studies of air pollution and health involve far fewer daily counts and while filtered Gaussian data is still Gaussian, this is not generally the case for other distributions, such as the Poisson.

Example 10.8. *Moving average processes*

The independent, random elements, w_t, in the AR models can represent shocks to a system such as a weather system that produces extreme cold.

These systems may take time to subside leading to moving average processes. In the case of the second order, MA(2) this takes the form

$$Y_t = W_t + \beta_1 W_{t-1} + \beta_2 W_{t-2}$$

Example 10.9. *ARMA processes*

An ARMA(p,q) process combines an AR(p) process and an MA(q) process. For example, an ARMA(1,1) would be

$$Y_t = \alpha_1 Y_{t-1} + W_t + \beta_1 W_{t-1}$$

Example 10.10. *Inference for ARMA processes*

Ergodicity suggests that for a weakly stationary, temporal process measured with additive measurement error, we can use the following method of moments estimators:

$$\hat{\mu}(t) \equiv \frac{\sum_{t=1}^{T} Z_t}{T}$$

$$\hat{C}(\tau) = \frac{\sum_{t=1}^{T-\tau}(Z_{t+\tau} - \hat{\mu})(Z_t - \hat{\mu})}{T}$$

Example 10.11. *Backshift operators*

The backshift operator is defined as $BW_t = Y_{t-1}$ for any stochastic process indexed by time $t \in T$. Thus for an AR(1) process we have $Y_t = \alpha B Y_t + w_t$ or $w_t = (1 - \alpha B)Y_t = \phi(B)Y_t$. Thus $Y_t = \phi(B)^{-1}w_t$ in terms of the white noise process w_t. Another example is the MA(1) moving average process of order 1 where now $Y_t = w_t + \beta w_{t-1} = (1 + \beta B)w_t = \theta(B)w_t$. Finally putting these things together gives us the ARMA model $Y_t = \phi(B)^{-1}\theta(B)w_t$, a combination of the AR and MA models. This is a common way of representing the autoregressive process in terms of the innovations process w_t.

Example 10.12. *Long memory processes*

A long memory process is one whose temporal auto-correlation decays more slowly over time than one for an autoregressive process (Craigmile, Guttorp, & Percival, 2005).

If $d = 2$, the standard binomial expansion of $(1 - B)$ where B is a backshift operator would be

$$
\begin{aligned}
(1 - B)^2 &= \binom{2}{0} + \binom{2}{1}(-B) + \binom{2}{2}(-B)^2 \\
&= \frac{2}{0!} + \frac{2 \cdot 1}{1!}(-B)\frac{2!}{2!}(-B)^2
\end{aligned}
\tag{10.4}
$$

The same expansion as expressed in Equation 10.4 can be used when we replace 2 by d. When d is a fraction then the series cannot terminate and has an infinite number of terms. This is what gives the process its very long memory.

Given the data, many approaches have been developed for determining and fitting process model including the method of moments, least squares, likelihood based and Bayesian methods. These have been implemented in the various software packages. The breadth of this subject is much too great for a comprehensive treatment here and instead the reader is referred to one of the many excellent books now available on time series.

10.5   The spectral representation theorem and Bochner's lemma

Many time series processes which are indexed by integer valued time points can be expressed in terms of the sum of sine and cosine terms. Bochner's lemma tells us the covariance function of weakly stationary series can be characterised by these trigonometric functions.

To state that theorem requires a brief review of some important basic concepts. We first recall the notion of positive definiteness for covariance matrices. Such matrices are symmetric and satisfy the inequality

$$
\sum_i \sum_j a_i a_j C(t_i - t_j) > 0
$$

for any vectors of constants $a_1 : a_m$ and time points $t_1 : t_m$, where $u_1 : u_m$ standards for $(u_1, \ldots, u_m)$ where in general the notation $a : b$ means $a, a+1, \ldots, b$. Thus $r_{a:b}$ means the vector $(r_a, r_{a+1}, \ldots, r_m)$ and so on.

We also need some basic ideas from the theory of complex numbers, starting with the imaginary number $i = \sqrt{-1}$. This is the number, which when squared, gives -1 and is the number on which the complex number system is built. A complex number $z = x + iy$ has both a real component x and an imaginary one y. To visualise things, you can plot z in the complex plane as a point (x, y) just like you would plot a point in R^2. A close relative of z is its complex conjugate $\bar{z} = x - iy$. Straightforward algebra shows that $z\bar{z} = x^2 + y^2$, the distance square from the origin $(0, 0)$ to (x, y) in the complex plane. The complex number represented by the point

$w = (cos(\theta), sin(\theta)) = cos(\theta) + isin(\theta)$, lies on the rim of a unit circle in the complex plane: $w\bar{w} = 1$. When θ goes from 0 radians to 2π radians, w goes through one revolution around the unit circle, i.e. one cycle or a frequency of 1. It turns out that w has a remarkable representation, $w = \exp\{i\theta\}$. In fact if $w_i = \theta_i$, $i = 1,2$ we can compute $w_1 w_2$ as $\exp\{i(\theta_1 + \theta_2)\}$, showing why the exponential function representation is so important. We can bring the temporal dynamic into this representation by letting $\theta = 2\pi\omega t$. Thus as time advances one unit, from t to $t+1$, the exponential is advanced around the unit circle in the complex plane from $\exp\{i2\pi\omega t\}$ $\exp\{i2\pi\omega t\}\exp\{i2\pi\omega 1\} = \exp\{i2\pi\omega(t+1)\}$.

This appealing idea also turns out to be of fundamental importance for the theory being considered here. We start by examining the *spectral representation theorem* for weakly stationary processes. A simple example might start with a completely deterministic process $\varepsilon_t = cos(2\pi\omega_0 t)$. Since in general the $sin(x)$ is an odd function, i.e. $sin(-x) = -sin(x)$, this process can be rewritten as

$$
\varepsilon_t = \frac{cos(2\pi[\omega_0]t) + isin(2\pi[\omega_0]t)}{2} + \frac{cos(2\pi[-\omega_0]t) + isin(2\pi[-\omega_0]t)}{2}
$$

$$
= \sum_{\omega=-\omega_0}^{\omega_0} \exp\{i(2\pi[\omega]t)\}dU(\omega) \tag{10.5}
$$

where U denotes the cumulative distribution function (CDF) that puts half of the probability on each of ω_0 and $-\omega_0$. This is a somewhat complicated way of expressing the cosine but it demonstrates how a function can be represented as the weighted sum of two processes that move on the unit circle in the complex plane. The key to the success of the representation was a discrete distribution whose probability mass function $dU(\omega) = u(\omega)$ is symmetric around $\omega = 0$ which results in the sine terms cancelling out, thus eliminating the imaginary number i. This could be extended to allow u to include additional frequencies.

$$
\varepsilon_t = \sum_{\omega=-\omega_0}^{\omega_M} \exp\{i(2\pi[\omega]t)\}dU(\omega) \tag{10.6}
$$

The previous paragraphs combined with Exercise 10.6 shows how we could construct a weakly stationary random process, in the case of a power spectrum with a finite number of frequencies or to include a random component.

A simple version of the spectral representation theory says that all weakly stationary processes over continuous time, ε_t, may be presented by

$$
\varepsilon_t = \int_{-\infty}^{\infty} e^{i\omega t} \, dU(\omega) \tag{10.7}
$$

where U is a complex valued process with orthogonal increments, meaning for example that if $\omega_1 < \omega_2 < \omega_3 < \omega_4$ then

$$
E[U(\omega_2) - U(\omega_1)]\overline{[U(\omega_4) - U(\omega_3)]} = 0
$$

In other words, any weakly stationary process can be represented as a sum of cosines and sines with random coefficients, i.e. amplitudes. A similar representation holds for discrete time and even when 'time' is replaced by 'space'.

10.5.1   The link between covariance and spectral analysis

Spectral analysis is the study of processes over their frequency domain, and it aims to determine the contribution of each frequency to the variance of the process. In other words, it looks for where the 'power' driving the process is concentrated.

As seen in Section 10.4.1 the covariance function characterises stationarity. Bochner's lemma provides a link between the spectral representation of a time series and the covariance.

Lemma 1. (Bochner) If C is a positive definite covariance function for a stationary process then there exists a spectral distribution function $F(\omega)$, $-1/2 \leq \omega \leq 1/2$ such that

$$C(\tau) = \int_{-1/2}^{1/2} \exp\{2\pi i \omega \tau\} dF(\omega), |\tau| = 0, 1, \ldots. \tag{10.8}$$

The spectral CDF in this lemma can have a probability density function called the spectral density function under conditions given in the following Corollary.

Corollary 1. $\sum_{\tau=-\infty}^{\tau=\infty} |C(\tau)| < \infty$ implies $dF(\omega) = f(\omega)d\omega$.

Note that as in the case of the spectral representation theorem, symmetry is required, i.e. f must be symmetric about 0 to avoid imaginary numbers.

Bochner's theorem extends to continuous time and spatial processes in a natural way. It is of practical value since it can be used to construct temporal (and spatial) covariance functions simply by specifying f, otherwise finding a legitimate covariance function is difficult. When F is discrete, being concentrated on a countable set $\{\omega_i\}$, the integral in Equation 10.8 becomes a sum of *sin* and *cos* terms for a weakly stationary process. and the spectral distribution relates to the random measure in the spectral representation theorem as follows:

$$E[dU(\omega)d\overline{U(\omega)}] = dF(\omega) \tag{10.9}$$

A large value of $dF(\omega)$ corresponds in an informal sense with the size of the amplitudes attached to the sines and cosines. A similar result is found when the process is indexed by discrete time, except the integral in Equation 10.7 is taken over $[-\pi, \pi]$ instead.

Spectral representation theory points to the need to estimate the spectral distribution in order to learn about recurring events and their frequency. Equation 10.6 points to a method for computing that estimate when a finite number of frequencies in the spectrum contribute most of the variability in the process as reflected in Bochner's lemma. Let us assume the process has been detrended and the resulting residuals ε_t

are weakly stationary. For simplicity, these are assumed to be observed without error. In this case, Equation 10.6 suggests the inversion of the relationship between ε_t and $U(d\omega)$ and after some refinement, this idea leads to the estimate of the spectral mass function $I(\omega)$ (Chatfield, 2013), which when plotted against ω is sometimes called the periodogram. Studying this estimate or Bayesian versions of it can indicate unanticipated frequencies, for example a twice a day twelve-hour cycle in urban ozone concentrations in addition to the obvious twenty-four hour cycle (Huerta et al., 2004). Such a discovery can then be turned into a sine–cosine temporal mean model for the ozone field.

10.6   Forecasting

The negative health impacts of environmental hazards, such as criteria air pollutants, has led to the need in some urban areas for forecasts of future levels of pollution. These are used to inform planning of activities for susceptible individuals, for example California's South Coast Air Quality Management District provides online a detailed map of forecasts in the region around Los Angeles.

This section introduces the classical theory of forecasting temporal processes. Two general approaches are described; the first uses available data to estimate coefficients in a forecasting model and then applies that model while the second exploits autocorrelation in the temporal process to forecast future values.

10.6.1   Exponential smoothing

Following Chatfield (2013), we begin by describing the exponential smoothing model. When there is no trend or seasonality the next value is predicted from the observations to date,

$$\hat{Y}_{t+1} = c_t \hat{Z}_t + \ldots + c_1 \hat{Z}_1 \qquad (10.10)$$

where $c_i = \alpha(1 - 1\alpha)^{i-1}$ with $0 < \alpha < 1$ which means that the most recent observation gets the most weight. Equation (10.10) implies

$$\hat{Y}_{t+1} = c_t \hat{Z}_t + (1 - \alpha)\hat{Y}_t \qquad (10.11)$$

An extension to this idea is at the heart of the Holt–Winters (HW) approach. Suppose $g(t)$, $t > 0$ is a differentiable function that is observed at unit intervals $t = t_0, t_1, \ldots$, with $t_{i+1} = t_i + 1$, $i = 1, \ldots, I$, where I is the total number of segments that the function is split into. Then

$$g(t_{i+h}) \approx g(t_i) + hg'(t_i) \qquad (10.12)$$

This gives a predictor for g h time steps ahead based on the present level $g(t_i)$ and the slope $g'(t_i)$. The HW approach extends Equation 10.11 by building upon the concept in Equation 10.12 and treating $\hat{Y}_t = S_t$ as the current 'level' of the process. So S_t is replaced by $S_t + T_t$ where T_t represents a fitted trend, $T_t = \rho(S_t - S_{t-1}) + (1-\rho)T_{t-1}$,

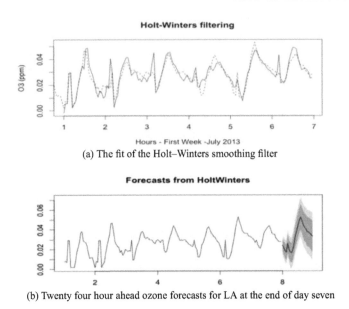

(a) The fit of the Holt–Winters smoothing filter

(b) Twenty four hour ahead ozone forecasts for LA at the end of day seven

Figure 10.4: Ozone concentration (ppm) levels at the Los Angeles site with ID 840060370113 for the first week of July 2013 with forecasts for following 24 hours.

which is a weighted combination of the previous trend and change in the level. The l step ahead forecast is then $\hat{Y}_{t+l} = S_t + lT_t$. Further extensions to the HW method can incorporate seasonality.

Example 10.13. *Forecasting ozone levels*

We return to Example 10.1 with the objective of forecasting ozone concentrations for the next twenty-four hours. These forecasts are based on measured concentrations in Los Angeles from the first week July 2013. We first fit the Holt–Winters model and see the results in the upper panel of Figure 10.4. The twenty-four hour ahead forecast on day eight is seen in the lower panel.

10.6.2 ARIMA models

The second approach to forecasting exploits auto-correlation in the temporal process to forecast future values. This approach requires that we first determine the autocorrelation in the process series. This can be assessed using the *autocorrelation function* (ACF), which is equal to ρ_τ. When plotted against lag τ, it is known as the correlogram. Recall that for stationary temporal processes Y_t, $t > 0$, the autocorrelation is $\rho_\tau = Corr(Y_t, Y_{t-\tau}) = Cov(Y_t, Y_{t-\tau})/Var(Y_t)$ for all $t > \tau$ and $\tau > 0$.

When the process is non-stationary, the correlogram can be used as a diagnostic tool (Chatfield, 2013). For example an increasing trend will be seen as a slowly declining ACF. A periodic series will induce a similar periodic pattern in the correlogram. This diagnostic role is an important one and should be part of any preliminary data analysis.

The *partial auto-correlation* (PACF) also plays a key role in the initial analysis of a temporal series. (PACF) Consider the autoregressive temporal process model as seen in Example 10.6; this is a Markov model in the sense that given Y_{t-1}, Y_t will be independent of all previous observations from Y_{t-2} to Y_1. In this case, the correlogram will show a lag 2 effect. This arises as the correlation between Y_{t-2} and Y_t comes from their mutual association with Y_{t-1}. The partial ACF eliminates this spurious correlation by computing the autocovariance function between Y_{t-2} and Y_t conditional on Y_{t-1} i.e. $E[(Y_{t-2} - \mu)(Y_t - \mu) \mid Y_{t-1} = y_{t-1}]$. Therefore, in practice both the ACF and PACF need to be studied as a part of a preliminary data analysis.

One way of incorporating autocorrelation in forecasting involves pre-filtering the process to remove the (estimated) regular components leaving a stationary process for the (estimated) residuals. An ARMA(p,q) process model might then be fit to these residuals and these models then used to forecast future, as yet unobserved residuals. These forecasts can be combined with the estimated future values of the regular components, i.e. the trend and seasonality, to get forecasts of the process.

However finding the regular components can often prove problematic. The Box–Jenkins provides an approach to this problem, which involves differencing the temporal series, until weak stationarity is obtained. The first difference of a series is $Y_t(1) = Y_t - Y_{t-1}$, the second $Y_t(2) = Y_t(1) - Y_{t-1}(1) = Y_t - 2Y_{t-1} + Y_{t-2}$, etc... An ARMA model is then fit to the result and forecasts made. These forecasts are then integrated back to the original non-stationary process. This is known as an integrated ARMA (ARIMA) process.

10.6.3 *Forecasting using ARMA models*

Consider the simple case where the process is given by $Y_t = \beta t + \varepsilon_t$, $t = 1, 2, \ldots, T$. An analyst recognising the trend could estimate β in order to get the series of estimates $\hat{\varepsilon}_t = Y_t - \hat{\beta}t$. If these represent a stationary process then an ARMA model could be fit giving the forecast $\hat{\varepsilon}_{T+1}$ and thus the one step ahead forecast, $\hat{Y}_{T+1} = \hat{\beta}(T + 1) + \hat{\varepsilon}_{T+1}$. However, if the analyst did not recognise the trend, they might begin by taking the first order difference $\nabla^1 Y_t = Y_t - Y_{t-1} = \beta + \varepsilon_t^*$, where $\varepsilon_t^* = \varepsilon_t - \varepsilon_{t-1}$, $t = 2, 3, \ldots, T$. The trend would now have been eliminated and if the ε_t^* comprise a stationary process, an ARMA model might be fit to get the series $\hat{\varepsilon}_t^*$, $t = 2, \ldots, T + 1$, which includes the forecast. Just as in the case of continuous time $g(t) = \int_{t-1}^{t} g'(u)du + g(1)$, and so here we have $\hat{Y}_2 = \hat{\varepsilon}_2^* + Y_1$ etc... The ARMA would give a model for the ε_t^* that includes both autoregressive as well as moving average components. Finally we can adapt this same idea to deal not only with trend as above, but also with seasonality.

In general, an ARIMA(p,d,q) model represents the dth difference in the process series as an ARMA process. The seasonal version is based on differences $\nabla^d Y_t$ that give d–step seasonal differences, resulting in a d-differenced series. This approach can deal with regular components, i.e. trends, seasonality, etc, without the need to model them explicitly. The ARMA parameters would be estimated for the differenced series and predictions made. Forecasts on the original scale can then be reconstructed by 'integrating' the differences.

Example 10.14. *Forecasting volcanic ash*

Volcanic ash is not a substance that is commonly encountered in environmental epidemiology, however it can be widely distributed by winds and it can be a significant health hazard. A study of the Mount St Helens volcanic eruption in May and June of 1980 reports thirty-five deaths due to the initial blast and landslide. Others are reported to have died from asphyxiation from ash inhalation (Baxter et al., 1981). The respirable portion of the ash was found to contain a small percentage of crystalline free silica, a potential pneumoconiosis hazard, and a number of acute health effects were found in those visiting emergency rooms including asthma, bronchitis and ash-related eye problems.

Hickling, Clements, Weinstein, and Woodward (1999) reported that the ash plume of the relatively small Mount Ruapehu eruption of June 1996 in New Zealand extended over several hundred kilometres. A comparison of rates of respiratory disease, stroke and ischaemic heart disease in the three month period following the blast with the same time period over the previous seven years showed evidence of acute health impacts. For example, a relative risk of RR = 1.44 was found for acute bronchitis.

Figure 10.5 shows the time series of volcanic ash and 10.6 the ACF and PACF. From the plot of the time series a complex pattern can be seen which is the result of eruptions occurring at random times. The large spikes suggest the possibility of moving average components, this is a way that the ARMA process has of incorporating shocks that abate over the period following their occurrence.

There are no obvious regular components in the series that would induce autocorrelation so we turn to analysis of the ACF and PACF. The correlogram in the left hand panel of Figure 10.6 points to a significant autocorrelation at lag three and the possibility of an ARMA(0,3) model to capture the persistence in the ash level following a shock and hence an MA(3) component

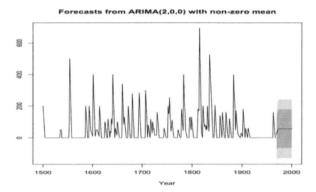

Figure 10.5: Forecasting future atmospheric levels of volcanic ash. The forecast using an ARIMA(2,0,0) is shown on the right hand side of the plot.

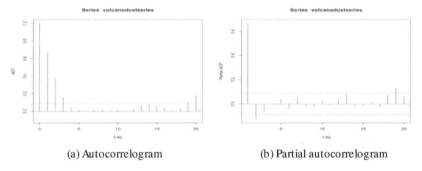

(a) Autocorrelogram (b) Partial autocorrelogram

Figure 10.6: The autocorrelogram (ACF) and partial autocorrelogram (PACF) plots for data on volcanic ash. Panel (a) shows the ACF and suggests an MA(3) model would capture the persistence of the ash in the atmosphere for three periods. Panel (b) suggests that some of that persistence might be spurious correlation in an AR(2) model.

or a mixture of an AR and MA model might be considered. The PACF plot in the right hand panel of the figure suggests a significant lag 2 effect suggesting an ARMA(2,0) model.

The lack of obvious regular components together with the correlogram plots, suggests the ARIMA approach differences may be used to achieve a stationary process. The auto.arima function in the R forecast library gave the following results (using the BIC criterion).

```
ARIMA(2,0,0) with non-zero mean

Coefficients:
ar1        ar2         intercept
  0.7533    -0.1268       57.5274
```

```
(0.0457)      (0.0458)        (8.5958)

sigma^2 estimated as 4870:   log likelihood=-2662.54
AIC=5333.09     AICc=5333.17    BIC=5349.7
```

This analysis suggests that differencing is not needed and that an AR(1) process model would be sufficient. The AIC criterion suggested an ARIMA(1,0,2) instead, this one including an MA(2) component to represent the shock. The earlier results seen in the ACF and PACF analyses point to the ARMA(2,0,0).

We can now forecast future ash levels using an ARIMA(2,0,0) model, the results of which can be seen in Figure 10.5.

10.7   State space models

The state space or dynamic linear modelling approach has its roots in the celebrated Kalman–Bucy filter. A Bayesian version of this was developed in the 1970s (Harrison & Stevens, 1971) and the method has been developed extensively since. It now provides a flexible and general tool for temporal and spatio–temporal modelling (Harrison, 1999).

State space models arise from the notion that a measured signal will be contaminated by noise. In the simplest case a state space model would consist of two equations, or levels: the observation equation and the system equation. In the observation equation, the 'signal' at time t is taken to be a combination of the variables in an underlying *state vector* also at time t, i.e. $Y_t = \mathbf{F}_t \mathbf{Z}_t + \mathbf{v}_t$, where $\mathbf{v}_t \sim N(0, \mathbf{V}_t)$. The state vector is the minimum set of information from the past data which contains all the information necessary for predicting future behaviour. The state vector therefore has the Markov property that the latest value is all that is required. Whilst it may not be possible to observe all (or even any) of the elements of the state vector, $\mathbf{Z}_t$, directly, assumptions can be made about how it changes in time, leading to the system equation, $\mathbf{Z}_t = \mathbf{G}_t \mathbf{Z}_{t-1} + \varepsilon_w$, where $\varepsilon_w \sim N(0, \sigma_w^2)$. Once data have been observed, inference can be made about this underlying state vector, which can often be thought of as the true underlying level of the signal. Due to the set up of the state space model, inference can be performed sequentially, updating the estimates of the parameters as more data become available over time.

10.7.1   *Normal Dynamic Linear Models (DLMs)*

Analogous to the idea of state space models is that of dynamic linear models (DLMs) (West, Harrison, & Migon, 1985). Here, the underlying process over times $t \in T$ is characterized by a sequence of random vectors $\mathbf{Z}_t$ which conditional on $\mathbf{Z}_{t-1}$ have a multivariate normal distribution with expectation $\mathbf{G}_t \mathbf{Z}_{t-1}$, and covariance $\mathbf{W}_t$. The $\{\mathbf{G}_t\}$ are called the transition matrices.

At each time $t \in T$, a univariate observation Y_t is made (the multivariate extension is analagous). Conditional on $\mathbf{Z}_t$ these have a univariate normal distribution with expectation $\mathbf{F}_t \mathbf{Z}_t$ and covariance matrix $\mathbf{V}_t$ for a matrix $\mathbf{F}_t$ that is usually taken to be known. If $\mathbf{F}_t, \mathbf{G}_t, \mathbf{V}_t, \mathbf{W}_t$ do not depend on t, the combination of observation and system equations yields a constant DLM.

$$
\begin{aligned}
\text{Observation equation: } Y_t | \mathbf{Z}_t &\sim N(\mathbf{F}_t \mathbf{Z}_t, \mathbf{V}_t) \\
\text{System equation: } \mathbf{Z}_t | \mathbf{Z}_{t-1} &\sim N(\mathbf{G}_t \mathbf{Z}_{t-1}, \mathbf{W}_t) \\
\mathbf{Z}_1 &\sim N(\mathbf{m}_1^*, \mathbf{c}_1^*)
\end{aligned}
\tag{10.13}
$$

This can also be presented in additive form, giving the traditional form of a state space model

$$
\begin{aligned}
\text{Observation equation, } Y_t &= \mathbf{F}_t \mathbf{Z}_t + \mathbf{v}_t \\
\text{System equation, } \mathbf{Z}_t &= \mathbf{G}_t \mathbf{Z}_{t-1} + \mathbf{w}_t.
\end{aligned}
\tag{10.14}
$$

Here $\mathbf{v}_t \sim N(0, \mathbf{V}_t)$, $\mathbf{w}_t \sim N(0, \mathbf{W}_t)$ are stochastically independent of $\mathbf{Z}_t, \mathbf{Z}_{t-1}$ respectively and of each other.

Estimating the parameters

Having constructed a normal DLM, the main responses of interest, $\mathbf{Z}_t$ can be predicted at each time, t. In the simplest case, $\mathbf{F}_t = 1$, $\mathbf{G}_t = \mathbf{I}$ (the identity matrix), $\mathbf{V_t} = V$ and $\mathbf{W}_t = W$, which is known as a *steady state model*. In cases where the two variances are known, the resulting estimate can be found easily using Bayes' Theorem. By considering the likelihood of the observed data and treating the underlying level as a 'prior' for the state vector, $\mathbf{Z}$, then the posterior distribution can also be seen to be normally distributed with the posterior mean, $E(\mathbf{Z}_t | Y^t = y^t)$ where in general Y^t denotes the vector consisting of all $Y_{t'}$ with all $t' \in T : t' \leq t$. That posterior mean is a weighted average of the observed data, y_t and the previous day's underlying level, Z_{t-1}. More precisely, in the case of univariate Z's,

$$
E(Z_t | Y^t = y^t) = E(Z_{t-1} | Y^t = y^t) + A_t [y_t - E(Z_{t-1} | Y^t = y^t)] \tag{10.15}
$$

where A_t is a weighting factor based on the ratio of two variances, $A_t = Var(Z | Y^t = y^t) / Var(y_t | Z)$.

This is recognisable as one of the updating formulas for the Kalman Filter (Meinhold & Singpurwalla, 1983; Harvey, 1993).

It is unlikely in practice that the variances will be known, in which case they too will have to be estimated from the data. One approach is to use a fully Bayesian approach, assigning probability distributions to all the unknown parameters, and then sample from their posterior distributions using MCMC techniques. Details of this approach can be found in Gamerman and Smith (1996) and Carter and Kohn

(1994) with a description of the general framework of dynamic hierarchical models in Gamerman and Migon (1993).

Example 10.15. *Dynamic GLMs in environmental epidemiology*

A dynamic generalised linear model (DGLM) extends a generalised linear model by allowing a subset of the regression parameters to evolve over time as an autoregressive process. Lee and Shaddick (2008) used DGLMs to model the changing effects of air pollution on health. Their model was of the form,

$$
\begin{aligned}
y_t &\sim \text{Poisson}(\mu_t) && \text{for } t = 1,\ldots,N_T \\
\ln(\mu_t) &= \mathbf{z}_t'\beta_t + \mathbf{x}_t'\beta_x \\
\beta_t &= F_1\beta_{t-1} + \ldots + F_p\beta_{t-p} + v_t && v_t \sim N(0,\Sigma_\beta) \\
\beta_0,\ldots,\beta_{-p+1} &\sim N(\mu_0,\Sigma_0) && (10.16) \\
\alpha &\sim N(\mu_\alpha,\Sigma_\alpha) \\
\Sigma_\beta &\sim \text{Inverse-Wishart}(n_\Sigma, S_\Sigma^{-1})
\end{aligned}
$$

Here, the vector of health counts is denoted by $\mathbf{y} = (y_1,\ldots,y_{N_T})'_{N_T \times 1}$, and the covariates include an $J \times 1$ vector $\mathbf{x}_t$, with fixed parameters $\beta_x = (\beta_{x1},\ldots,\beta_{xJ})'_{J \times 1}$, and a $q \times 1$ vector $\mathbf{x}_t$, with dynamic parameters $\beta_t = (\beta_{t1},\ldots,\beta_{tq})'_{q \times 1}$. The dynamic parameters are assigned an autoregressive prior of order p with the variability in the process controlled by a $q \times q$ variance matrix Σ_β, which is assigned a conjugate inverse-Wishart prior. For univariate processes Σ_β is scalar, and the conjugate prior simplifies to an inverse-gamma distribution. The evolution and stationarity of this process are determined by Σ_β and the $q \times q$ autoregressive matrices $F = \{F_1,\ldots,F_p\}$, the latter of which may contain unknown parameters or known constants. For example, a univariate first-order autoregressive process is stationary if $|F_1| < 1$.

10.8 A hierarchical model for temporally varying exposures

In this section we give details of a hierarchical model described by Shaddick and Wakefield (2002) and used for a spatial process in Chapter 9, Section 9.12. Here, the process model is a temporal one and constitutes an AR process. As described in Section 9.12 there are three stages to the model: (i) the observation, or data, model; (ii) the process model which in this case now describes the form of the underlying temporal process and (iii) assigning prior distributions to the unknown parameters. For clarity, the material presented within this chapter on classical time series methodology was presented as though there was no measurement error present, i.e. $Y_t = Z_t$ in the formulation seen in Equation 10.1 but here we revert to the full measurement-process setup.

Stage 1 - Observation model

At the first stage of the model, the observed data is related to an underlying temporal process which is not observable but may be measured, possibly with error.

$$y_t = z_t + v_t$$

where y_t denotes the observed level of the exposure at time t for $t = 1,..,N_T$ where N_T is the number of time points. In this model u_t represents measurement error which is assumed to be independent and identically distributed, $N(0,\sigma_v^2)$ or equivalently $\mathbf{v} \sim MVN_S(0_{N_T}, \Sigma_v)$ where Σ_v is a diagonal matrix.

Stage 2 - Process model

In this stage the underlying levels of the exposure, Z_t, are assumed to comprise an underlying trend, μ_t together with auto-regressive process, γ_t.

$$\begin{aligned} z_t &= \mu_t + \gamma_t \\ \gamma_t &= \alpha\gamma_{t-1} + w_t \end{aligned} \qquad (10.17)$$

where μ_t is a temporal mean term, $\mu_t = \sum_{j=1}^J \beta_j X_{jt}$, which might contain (functions of) time as explanatory variables, and γ_t is a spatial process with w_t normally distributed with zero mean and variance, σ_w^2, i.e. $w_t \sim N(0,\sigma_w^2)$ for all t. For $\alpha \in (0,1]$ we have an AR(1) process while for $\alpha \geq 1$ we have a non-stationary random walk, RW(1).

Stage 3 - Hyperpriors

The prior distribution for the vector of regression coefficients, β is assumed to be a $N(\mu_\beta, \sigma_\beta^2)$. For the precision parameters a Gamma prior is selected, for instance $\sigma_u^{-2} \sim Gam(a_u, b_u)$ and $\sigma_w^{-2} \sim Gam(a_w, b_w)$. The prior for the AR parameter, α, needs to allow the constraint that it be $\alpha \in (0,1]$ which may be achieved using a logistic transformation of a continuous distribution, such as the normal of beta, or by using a uniform.

The joint posterior distribution is of the form

$$\begin{aligned} p(z,\beta,\sigma_v^2,\sigma_w^2|y) &= p(y)^{-1} \left\{ \prod_{t=1}^T p(y_t|z_t,\beta,\sigma_v^2) \right\} \\ &\times \left\{ \prod_{t=2}^T p(z_t|z_{t-1},\sigma_w^2,\alpha) \right\} p(\beta)p(\sigma_v^2)p(\sigma_w^2)p(\alpha) \end{aligned}$$

It is difficult, if not impossible, to deal with such a distribution analytically but obtaining samples from it can be relatively straightforward using MCMC. Alternatively R–INLA could be used to perform Bayesian inference using this model.

Example 10.16. *Implementation in WinBUGS*

Following on from the material in Chapter 5 and the implementation of the example shown in in Chapter 9, Section 9.12 we discuss the implementation of the model using WinBUGS. We concentrate on the case where the temporal process is a random walk.

The WinBUGS code for fitting a random walk model, where $\alpha = 1$, is as follows (ignoring covariates for simplicity):

```
model {
        for (t in 2:(NT-1)) {
# observation model
                y[t] ~ dnorm(gamma[t],tau.v)

}  # t loop
y[1]~dnorm(gamma[1],tau.v)
y[NT]~dnorm(gamma[n],tau.v)

    tau.v ~ dgamma(1,0.01)
}


        for (t in 2:(NT-1)) {

# system model
                tmp.gamma[t] <- (gamma[t-1]+gamma[t+1])/2
                gamma[t] ~ dnorm(tmp.gamma[t],tau.w2)

} # t loop

gamma[1]~dnorm(gamma[2],tau.w)
gamma[NT]~dnorm(gamma[NT-1],tau.w)

tau.w ~ dgamma(r.w,d.w)
sigma.w <- 1 / sqrt(tau.w)

} # end of model
```

Note that because we are dealing with a cyclical graph at this stage, unless we make specific allowance there will be double counting of the likelihood terms (where for example γ_t will appear as both a parent of γ_{t-1} and as a child of γ_{t+1}) and so we have to explicitly specify some of the full conditional distributions (using the RW structure). It is possible to do this in WinBUGS, although not widely documented. We need to explicitly find the contribution of the likelihood (the data) to the posterior for σ_w^2.

The gamma prior with the normal likelihood combine to give a gamma posterior $p(\gamma|\tau_w) \sim N(\gamma_{t-1},\tau_w)$. Note our use of $\tau_w = 1/\sigma_w^2$. $p(\tau_w) \sim Ga(r,d)$

$$p(\tau_w|\gamma) \quad \propto \quad d^r \tau_w^{(r-1)} \exp(-d\tau_w)$$

$$\times \quad \tau_w^{(n/2)} \exp\{\tau/2 \sum_{t=2}^{N_T} (\gamma_t - \gamma_{t-1})^2\}$$

$$\propto \quad d^r \tau_w^{(r+n/2-1)} \exp(\tau_w\{d + \sum_{t=2}^{N_T} (\gamma_t - \gamma_{t-1})^2)\}$$

Therefore the posterior $p(\tau_w|\gamma) \sim Ga(r+n/2, d + \sum_{t=2}^{N_T}(\gamma_t - \gamma_{t-1})^2/2)$.

We need to calculate the contribution of the likelihood ourselves and then combine this with the prior to give the posterior.

```
model {
        for (t in 2:(NT-1)) {

# calculate the contribution to the likelihood for
# full conditionals
tau.w.like[t] <-pow((gamma[t]-tmp.gamma[t]),2)

} # t loop

tau.w.like[1] <-    0
tau.w.like[T] <-    pow((gamma[T]-gamma[T-1]),2)


tau.w2 <- tau.w*2
        d <-1
        r <- 0.01
    d.w <- d+sum(tau.w.like[])/2
    r.w <- r + n/2
    tau.w ~ dgamma(r.w,d.w)


} # model
```

Note this uses a prior of $Ga(1,0.01)$ for τ_w which is 'hard–wired' into the code at this point, the values of r and d could also be an input to the model in the form of data.

The full code is included in the online resources. It is noted that this example is partly to show how full conditionals can be specified by the user. In practice it might be simpler in this case to draw on the equivalence with an intrinsic CAR (conditionally autoregressive) model and to use the in-built WinBUGS function for this (see Example 8.3).

The result of the analysis can be seen in Figure 10.7, which shows the fitted values, the medians of the posterior distribution for each day, from the RW process together with 95% credible intervals. Recall from Chapter 4 that missing data are treated as unknown parameters and samples from their posterior distributions can be obtained. Figure 10.7 shows the estimates of PM_{10} on

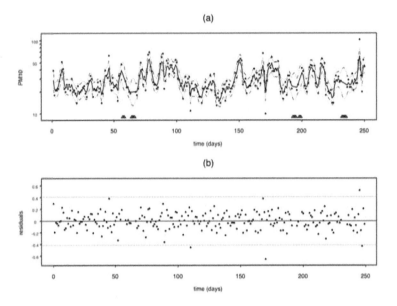

Figure 10.7: Time series of 250 days of observed and estimate levels (together with their differences) of particulate matter (PM$_{10}$) at the Bloomsbury monitoring site in London.

days where there was no monitoring data. It can clearly be seen that the uncertainty associated with periods of missing data is increased when compared to times for which data is available.

Example 10.17. *Implementation in R–INLA*

Implementing the model using R–INLA is somewhat simpler as there are pre-defined RW and AR latent process models. The code for this model, using a RW and including the covariates for the temporal mean term, would be of the form,

```
formula = y ~ x + f(t, "rw")
model = inla(formula, data=data)
```

10.9   Summary

The chapter contains the theory required for handling time series data and the reader will have gained an understanding of the following topics:

- That a temporal process consists of both low and high frequency components, the former playing a key role in determining long-term trends while the latter may be associated with shorter-term changes.

- Techniques for the exploratory analysis of the data generated by the temporal process, including the ACF (correlogram) and PACF (periodogram).

- Models for irregular (high frequency) components after the regular components (trend) have been removed.

- Methods for forecasting, including exponential smoothing and ARIMA modelling.

- The state space modelling approach, which sits naturally within a Bayesian setting and provides a general framework within which a wide class of models, including many classical time series models, can be expressed.

- Implementing time series processes within a Bayesian hierarchical framework.

Exercises

Exercise 10.1. Give a direct proof that the random process in Equation 10.5 is weakly stationary.

Exercise 10.2. Find a recursive relation for the autocorrelation function of an AR(2) process.

Exercise 10.3. For a white noise process indexed by discrete time, find the spectral density function.

Exercise 10.4. For a random walk process, find the mean and variance as a function of time.

Exercise 10.5. This exercise explores the effect of the regular component of a temporal process if hidden in the irregular component.

(i) A statistician observes without error a process $Y_t = a + bt, t = 0, 1, 2, \ldots T$ without realizing it is deterministic. Find a theoretical formula for an estimate of the autocorrelation for this process. In particular, show how the sum of the geometric series $\sum_{i=t}^{n} x^i$ can be used to find the first and second 'sample' moments you need for this exercise.

(ii) What conclusions can you draw from this exercise about the possible effects of the deterministic component on the computed autocorrelation of a time series?

Exercise 10.6. Perform the same kind of analysis as in Exercise 10.5 for another deterministic function for example, a periodic function.

Exercise 10.7. Given a stationary series, prove that if the autocovariances are all positive then the mean of the process will be estimated with greater variance than if all the autocovariances are null.

Exercise 10.8. Obtain the autocorrelation function of an ARMA(1,1) process, writing it as an MA(∞).

Exercise 10.9. Show how an AR(2) process can be represented in state space modelling form.

Exercise 10.10. We will call a stationary temporal process Y_t, $-\infty < t, \infty$ L2 continuous if for every t, $\lim_{h \to 0} E[Y_{t+h} - Y_t]^2 = 0$. Show that Y_t is L2 continuous if and only if the covariance function $C(t)$ is continuous at $t = 0$. Extend this result to L2 differentiability. This exercise shows that the "smoothness" of a process is determined by the smoothness of the covariance function.

Exercise 10.11. Show that a random walk process of order 1 can be expressed in terms of an intrinsic CAR model, i.e. if $p(\gamma_t | \gamma_{t-1}) \sim N(\gamma_{t-1}, \sigma_w^2)$ then

$$
p(\gamma_t | \gamma_{-t}, \sigma_w^2) \sim
\begin{cases}
N(\gamma_{t+1}, \sigma_w^2) & \text{for } t = 1 \\
N\left(\frac{\gamma_{t-1} + \gamma_{t+1}}{2}, \frac{\sigma_w^2}{2}\right) & \text{for } t = 2, ..., N_T - 1 \\
N(\gamma_{t-1}, \sigma_w^2) & \text{for } t = N_T
\end{cases}
$$

where m_{-t} represents the vector of m's with m_t removed. Pay particular attention to any assumptions that need to be made when $t = 1$ and $t = T$.

Exercise 10.12. Show that a Gamma prior, τ_w $Ga(a, b)$ combines with normal likelihood, $[\gamma_t | \gamma_{t-1}, \tau_2] \sim N(\gamma_{t-1}, \tau_w)$, to give a Gamma posterior, paying particular attention to the form of the updated parameters.

Exercise 10.13. Using WinBUGS
The models, data and initial values for the London PM_{10} example can be found on the website.

(i) Open model1.odc and load the data (model1-data.odc) and compile the model with two chains. Initial values can be found for two chains in model1-inits1.odc and model2-inits2.odc.

(ii) Run the model for a suitable number of iterations and calculate summary statistics for the posterior distributions of m, sigma.v and sigma.w.

(iii) In R plot the estimated values of gamma against the observed data, y. What do you conclude? Note that you may have to deal with the different lengths of the two series, and remember that theta has no missing values in it.

(iv) Plot a suitable summary of the posterior values of gamma (including their uncertainty) against time. What do you conclude about the uncertainty in the values of gamma when the original data is missing?

Exercise 10.14. Using R–INLA
(i) Perform the analysis in Exercise 10.14 using R–INLA and compare the results to what you obtained using WinBUGS.

(ii) Repeat the analysis using different models for the underlying temporal process: (i) RW(2), (ii) AR(1) and (iii) AR(2). What do you conclude about the appropriateness of these models for representing the temporal structure of this data? Which one would you choose and why?

Chapter 11

The interplay between space and time in exposure assessment

11.1 Overview

In recent years there has been an explosion of interest in spatio–temporal modelling and there have been a number of noteworthy publications in this field, including Le and Zidek (2006), Cressie and Wikle (2011) and Banerjee et al. (2015). However none of these are specifically concerned with environmental epidemiology, where interest is in the relationship between human health and spatio–temporal processes of exposures to harmful agents. This might be, for example, the relationship between deaths and air pollution concentrations or future climate simulations, the latter of which may involve 1000's of monitoring sites that gather data about the underlying multivariate spatio–temporal field of precipitation and temperature. Although the main concern may be the effects on human health there may be other effects, for example acid precipitation and its negative impact on the flora and fauna in addition to those related to human health.

11.2 Strategies

There are many ways in which space and time can be incorporated into a statistical model and we now consider a selection. One must first choose the model's space–time domain. Is it to be a continuum in both space and time? Or a discrete space with a finite number of locations at which measurements may be made?

Time is obviously different than space. For one thing, it is directed, whereas any approach to adding direction in space is bound to be artificial. A major challenge in the development of spatio–temporal theory has been combining these fundamentally different fields in a single modelling framework. Much progress has been made in this area over the last three or four decades to meet the growing need in applications of societal importance, including those in epidemiology.

As seen in Chapter 10 there are competing advantages to using finite (discrete) and continuous domains. Indeed a theory may be easier to formulate over a continuous domain, but practical use may entail projecting them onto a discrete domain. Time is regarded as discrete because measurements are made at specified, commonly equally spaced, time points. The precise methodology will be determined by the nature of the data that is available over space, for example is it point-referenced or collected on a lattice?

Some general approaches to incorporating time are as follows:

Approach 1: Treat continuous time as another spatial dimension, e.g. spatio–temporal kriging (Bodnar & Schmid, 2010). There is extra complexity in constructing covariance models compared to purely spatial process modelling (Fuentes, Chen, & Davis, 2008) and possible reductions in the complexity based on time having a natural ordering (unlike space) are not realised.

Approach 2: Represent the spatial fields represented as vectors $Z_t : N_S \times 1$, and combine them across time to get a multivariate time series.

Approach 3: Represent the time series as vectors, $Z_s : 1 \times N_T$, and use multivariate spatial methods e.g. co-kriging.

Approach 4: Build a statistical framework based on deterministic models that describe the evolution of processes over space and time.

Approach 1 may appeal to people used to working in a geostatistical framework. Approach 2 may be best where temporal forecasting is the inferential objective while Approach 3 may be best for spatial prediction of unmeasured responses. Approach 4 is an important new direction that has promise because it includes background knowledge through numerical computer models. For further details of this approach see Chapter 14, Section 14.3.

If the primary aim is spatial prediction then you would want to preserve the structure of the spatial field. However if the primary interest is in forecasting this would lead to an emphasis in building time series models at each spatial location. The exact strategy for constructing a spatio–temporal model will also depend on the purpose of the analysis. A regulator may need to estimate the effect of ambient levels of an environmental hazard on human health. This would be expressed as a concentration response function (CRF) (see Section 12.5.1) that relates the levels of ambient concentrations to health outcomes. That function would indicate the potential beneficial effect of reductions in ambient levels. An environmental epidemiologist might wish to plug a process model into an exposure response function (ERF) where it is the predicted exposures rather than ambient levels. Predicting personal exposures means that both ambient and indoor sources e.g. carbon monoxide from a poorly maintained gas heater, would be represented and thus give a truer exposure profile (Shaddick, Lee, Zidek, & Salway, 2008; Zidek, Meloche, Shaddick, Chatfield, & White, 2003; Zidek, Shaddick, White, Meloche, & Chatfield, 2005). Interest may lie in forecasting an ambient measurement twenty-four hours ahead of time. Or to spatially predict such levels at unmonitored sites to get a better idea of the exposure of susceptible school children in a school far from the nearest ambient monitor. In deciding how to expand or contract an existing network of monitoring sites in order to improve prediction accuracy or to save resources, a spatio–temporal model will be required together with a criterion on which to evaluate the changes you recommend. This last topic is the subject of Chapter 13.

11.3 Spatio–temporal models

A spatial–temporal random field, $Z_{st}, s \in \mathscr{S}, t \in \mathscr{T}$, is a stochastic process over a region and time period. This underlying process is not directly measurable, but realisations of it can be obtained by taking measurements, possibly with error. Monitoring will only report results at N_T discrete points in time, $T \in \mathscr{T}$ where these points are labelled $T = \{t_1, \ldots, t_{N_T}\}$. The same will be true over space leading to a discrete set of N_S locations $S \in \mathscr{S}$ with corresponding labelling, $S = \{s_1, \ldots, s_{N_T}\}$.

If the temporal aspect is incorporated into the structure of Equation 9.1, then Z_{st} can be represented as a space–time random field, which can again be expressed in terms of a hierarchical model such as that introduced in Section 1.5 for the measurement and process models

$$
\begin{aligned}
Y_{st} &= Z_{st} + v_{st} \\
Z_{st} &= \mu_{st} + \omega_{st}
\end{aligned}
\tag{11.1}
$$

where v_{st} represents independent random measurement error. The term μ_{st} is a spatio–temporal mean field (trend) that is often represented by a model of the form $\mu_{st} = x_{st}\beta_{st}$. For many processes this mean term represents the largest source of variation in the responses. Over a broad scale it might be considered as deterministic if it can be accurately estimated, e.g. as an average of the process over a very broad geographical area. However where there is error in modelling μ_{st} the residuals ω_{st} play a vital role in capturing the spatial and temporal dependence of the process.

The spatio–temporal process modelled by ω can be broken down into separate components representing space, m, time, γ and the interaction between the two, κ (Cressie & Wikle, 2011).

$$
\omega_{st} = m_s + \gamma_t + \kappa_{st}
\tag{11.2}
$$

The first two of these can be modelled using models seen in Chapters 9 (space) and 10 (time). In this example, **m** would be a collection of zero mean, site-specific deviations (spatial random effects) from the overall mean, μ_{st} that are common to all times. For time, γ would be a set of zero mean time-specific deviations (temporal random effects) common to all sites. The third term κ_{st} in Equation 11.2 represents the stochastic interaction between space and time. For example, the effect of latitude on temperature depends on the time of year. The mean term, μ_{st} may constitute a function of both time and space but the interaction between the two would also be manifest in κ_{st}. This would capture the varying intensity of the stochastic variation in the temperature field over sites which might also vary over time. In a place such as

California the temperature field might be quite flat in summer but there will be great variation in winter. It is likely that there will be interaction acting both through the mean and covariances of the model.

Many variations of the models in Equations 11.1 and 11.2 have been published for specific applications including ones that take a multi-resolution approach where terms corresponding to stochastic subregional blocks of medium level resolution and then micro-level models for local levels (Reinsel, Tiao, Wang, Lewis, & Nychka, 1981; Sahu et al., 2006; Caselton & Zidek, 1984; Schumacher & Zidek, 1993).

Example 11.1. *Effect of wildcat drilling in Harrison Bay, Alaska*

Here $N_T = 2$ and $t = 1, 2$ represent times before and after the startup of exploratory drilling in Harrison Bay, Alaska, the Beaufort Sea oil field having already been established. Interest in this case was on human welfare rather than human health, namely on the effect of such drilling on the food chain of the indigenous people who lived in that area. Clearly the risk of this drilling would depend on how wind and sea currents carried the plume of drilling mud which is used to lubricate the drill stem as it digs into the earth. Experts were asked to independently draw boundaries of what they saw to be the zones of equitable risk. There was surprising agreement amongst the experts. This led in the end to a model of the form

$$Y_{st} \quad = \quad Z_{st} + v_{st}, \; t = 1, 2, \; s \in S \qquad (11.3)$$
$$Z_{st} \quad = \quad \mu_{st} + \omega_{st}, \qquad (11.4)$$
$$\mu_{st} \quad = \quad \mu + \beta_s + \gamma_t x_t$$
$$\beta_s \quad ind \sim \quad N(0, \sigma_\beta^2)$$

where the dummy variable is $x_t = I\{t = 2\}$ and b_s puts s into its zone of equitable risk. This simple model was chosen, in part because of its simplicity; it resulted in a paired t–test like analysis to detect change. However to allow fully for uncertainty, random effects are assigned to the risk zones. The spatial domain S consisted of a geographic grid superimposed on the risk zones. The eventual design was based on that knowledge that the National Oceanic and Atmospheric Agency (NOAA), which was overseeing the project, would prefer a simple method of analysis for assessing the impact. The Bayesian elements were used to find the expected value of the uncertain non-centrality parameter for the test. This depended on the subset S that was to be selected and so was optimised to find the optimal design. That led to maximising the contrast in the field, with the optimal sites distributed between low and high risk zones. That in turn led to a theoretical paper that generalised this approach (Schumacher & Zidek, 1993).

Example 11.2. *Modelling pollution fields*

It has long been recognised that particulate air pollution is associated with adverse health impacts in humans. Thus it is now a criteria air pollutant that is regulated to ensure air quality. Of particular concern is $PM_{2.5}$, consisting of small particles formed from gaseous emissions, for example, from the burning of wood. Both their mass μgm^{-3} as well as their counts *ppm* are considered important since a large number of tiny particles in the $PM_{2.5}$ mix, for example those of size less than 1 micron in diameter, can penetrate deeply into the lung.

Primary interest lies in the spatial prediction of the $PM_{2.5}$ field. However, in general a spatio–temporal modelling approach is preferable since the quasi replicates of the spatial field over time enables better parameter estimation. Sahu et al. (2006) used the following model for the underlying spatio–temporal mean term,

$$\mu_{st} = \beta_0 + \beta_1 p_s + \beta \alpha_s \times p_s + \sum_{l=2}^{12} \zeta_l u_{tl} \qquad (11.5)$$

where the dummy $u = I\{t$ is in month m$\}$ tells us the month in which the time (week) $t = 1, \ldots, 52$ is located. Although in some ways an appealing and simple way of handling seasonality when temporal replicates are available it comes at a cost of eleven degrees of freedom. These are used in estimating the set of the $\{\zeta_l\}$ coefficients. Here p_s denotes human population density while α_s is an rural–urban indicator function. The process model used by Sahu et al. (2006) was

$$Z_{st} = \mu_{st} + \omega_{st} + p_s v_{st} \qquad (11.6)$$

where the the sum of the last two terms is thought of as representing a spatially varying temporal trend. To complete the model description for this example, we need a covariance structure and here Sahu et al. (2006) assumed that space and time are separable (see Subsection 11.3.1) and they used the exponential covariance functions for both.

11.3.1   *Separable models*

In most applications, modelling the entire spatial–temporal structure will be impractical because of high dimensionality. A number of approaches have been suggested to deal with this directly and we now discuss the most common of these, that of assuming that space and time are *separable* (Gneiting, Genton, & Guttorp, 2006). This is in contrast to cases where the space–time structure is modelled jointly which are known as *non-separable* models.

Separable models impose a particular type of independence between space and time components. It is assumed the correlation between Z_{st} and $Z_{s't}$ is $\rho_{ss'}$ at every time point t while the correlation between Z_{st} and $Z_{t's}$ is $\rho_{tt'}$ at all spatial time points s.

The covariance for a separable process is therefore defined as

$$Cov(Z_{st}, Z_{s't'}) = \sigma^2 \rho_{ss'} \rho_{tt'}$$

for all $(s,t), (s',t') \in \mathscr{S} \times \mathscr{T}$.

Expressed in matrix form, for Gaussian processes, we get the Kronecker product for the covariance matrix,

$$\Sigma^{Tp \times Tp} = \sigma^2 \rho_T^{N_T \times N_T} \otimes \rho_S^{N_S \times N_S}. \tag{11.7}$$

where ρ_1 is the between row temporal autocorrelations and ρ_2 is the between column spatial correlations.

Example 11.3. *Kronecker products*

The Kronecker product is an operation on two matrices of arbitrary size resulting in a block matrix. It is a generalisation of the outer product from vectors to matrices.

For matrices $\mathbf{A} : n \times m$ and $\mathbf{B} : p \times q$ their Kronecker product is defined as $\mathbf{A} \otimes \mathbf{B}$ as the linear operator acting on $\{Z : m \times q\}$ as follows:

$$(\mathbf{A} \otimes \mathbf{B})\mathbf{Z} = \mathbf{A} \, \mathbf{Z} \, \mathbf{B}' \tag{11.8}$$

From this definition, properties of the product can be easily proven. The following are examples:

$$
\begin{aligned}
(\mathbf{A} \otimes \mathbf{B})(\mathbf{C} \otimes \mathbf{D}) &= (\mathbf{AC} \otimes \mathbf{BD}) & (11.9)\\
(\mathbf{A} \otimes \mathbf{B})' &= \mathbf{A}' \otimes \mathbf{B}' & (11.10)\\
(\mathbf{A} \otimes \mathbf{B})^{-1} &= \mathbf{A}^{-1} \otimes \mathbf{B}^{-1}, & (11.11)
\end{aligned}
$$

if $\mathbf{A}$ and $\mathbf{B}$ are nonsingular

The model shown in Equation 11.14 assumes temporal correlations are the same at every site. Likewise, the spatial correlations are the same at every point in time. These are strong assumptions which greatly simplify things but they do seem to be reasonable in a lot of applications, for example through cross-validation yields good results (Le & Zidek, 2006). Due to the reduction in computational burden that comes with this approach, the majority of work on space–time modelling tends to be based on analysing the temporal and spatial aspects separably, and then to combine the chosen models in a single separable model.

Example 11.4. *Spatial prediction for daily levels of particulate matter*

Sun, Zidek, Le, and Ozkaynak (2000) use the approach suggested by Le, Sun, and Zidek (1997) to develop a spatial predictive distribution for the space–time field of daily ambient PM_{10} in Vancouver, Canada. For simplicity, they analysed each of the monitoring sites separately and chose a single, AR(1), model to represent the temporal structure. They recognize the possibility that spatial correlation between sites might 'leak' into the lagged values of the series, due to modelling each site univariately. This would not have happened if a multivariate auto-regression was used, but such an approach may be infeasible with a large number of monitored and unmonitored sites. They then used the de-trended residuals to obtain posterior distributions of the covariance matrix, which was then extended to include unmonitored sites using a semi-parametric approach for the spatial covariances of hourly ozone levels (Sampson & Guttorp, 1992), and allowed the parameters of the model to vary as functions of time of day.

Example 11.5. *Modelling wind velocity*

Haslett and Raftery (1989) modelled wind velocity measurements in Ireland using a combination of kriging and ARMA time series models to predict values at gauged sites that have short runs of data. They use kriging estimates of the variance of the random field, using an exponential model for the covariance–distance relationship. They built space–time models by linking models for the observed data, Y, directly at each of the sites through a spatially correlated set of innovations whose covariance structure was derived from the underlying continuous spatial process, which was assumed to be realized independently at each t. They then combine the spatial and temporal aspects in a model of the form

$$Y_{st} = \mu_s + \nabla^{-d}\phi(B)^{-1}\theta(B)\omega_{st} \tag{11.12}$$
$$\mathbf{m}_t = (m_{t1}, ..., m_{tN_S})' \sim N_{N_S}(0, \sigma_m^2\Sigma_s)$$

where the matrix Σ_s represents the spatial correlation, and does not depend on time. ∇^{-d} is the Binomial expansion of $(1 - B)^{-d}$, where B is the backshift operator as described in Chapter 10, Section 10.4.2, and d can have non-integer value, leading to fractional differencing and long memory processes. It is noted that in Equation 11.1, the observed data is modelled directly, i.e. there is measurement error term in level one.

Example 11.6. *Winter temperature*

In an early Bayesian application Handcock and Wallis (1994) consider the spatio–temporal modelling of winter temperature data but their approach was to carry out separate spatial analyses in each year using a Gaussian random field model. The mean and covariance parameters of this model were then examined temporally and found to be stable, which they concluded was justification for the use of a simple model that essentially treated spatial and temporal aspects separately. They developed a stationary spatial–temporal random field for mean temperature using eighty-eight weather stations in the northern U.S. They first developed a Gaussian random field map using a Bayesian kriging approach as a function of latitude, longitude and elevation, examining variograms in different directions to look for anisotropy. Having produced a spatial structure, they consider the yearly temperatures by examining each site separately, fitting AR(1) models and finding that only three out of the eighty-eight were significant, leading to the conclusion that there was little evidence of short-term temporal dependency. They then examined long-term dependency using ARIMA models with non-integral or fractional differencing, as discussed in Example 11.5, finding little evidence of long memory dependence and also, that the spatial structure was changing over time.

Example 11.7. *Models with cyclical variance*

Recall that the general hierarchical model involves first and foremost a process model. The relationship between measurements and the process is described in the measurement model. In addition there is the parameter model, which in a Bayesian framework will contain the prior distributions assigned to all unknown parameters. Hence, the model parameters are random quantities and can be assigned spatial and temporal distributions. For example consider the process model

$$Z_{st} = \mu + \beta_s cos(2\pi\omega t) + v_{st}, \; s \in S, \; t = 1, \ldots N_T \qquad (11.13)$$

with μ constant, $v_{st} \; ind \sim N(0, \sigma_v^2)$ and $\beta = (\beta_1, \ldots, \beta_p) \sim N_{N_s}(0, \Sigma_\beta)$. It has the unusual property that the process's marginal variance $Var(Z_{st}) = \Sigma_\beta cos^2(2\pi\omega t) + \sigma_v^2$ variance is cyclical. Huerta et al. (2004) use a more complex version of such a model that has been used and criticised for a variety of reasons (Dou, Le, & Zidek, 2007). Amongst other things, models such as these have wiggly credibility bands around the path of the process over time $Z_{st}, \; t = 0, 1, \ldots, T$ for a fixed s. Such a band may seem unnatural depending on the context and nature of the prior distribution on the process amplitudes β. That distribution could represent the modeller's epistemic uncertainty (see Chapter 3) about the size of β. In that case it would be expected to shrink to zero in the future as increasing amounts of data become available over time and knowledge about β increases to certainty. More likely β would also include aleatory uncertainty in nature to allow for things like fluctuating wind directions for example. In any case, careful thought needs to be given in selecting the parameter model in a Bayesian context to insure that the aleatory uncertainty about a process is preserved as the epistemic uncertainty is resolved (see Chapter 3).

11.3.2   Non-separable processes

Non-separable processes will often be more difficult to understand than when separation processes can be assumed for space and time and as a consequence modelling is often complex. In particular, dealing with the Kronecker products (see Example 11.3) that define covariances poses technical challenges if the wrong approach is taken. To illustrate, consider the simple problem of showing that $(\mathbf{A} \otimes \mathbf{B})^{-1} = \mathbf{A}^{-1} \otimes \mathbf{B}^{-1}$. This problem proves to be very difficult if we ignore the algebraic roots of the Kronecker product as a linear operator (see below) and instead use the matrix definition which for simplicity in the case of 2×2 matrices is the 4×4 matrix given by:

$$\begin{pmatrix} a_{11} & a_{12} \\ a_{21} & a_{22} \end{pmatrix} \otimes \begin{pmatrix} b_{11} & b_{12} \\ b_{21} & b_{22} \end{pmatrix} = \begin{pmatrix} a_{11}\mathbf{B} & a_{12}\mathbf{B} \\ a_{21}\mathbf{B} & a_{22}\mathbf{B} \end{pmatrix} \qquad (11.14)$$

Another example, also very relevant to the development of spatio–temporal models, is the separable case for Gaussian processes. This is based on matrix multiplication.

Suppose that

$$\mathbf{Z} = \begin{pmatrix} z_{11} & z_{12} \\ z_{21} & z_{22} \end{pmatrix} = \begin{pmatrix} \mathbf{z}_1 \\ \mathbf{z}_2 \end{pmatrix}$$

In this case, what is $(\mathbf{A} \otimes \mathbf{B})\mathbf{Z}$? The answer obtained by using the matrix definition of the product employs the idea of vectorizing the matrix $\mathbf{Z}$ more precisely to define

$$vec(\mathbf{Z})^{4 \times 1} = \begin{pmatrix} \mathbf{z}_1^T \\ \mathbf{z}_2^T \end{pmatrix}$$

Then we can show that

$$(\mathbf{A} \otimes \mathbf{B})vec(\mathbf{Z}) = vec(\mathbf{A}\mathbf{Z}\mathbf{B}') \qquad (11.15)$$

However even in this simple case the technicalities are relatively complex. A much better approach that avoids use of the vec operator treats the Kronecker product as a linear operator.

So why is all this important for modelling spatio–temporal Gaussian processes? There, the domain where measurements will be taken is $(s,t) \in \mathscr{S} \times \mathscr{T}$ where

$\mathscr{S} \times \mathscr{T}$ denotes what is called the 'product space' of $\mathscr{S}$ and $\mathscr{T}$. Over that domain responses for a separable Gaussian process can be represented by a random matrix with a matric normal distribution (see Appendix 14.5):

$$\mathbf{Z}^{N_S \times N_T} \sim N_{N_S \times N_T}[\mu, \sigma^2 \rho_S \otimes \rho_T]$$

So if the temporal auto correlation matrix were known, we could easily reduce the process to another with independent replicates over time as follows.

$$\mathbf{Z}^* = (\mathbf{I} \otimes \rho_T^{-1/2})\mathbf{Z} \sim N_{N_S \times N_T}[(\mathbf{I} \otimes \rho_T^{-1/2})\mu, \sigma^2 \rho_S \otimes \mathbf{I})]$$

Even if ρ_T is unknown, in some cases it may be possible to estimate it well, for example when it has a simple parametric form and there are many time points.

11.4   Dynamic linear models for space and time

Here we extend the state space model for the temporal setting introduced in Section 10.7 to the spatio–temporal setting. As before the key elements are the measurement and process models in the linear Gaussian situation where for $t \in T$ $\mathbf{Y}_t : N_S \times 1$ denotes the measurement sequence and $\mathbf{Z}_t : N_S \times 1$, the process sequence:

$$\begin{aligned}
\mathbf{Y}_t &= \mathbf{F}_t^T \mathbf{Z}_t + v_t, v_t \sim N[0, V_t] \\
\mathbf{Z}_t &= \mathbf{G}_t \mathbf{Z}_{t-1} + w_t, \quad w_t \sim N[\mathbf{0}, \mathbf{W}_t]
\end{aligned}$$

where $\mathbf{F}_t : p \times N_S$, $\mathbf{G}_t : p \times p$, $V_t : N_S \times N_S$, and $\mathbf{W}_t : p \times p$ are known. Generally $\mathbf{F}_t$ is called the 'design matrix', v_t the observational error, $\mathbf{G}_t$ the state matrix and ω_t, the evolution error with evolution covariance matrix $\mathbf{W}_t$. To finish the DLM's specification we need

$$[\mathbf{Z}_0 | \mathbf{Y}_0] \quad \sim \quad N[\mathbf{m}_0, \mathbf{C}_0]$$

The model is implemented by applying the forward filtering–backward sampling algorithm (Harrison, 1999).

Example 11.8. *Modelling hourly ozone concentrations using DLMs*

The DLM has been applied to model hourly ozone concentrations in both Mexico City and Chicago (Huerta et al., 2004; Dou, Le, Zidek, et al., 2010; Dou et al., 2012). We now describe a version of the DLM that was developed to represent the daily cycles in the levels of ozone concentrations in urban areas (Huerta et al., 2004). The model recognizes the need to reflect both aleatory and epistemic uncertainty as discussed in Example 3.5 and its sequel. It does this by incorporating in the measurement model's random residual terms m_{st} spatial (but not temporal) correlation. More precisely

$Cov(\mathbf{m}_t) = \sigma_y^2 \exp(-\mathbf{D}/\lambda_y)$, where $\mathbf{m}_t = (m_{s_1t}, \dots, m_{s_{N_s}t})$ and $\mathbf{D}$ denotes the intersite distance matrix. The data and process models for the DLM are then defined by

$$\text{Data model: } Y_{st} = Z_{1t} + S_{1t}Z_{2st} + S_{2t}Z_{3st} + v_{st}$$

Process models:

$$Z_{1t} = Z_{1(t-1)} + \gamma_t^1$$

$$Z_{jst} = Z_{js(t-1)} + \gamma_{jst}^2, \quad j = 2,3$$

The twelve and twenty-four hour ozone cycles are captured by $S_{jt}(a_j) = cos(\pi jt/12) + a_j sin(\pi jt/12)$, $j = 1,2$ while the Z_{jst}, $j = 2,3$ are random amplitudes. The model parameters are the processes in the model. The first Z_{1t} depends only on time and plays the key role of establishing the baseline level. It evolves according to a random walk model and with an innovation term distributed as $\gamma_t^1 \sim N(0, \sigma_y^2 \tau_y^2)$. The remaining two coordinates of the trivariate vector valued process $\mathbf{Z}_t$, i.e. the amplitudes Z_{jst}, $j = 2,3$, evolve according to a random walk like the first coordinate. Their innovation terms are allowed to be spatially dependent, i.e. letting $\gamma_{jt}^2 = (\gamma_{j1t}, \dots, \gamma_{js_{N_s}t}^2)$, $Cov(\gamma_{jt}^2) = \sigma_y^2 \tau_j^2 \exp(-\mathbf{D}/\lambda_j)$. Finally the parameters $\tau_j^2, \tau_y^2, \lambda_j$ are specified in advance. We could have added an additional term to capture the effect of covariates such as temperature but for simplicity that is ignored here.

The model was initially applied over Mexico City (Huerta et al., 2004) to model a sequence of measurements made at ten sites. Both spatial prediction as well as temporal forecasting were performed.

The promise of the DLM approach led to its application (by the second author and his co-investigators) in regions in the eastern United States. Approximately 300 sites were involved and measurements were made over the so-called ozone season of about 120 days (2880 hours) during the summer. Interest was in spatial prediction to rural areas which have few ozone sites, despite the importance of the pollutant's effects on human welfare, crops, forests and both flora and fauna in general. It soon became obvious that the extension of the DLM developed for Mexico City would not work in our application. There the models involved nearly 1.7 million parameters due to the large number of monitoring sites and measurements over time. We found that with even substantial computational power, only a maximum of ten sites could be handled for data covering the entire ozone season.

Thus in the end an investigation was undertaken for urban areas with ten sites to explore the possibility of using the method in those domains. There the method worked quite well, although we did discover some unusual features, one of which has been previously discussed involving Equation 11.13, that the sinusoidal character of the mean function induces wiggly posterior credibility bands around the inferred process due to the random amplitudes in the model. In other words the degree of uncertainty expressed by those bands varies cyclically over time. That uncertainty would be a combination of aleatory and epistemic uncertainty so this effect would be muted as the amount of available data increased. Finally the random walk (non-stationary) model used in this model to reduce the overall number of parameters in the model leads to posterior credibility bands that increase in width as time increases—even when conditioning on the full set of measurement collected over all those time points. This would seem to be an argument against use of the random walk model in Bayesian dynamic modelling (Dou et al., 2012).

11.5   An empirical Bayes approach

Dealing with large amounts of spatio–temporal data will often mean that implementing DLMs can be computationally prohibitive. One possible solution to this difficulty is to consider an empirical Bayes approach, known as the Bayesian spatial predictor (BSP) (Le & Zidek, 2006). An empirical comparison of the BSP and DLM is given in Example 11.9 but first we review the theory behind the approach.

To align our description of the BSP model with that in Le and Zidek (2006) entails a slight modification in our notation. In the BSP model the responses are represented as elements of a random matrix where the columns rather than rows represent spatial sites. Each gauged site (a location at which there is a monitor) $s \in S$ has a vector of k responses corresponding to number of different exposures meaning this is a multivariate model. A gauged site therefore has k 'quasi-sites' where measurements could be made but gauges may not have been installed to measure all responses at that site. Therefore the number of columns in the response matrix would be $N_S \times k$ where the number of rows would be N_T.

For expository simplicity we will describe the univariate case where $k = 1$. However, we emphasise the importance of multivariate responses There may be several responses of interest at each site and even when interest focuses on a single response, it can be advantageous to incorporate other responses, so that strength can be borrowed from them through their stochastic dependence. Also, even with a univariate response it can sometimes be desirable to group its successive realisations over time in blocks, thus creating a multivariate Gaussian field. For instance we could put hourly responses into blocks of length 24 hours and make the t's represent days. This strategy means avoiding the challenging issue of modelling fine scale temporal dependence.

In applications some sites will not have been monitored or 'gauged' in the terminology of Le and Zidek (2006), for the entire period $t \in T$ and they will be labelled with a u, standing for 'ungauged'. Others will have had monitors from the beginning when $t = 1$ and they are labelled with a g. In general startup times will be staggered with some starting very recently. Commonly even after a site is gauged, it does not monitor all the responses of interest, leading to a complex pattern of nonrandom blocks of missing data in both ungauged and gauged sites. Therefore we need notation to distinguish between the observed and missing responses at gauged sites leading to labels g_o and g_m.

With that background we may now describe the process as through a response matrix:

$$Z^{N_T \times N_S} = \left[Z^{[u]}, \left(\begin{array}{c} Z^{[g_1^m]} \\ Z^{[g_1^o]} \end{array} \right), \cdots, \left(\begin{array}{c} Z^{[g_k^m]} \\ Z^{[g_k^o]} \end{array} \right) \right] \qquad (11.16)$$

where:

- m means no measurement for this response;
- g means the site is gauged and it also stands for the total number of ungauged sites to reduce the notational burden;
- u means an ungauged site and also the total number of ungauged sites;
- $N_S = u + g$.

Equation 11.16 shows the staircase pattern in the data matrix once the gauged sites have been reordered from newest to oldest (from left to right).

Both the measurement or process models may incorporate factors and covariates. We will call factors things the experimenter controls, for example rural–urban in the selection of sites. Covariates would be the predictors or potential confounders that may co-vary with the response of interest.

The design matrix X in the BSP model contains just those covariates that are common to all sites and they may be discrete or continous. A discrete example would be a binary (present or absent) and continuous ones would be things like 'temperature' measured at the airport, or 'degree of visibility', when the domain of interest is an urban area. The BSP conditions on the design matrix and treats it as a constant.

But there is a second type of covariate that is site-specific, for example hourly concentrations NO_x (oxides of nitrogen) X_{st} when hourly ozone concentrations $Y_{st} = Z_{st}$ (assuming no measurement error) are the responses of interest. The BSP can handle this case because it is a multivariate model that would first model the joint distribution $[Y_{st}, X_{st}]$ and then use the conditional distribution $[Y_{st} \mid X_{st} = x_{st}]$ in the analysis of the health effects of ozone. BSP does not explicitly allow for site-specific factors such as elevation for example, but it can be adapted for use in that context.

For expository simplicity, assume just one covariate X_t over the domain of interest, one that is treated as a constant $X_t = x_t$. The BSP measurement model is then for

$s \in S$ and $t \in T$

$$Z_{st} = x_t \beta_s + \omega_{st} = x_t \hat{\beta}_0 + x_t \tilde{\beta}_s + \omega_{st} \qquad (11.17)$$

for time t and site s. Here the non site-specific $\tilde{\beta}_0$ can be fitted by classical methods when there are lots of data so that the estimate has a small standard error. Removing the effect of $x_t \hat{\beta}_0$ or 'prefiltering' in the parlance of BSP theory, is thus like removing a constant from the random response. We can rewrite the model as

$$Z_{st}^* = Z_{st} - x_t \hat{\beta}_0 = x_t \tilde{\beta}_s + \omega_{st}$$

leaving the uncertain site-specific coefficient, $\tilde{\beta}(s)$, as a random site effect from the point of view of a frequentist or an uncertain fixed parameter from that of a Bayesian. For really smooth random fields, it can nearly be zero for all sites. In vector form we have $\mathbf{Z}_t^* = x_t \, [\tilde{\beta}_{s_1}, \ldots, \tilde{\beta}_{s_{N_s}}] + \omega_t$.

The BSP's distributional assumptions are:

$$\begin{cases} Z \mid \beta, \Sigma \sim N(\mathbf{X}\beta, \mathbf{A} \otimes \Sigma) \\[2mm] \beta \mid \Sigma, \beta_0, F \sim N(\beta_0, F^{-1} \otimes \Sigma) \\[2mm] \Sigma \sim GIW(\Theta, \delta) : \text{Generalized inverted Wishart} \end{cases} \qquad (11.18)$$

Here $\mathbf{A}$ is assumed to be known, so that by a simple transformation, it can without loss of generality be taken to be the identity matrix. In practice, a plug-in estimate may be used during the pre-filtering stage of the analysis. For instance for long temporal aggregates such as monthly averages: $\mathbf{A} \approx I_n$ $\mathbf{A} \approx I_n$ sometimes achieved by removing a single low frequency temporal component across all sites. The generalized inverted Wishart (GIW), which allows different degrees of freedom for different steps in the staircase, extends the standard Wishart distribution $\Sigma \sim IW(\Psi, \delta)$. The latter, which is the multivariate version of the σ^2 scaled chi-squared distribution, can be used in the absence of a staircase data pattern. However the GIW allows much more flexibility by allowing the response vector coordinates to be grouped into blocks with different degrees of uncertainty to represent different levels of uncertainty.

With g denoting the number of gauged or partially gauged sites we get the following posterior distribution:

$$[\mathbf{Z}^u \mid D, \mathscr{H}] \sim \left[Y^{[u]} \mid \mathbf{Z}^{[g_1^m, \ldots, g_g^m]}, D, \mathscr{H} \right] \times$$

$$\prod_{j=1}^{g-1} \left[\mathbf{Z}^{[g_j^m]} \mid \mathbf{Z}^{[g_{j+1}^m, \ldots, g_g^m]}, D, \mathscr{H} \right] \times \left[\mathbf{Z}^{[g_g^m]} \mid D, \mathscr{H} \right]$$

Each factor is a matric-t distribution (see Appendix 14.5). The mean, covariance, and degrees of freedom are functions of $\mathscr{H}$ and D. Thus the posterior is completely characterised given $\mathscr{H}$ since all the hyperparameters that determine the priors are in $\mathcal{H}$. The EnviroStat package uses an empirical Bayes approach and estimates those hyperparameters through a maximum likelihood approach.

Example 11.9. *A comparison of the DLM and BSP approaches*

In this example we compare the results of using the DLM and BSP approaches when modelling ozone concentrations in the Chicago area (Dou, 2007).

This example involves two different approaches to spatio–temporal modelling, the first is the dynamic linear modelling (DLM) approach where coefficients can change over time and the second is the Bayesian spatial predictor (BSP) approach which uses an empirical Bayes within Bayes approach to minimise computation times where a large number of sites are involved (Le & Zidek, 2006).

A total of twenty-four hourly ozone concentration monitoring sites were selected in the Chicago area, treating fourteen of them as gauged and keeping ten as validation (ungauged) sites. These can be seen in Figure 11.1. Application of the BSP in compact geographical regions like this case study, usually begins by pre-filtering; that is fitting a regional (not site-specific) trend (Li et al., 1999). More precisely, that would mean fitting $\tilde{\beta}_0$ in Equation 11.17 and subtracting the estimated trend from the series of measurements

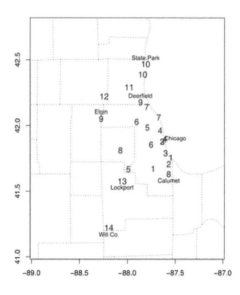

Figure 11.1: Locations of fourteen gauged and ten ungauged validation hourly ozone concentration monitoring sites in the Chicago area whose data were used in a comparison of the DLM and BSP approaches.

to get detrended residuals. Pre-whitening those residuals by fitting a time series model results in a second set of residuals. This could mean for example fitting a stationary time series model such as AR(2), in this case with coefficients that are constant across the region. Often the temporal dependence in the resulting residuals are negligible and the A in Equation 11.18 is approximately equal to the identity.

However that does not work in this case since space and time are not separable. Here fitting factors 'month' and 'weekday–weekend' followed by the fitting of a AR(2) model may yield residuals with greatly reduced spatial correlation compared with the original process (Dou et al., 2010). This has been referred to as 'correlation leakage' (Li et al., 1999) and it makes spatial prediction difficult. So an alternative approach was used, one that takes advantage of the multivariate model built into the BSP approach. Vectors of consecutive hourly concentrations in each day were used as the responses and a multivariate model was used for these vectors with 'month' and 'weekday–weekend' as the explanatory factors. Preliminary analysis pointed to the use of just two consecutive hours per day, meaning that the resulting vectors were separated by twenty-two hours and hence are uncorrelated. For example if we wished to predict the 11AM hourly concentration at an ungauged site based on the measurements at the gauged sites, we would put the 10AM and 11AM responses in the response bivariate vectors and make spatial predictions based on both 10AM and 11AM responses at the gauged sites, thus borrowing strength across both time and space for spatial prediction.

The DLM and BSP were empirically compared in two different contexts, the first using two hour vector responses, and the second five (Dou et al., 2010). In these contexts the BSP yields superior empirical spatial prediction performance in terms of the mean squared prediction error (MSPE). Table 11.1 shows results obtained by Dou (2007).

Ungauged Site	MSPE (DLM)	MSPE (BSP)
1	1.79	1.16
2	1.55	1.61
3	1.80	1.30
4	1.46	0.94
5	1.92	1.03
6	1.85	0.99
7	1.60	0.97
8	2.62	2.67
9	1.63	1.01
10	0.87	0.38

Table 11.1: The mean square prediction error $(\sqrt{ppb})^2$ at Chicago's ungauged sites comparing the multivariate BSP and the DLM. In all cases but that of the eighth ungauged site, the BSP predictions are more accurate than those of the DLM.

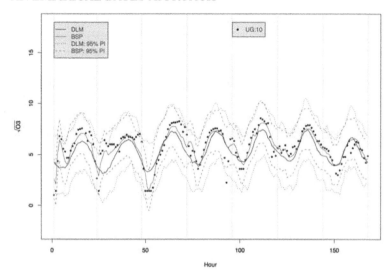

Figure 11.2: Spatial predictions of hourly ozone levels at the validation Site 10. The DLM and BSP approaches were used and their 95% credibility intervals are shown. The grey solid line represents the latter, which tends to track the observations at the validation site rather better than the DLM method. Note that the validation site lies in close proximity to the gauged sites. Both of the credibility bands tend to be too wide, the dashed 95% band for the BSP tending to be the narrower of the two, as can be seen near hour 50.

In Figure 11.2 we show the spatial predictions of values at Site 10 made by the two methods in Figure 11.2. As Site 10 is in close proximity to gauged sites, both predictors do quite well. These find that overall, the BSP is more accurate and its 95% credibility bands are narrower than those of the DLM, again reflecting the findings reported elsewhere (Dou et al., 2010).

Overall the BSP enjoys computational advantages over the DLM and it outperforms the latter based on empirical assessments made in this example and elsewhere (Dou et al., 2010). While the DLM was originally designed for temporal modelling, the BSP can also be used as a temporal forecasting methodology. Another empirical comparison of the two methods (Dou et al., 2012), also comes out in favour of the BSP for forecasting hourly ozone concentrations at least in the near future. However the BSP does require preliminary data analysis in the form of pre-filtering and pre-whitening, or equivalently the estimation of the covariance $\mathbf{A}$ in the distribution model. It also requires the assumption of separability of time and space, although that assumption can be side-stepped as seen in this example by resorting to multivariate response vector modeling.

The DLM is powerful, flexible and intuitive. However finding good hyperparameter estimates can be challenging. Computation times are long. It took 10 days for 3000 MCMC iterations running C code on dual processors whereas BSP took about four hours for the same sort of calculation. The posterior variance conditional on all the data increases over time due to random walk model used for the model coefficients and the resulting credibility bands tend to have a wobble induced by the random amplitude coefficients multiplying the sines and cosines in the model. It is difficult to assess the distribution of aleatory and epistemic uncertainty (see Chapter 3) as the amount of data increase, nor can we assess its degree of correlation leakage if any.

11.6   A hierarchical model for spatio–temporal exposure data

In this section we give details of a hierarchical model described by Shaddick and Wakefield (2002) and used to model a spatial process in Chapter 9, Section 9.12 and a temporal process in Chapter 10, Section 10.8. In addition we show how the temporal process described in Section 10.8 can be generalised to a multivariate case.

As described in the sections describing the spatial and temporal models in the previous chapters, there are three stages to the model: (i) the observation, or data, model; (ii) the process model which in this case now describes the form of the underlying spatial and temporal processes and (iii) assigning prior distributions to the unknown parameters.

The model is designed for cases where there are multiple pollutants being measured at a number of monitoring sites. The model allows for a temporal–pollutant interaction and a spatial–pollutant interaction, with the spatial model being constant across time, isotropic and stationary. The model and its assumptions are now described.

Stage 1 - Observation model

At the first stage, the measurements of each pollutant $(p = 1,...,P)$ over time $(t = 1,...,T)$ at each monitoring site $(s = 1,...,S)$ are modelled as a function of the true underlying level of the pollutant with a site adjustment and a pollutant-site specific error term.

$$y_{stp} = z_{stp} + v_{stp}$$

where y_{stp} denotes the observed level of the pollutant p, $p = 1,...,N_P$ at time t and location s for $t = 1,..,N_T$ where N_T is the number of time points and $s = 1,...,N_S$ where N_S is the number of monitoring sites. In this model v_{spt} represents the measurement errors which are assumed i.i.d. $N(0, \sigma_{sp}^2)$.

Stage 2 - Process model

In this stage the underlying levels of the exposure, Z_{stp}, are assumed to comprise an underlying trend, μ_{stp} together with a separable spatio–temporal process, ω_p for each pollutant.

$$
\begin{aligned}
z_{stp} &= \mu_{stp} + \omega_{stp} \\
\omega_{stp} &= m_{sp} + \gamma_{tp} \\
\gamma_t &= \alpha\gamma_{t-1} + w_{stp}
\end{aligned}
$$

$$(11.19)$$

Here $\mu_{stp} = \beta X$ where β is a vector of regression coefficients and X_{spt} represents an explanatory variables that may change temporally (for example, temperature), and spatially. The latter may represent, for example, spatial characteristics of the site that may be constant across time such as latitude and longitude (which could be used to remove any trend), or characteristics of the monitor, for example roadside or elevation. The subscript p allows these effects to be pollutant specific. and γ_t is a multivariate temporal process that induces temporal and pollutant dependence and m_{sp} represents the spatial effect of being at site s (for pollutant p).

Stage 2 (a) - Spatial/pollutant model

The collection of random effects $m_p = (m_{p1}, ..., m_{pN_S})'$, $p = 1, ..., P$, is assumed to arise from the multivariate normal distribution

$$
m_p \sim MVN(0_{N_S}, \sigma_{pm}^2 \Sigma_{pm})
$$

$$(11.20)$$

where 0_{N_S} is an $N_S \times 1$ vector of zeros, σ_{pm}^2 the between-site variance for pollutant p and Σ_{pm} is the $N_S \times N_S$ correlation matrix, in which element (s, s') represents the correlation between sites s and s', $s, s' = 1, ..., N_S$, for pollutant p.

Stage 2 (b) - Temporal/pollutant model

$$
\gamma_{pt} = \gamma_{p,t-1} + w_{pt}
$$

$$(11.21)$$

for $p = 1, ..., N_P$. Here $w_t = (w_{1t}, ..., w_{N_{P_t}})'$ are i.i.d. multivariate normal random variables with zero mean and variance-covariance matrix Σ_{N_P}. This matrix contains variances σ_{wp}^2 thus allowing different pollutants to have different amounts of temporal dependence, and $N_P(N_P - 1)/2$ covariance terms reflecting the dependence (more precisely the covariance) between each of the pollutants, conditional on the previous values.

Stage 3 - Hyperpriors

A normal prior $N(c,C)$ is assumed for β, where c is a $q \times 1$ vector and C a $q \times q$ variance-covariance matrix. Gamma priors are specified for the precisions, specifically $\sigma_{sp}^{-2} \sim Ga(a_v, b_v)$. The variance-covariance matrix, $\Sigma_P^{-1} \sim W_P(D,d)$ where $W_P(D,d)$ denotes a P–dimensional Wishart distribution with mean D and precision parameter d. Unless there is specific information to the contrary, i.e. that a monitor with different characteristics is used at a particular site, it is assumed $\sigma_{vs}^{-2} \sim Ga(a_v, b_v)$, $s = 1, ..., S$. A uniform prior may be used for ϕ_p, with the limits being based on beliefs about the relationship between correlation and distance (see Chapter 9, Section 9.12.1 for a discussion of issues related to sampling from the posterior distribution of this parameter when using MCMC).

Model assumptions

The assumptions of the model include the following:

- The measurement error variance σ_{sp}^2 does not depend on time, although the model is easily extendable to situations in which the measurement error may change as a function of t.

- The relationship between the pollutants is constant over time.

- The relationship between the pollutants is spatially constant.

- The temporal and spatial components are independent.

11.7 Approaches to modelling non-separable processes

The complexity of non-separable spatio–temporal processes often combined with computational issues has resulted in the development of a number of different approaches to modelling them. Here we provide a brief description of a selection of the available approaches.

The problems of high-dimensionality in modelling the entire space–time structure were addressed by Mardia, Goodall, Redfern, and Alonso (1998) as well as by Wikle and Cressie (1999) who reduce the dimensionality of the mean term in the same fashion as in principal components analysis (Chatfield & Collins, 1980), to an orthonormal sequence of deterministic spatial functions. The former describe these functions as 'principal' and 'trend fields', with the trend fields being functions of the co-ordinates of the sites and the principal fields selected from kriging estimates for a set of 'normative' sites. The hierarchical model (11.1) is now expressed as,

$$
\begin{aligned}
Y_{st} &= Z_{st} + v_{st} \\
Z_{st} &= \sum_{j=1}^{p} A_{tj} \phi_j(s) + m_s \\
A_t &= H A_{t-1} + J \eta_t
\end{aligned}
\tag{11.22}
$$

where either H and/or J must be non-diagonal in order to incorporate spatial structure into the temporal component. Mardia et al. (1998) call this method the 'kriged

Kalman filter' (KKF) and outline a likelihood-based estimation strategy while Wikle and Cressie (1999) give an (approximate) Bayesian estimation approach. In addition to the estimation techniques, the difference between the two approaches is that there is no measurement error term (v_{st}) in the first level of the KKF. This results in an error term consisting of two components, $v_{st} + m_s$, and as there is no attempt to incorporate the inherent variability in predicting the underlying process from the data, introducing the possibility of oversmoothed estimates (Cressie & Wikle, 1998). The KKF computes $E(\sum_{j=1}^{p} a_{tj}\phi_j(s)|Z_{st})$ whereas Wikle and Cressie (1999) compute $E(Y_{st}|Z_{st})$.

Wikle, Berliner, and Cressie (1998) present a fully Bayesian hierarchical model in which they approach the problems of high dimensionality by explicitly modelling parameters lower in the hierarchy. They expand the second level of Equation 11.1 to include long-term temporal patterns, $\beta_{1s}\cos(\Omega t) + \beta_{2s}\sin(\Omega t)$. They use a vector autoregressive process (VAR) (Chatfield, 2013; Cressie, 1993) to model the dynamic aspect of the space–time interaction, A_{st}, i.e. $A_t = HA_{t-1} + \eta_{st}$. The updating parameter, H_t, was simplified by using a 'nearest neighbour' VAR (Cressie, 1993) in which $A_{st} = a_{st}A_{(st-1)} + bA_{s^1(t-1)} + cA_{s^2(t-1)} + dA_{s^3(t-1)} + eA_{s^4(t-1)} + \eta_{st}$, where $|s^i - s| = 1$, the four locations within distance one of s. In this form, the autoregressive parameter for the location in question, a_{st}, varies spatially, but b, c, d and e do not. Consequentially, the model has no spatial interaction between A_{st} and $A_{s't} : s \neq s'$. Estimation was achieved using MCMC which was simplified to some extent by using Gaussian error terms at each level, except for the variance hyperparameters, which were inverse Gamma.

Brown, Diggle, Lord, and Young (2001) considered the spatio–temporal modelling of radar-rainfall data in Lancashire, UK. They consider both separable and non-separable models. In the former, they describe a state space model with observation equation $X_t = A + B_t X_t + v_t$ and second level being a first-order vector autoregressive progress, VAR(1) (Chatfield, 2013; Cressie, 1993), $A_{st} - \mu_{st} = H(A_{t-1} - \mu_{st}) + J\eta_{st}$. The covariance structure can be expressed as $\Gamma_k = H^k(H\Gamma_0 H' + JJ')$, with J influencing the purely spatial covariance, and H describing how the spatial covariance decays at larger time lags. They also fit a Gaussian density to the autoregressive parameters, A_{st}, which 'blurs' the field, allowing interaction between different sites at different time points (Brown, Karesen, Roberts, & Tonellato, 2000). If two locations, s and s' are close in space, then the values of A_{st} and $A_{s't}$ will be calculated using values from overlapping neighbourhoods, leading to the outcomes being highly correlated. Fitting the model, which they achieve using the MLE, requires the covariance matrix of all the grid points not just the gauged points, making the Kalman filter difficult. Instead they construct a complete multivariate model with the likelihood evaluated by inverting the covariance matrix. As previously mentioned, in most applications the size of the covariance matrix will prohibit using it directly, initiating the use of data reduction methods. In this particular application, the problems of dimensionality were lessened by being able to delete data points where there was no rainfall, as they added no information. The resulting covariance matrix, whilst still large (2000×2000) was then invertible.

A spatio–temporal model for hourly ozone measurements was developed by Carroll et al. (1997). The model, $Z_{st} = \mu_t + \omega_{st}$ combines a trend term incorporating temperature and hourly/monthly effects, $\mu_t = \alpha_{hour} + \beta_{month} + \beta_1 temp_t + \beta_2 temp_t^2$, which is constant over space, and an error model in which the correlation in the residuals was a nonlinear function of time and space. In particular the spatial structure was a function of the lag between observations, $COV(v_{st}, v_{s't'}) = \sigma^2 \rho(d, v)$, where d is the distance between sites and $v = |t' - t'|$ is the time difference, with the correlation being given by

$$\rho(d, v) = \begin{cases} 1 & d = v = 0 \\ \phi_v^d \psi_v & d \text{ otherwise} \end{cases}$$

where $\log(\psi_v) = a_0 + a_1 v + a_2 v^2$ and $\log(\phi_v) = b_0 + b_1 v + b_2 v^2$. The correlation of the random field is thus a product of two factors, the first, ψ_v^d depends on both the time and space, the second only on the time difference. Unfortunately, as Cressie (1997) pointed out, this correlation function is not positive definite. Using results from the model, there were occasions when $Cov(Z_{st}, Z_{s't}) > Cov(Z_{st}, Z_{st})$. This highlights a genuine lack of a rich set of functions that can be used as space–time correlation functions.

11.8 Summary

In this chapter we have seen the many ways in which the time can be added to space in order to characterise random exposure fields. In particular the reader will have gained an understanding of the following topics:

- Additional power that can be gained in an epidemiological study by combining the contrasts in the process over both time and space while characterising the stochastic dependencies across both space and time for inferential analysis.

- Criteria that good approaches to spatio–temporal modelling should satisfy.

- General strategies for developing such approaches.

- Separability and non-separability in spatio–temporal models, and how these could be characterised using the Kronecker product of correlation matrices.

- Examples of the use of spatio–temporal models in modelling environmental exposures.

Exercises

Exercise 11.1. (i) Verify Equation 11.15 for the 2×2 case and show your work.

(ii) Repeat (i) for the general case.

Exercise 11.2. Verify Equation 3.5 and show your work.

Exercise 11.3. Verify Equations 11.9–11.11 and show your work.

Exercise 11.4. A simple but sometimes useful spatio–temporal, multiresolution model is built on the idea that a location s falls into the region comprised of a number of relatively homogeneous subregions. The process is determined by a deterministic regional effect plus random subregional effects. Within subregions any residual effects can be considered white noise.

(i) Build a spatio–temporal model based on that information, assuming the space–time covariance is separable.

(ii) Determine that covariance in terms of the parameters of your model.

Exercise 11.5. Returning to Example 11.2

(i) Compute the process covariance matrix for the process at time $t = 2$.

(ii) Find a linear transformation of the process column vector $\mathbf{Y}_2 = (Z_{21}, \ldots Z_{21})'$ that makes its coordinates independent.

(iii) Determine the power function for the test of the null hypothesis of no change between time $t = 1, 2$ and give an expression for its expected value under the Bayesian model in that example.

Exercise 11.6. Add to the process model in Equation 11.13 the measurement model $Y_{st} = Z_{st} + v_{st}$, $s = 1, \ldots, N_S$, $t = 1, \ldots, N_T$.

(i) Find the posterior predictive distributive distribution of $Z_{(n+1)s}$ for a given site s and determine what happens as n, the amount of data available increases.

(ii) What deficiencies do you see in the proposed process say where Z represents hourly temperature, t the hour of the day, and $t = 0$ is set at the time when the hourly temperature is expected to be at its maximum? What is the frequency ω in this case?

Exercise 11.7. Using WinBUGS
The models, data and initial values for an implementation of the model in Section 11.6 are included in the online resources. The data consists of daily measurements of four pollutants measured daily in London.

(i) Open `model4.odc` and load the data (`model4-data.odc`) and compile the model with two chains. Initial values can be found for two chains in `model4-inits1.odc` and `model4-inits2.odc`.

(ii) Run the model for a suitable number of iterations and calculate summary statistics for the posterior distributions of `m`, `sigma.v` and `sigma.w`.

(iii) In R plot the estimated values of `gamma` against the observed data for the four pollutants, `y`. What do you conclude? Note that you may have to deal with the different lengths of the two series, remember that `gamma` has no missing values in it.

(iv) Plot a suitable summary of the posterior values of `gamma` (including their uncertainty) against time. What do you conclude about the uncertainty in the values of `gamma` when the original data is missing?

Exercise 11.8. Using R–INLA

The data for this question is the same as for Exercise 11.7 and are included in the online resources.

(i) Perform the analysis in Exercise 11.7 using R–INLA for a single pollutant, i.e. using a univariate AR process, measured over space and time. and compare the results to what you obtained using WinBUGS. Hint: you may want to consider using a model similar to that shown in (Cameletti et al., 2011).

(ii) Perform the same analysis on your chosen pollutant using WinBUGS and compare the results with those you obtained using R–INLA.

Exercise 11.9. Research question: Geographical site specific features such as elevation are not covariates in a formal sense, but they are explanatory factors. How can such features be incorporated in EnviroStat?

Chapter 12

Roadblocks on the way to causality: exposure pathways, aggregation and other sources of bias

12.1 Overview

Often the aim of a study is to determine the *cause* of a disease so that we may devise treatment or prevention. There are two broad types of study in medical research: experimental and observational. In an experiment or clinical trial we make some intervention and observe the result. In an observational study we observe the existing situation and try to understand what is happening.

When examining the association between two variables, bias may be introduced by confounding. A confounder is a variable which is strongly associated with the response and effect of interest which can result in spurious associations being observed between the variable of interest and a health outcome. In a randomised controlled trial the randomisation of treatment allocation is designed to eliminate, as far as possible, the effects of confounding variables. The interpretation of observational studies is more difficult than a randomised trial as bias due to confounding may influence the measure of interest. If careful consideration is taken beforehand to identify and measure important confounders that may differ between exposure groups, these can be incorporated in the analysis and the groups can be adjusted to take account of differences in these baseline characteristics. Of course, if further factors are present and not included these will bias the results.

Causality cannot be inferred from observational studies. Only in very rare cases have controlled experiments been used to assess the impact of an environmental hazard on human subjects, e.g. randomly selected, healthy subjects have jogged in exposure chambers into which varying concentrations of ozone are pumped with the outcome being a decrement in the subject's lung function. For the most part controlled experiments of this type are not possible due to ethical and practical reasons. Therefore observational studies are used in which the contrasts between high and low levels are used as a proxy for the treatment and non-treatment groups that would be used in a controlled trial.

The effect of an environmental hazard can be estimated using an observational study and the risk of exposure to the hazard is often expressed using the relative risk. This shows the increase in the health outcomes that is associated with a one-unit increase in the level of an environmental hazard. This has been interpreted as predicting the degree to which lowering the level of the hazard by say one unit would reduce its health effect. For example in their 2007 health risk assessment report, the US EPA staff used the results of published epidemiological studies of the effects when computing the expected decrease in O_3 related non-accidental mortality that would result if the ozone standards were reduced from the current levels. The implication is that there is a causal link between ozone and mortality as defined by the 'counterfactual' approach: if levels of O_3 were those specified by the proposed standard then there would be a specified number of deaths. The argument would be compelling if it had been demonstrated that O_3 caused the deaths assumed in the calculation. However, this is not usually possible using results from observational studies. This chapter will review the roadblocks encountered in establishing that casualty.

12.2 Causality

Causality can be a difficult concept to define although much has been written about the topic stretching at least as far back as Hume's 1753 treatise (Hume, 2011) where he states:

> "We may define a cause to be an object followed by another, and where all the objects, similar to the first, are followed by objects similar to the second. Or, in other words, where, if the first object had not been, the second never had existed."

In fact, Hume has mixed up two definitions of causation here as noted in the Stanford Encyclopedia of Philosophy's entry entitled "Counterfactual theories of causation." It is the second one that has been developed in succeeding years and would support the interpretation of the concentration response function above. Elaborate theories of causality now exist and include stochastic versions with applications in the health sciences (Robins & Greenland, 1989; VanderWeele & Robins, 2012).

Suffice it to say that the counterfactual definition cannot give us a basis for arguing that high levels of O_3 cause ill health although we have a *prima facie* case based on the fact of the observational studies used in its assessment. Given that we cannot claim a causative association between an environmental hazard and increases in adverse health outcomes, must we settle for mere association? An answer is provided by the Bradford–Hill criteria which are a group of minimal conditions necessary to provide adequate evidence of a causal relationship, although it should be noted that none of the criteria are sufficient on their own (Hill, 1965).

Strength of association: The stronger the association between a risk factor and outcome, the more likely the relationship is to be causal.

Specificity: There must be a one-to-one relationship between cause and outcome.

Biological plausibility: There should be a plausible biological mechanism which explains why exposures might cause adverse health effects.

Biological gradient: Change in disease rates should follow from corresponding changes in exposure (dose–response) with greater exposure leading to increase in adverse health effects.

Temporal gradient: Exposure must precede outcome.

Consistency of results: The same effect should be seen among different populations, when using different study designs and at different times.

Coherence: Does the relationship agree with the current knowledge of the natural history/biology of the disease?

Experimental evidence: Is there evidence that removal of the exposure alters the frequency of the outcome?

Analogy: Similar findings in another context may support the claim of causality in this setting.

Example 12.1. *Transfer of causality*

In this example, we see how even if causality were present it may be difficult to attribute it to a particular exposure when there are variables which are highly collinear. This phenomenon is called the transfer of causality. This can be described by a hypothetical example taken from Zidek, Wong, Le, and Burnett (1996).

A health count, Y, is Poisson distributed with mean $\exp\{\alpha_0 + \alpha_1 z\}$ conditional on an exposure $Z = z$ with $\alpha_0 = 0$ and $\alpha_1 = 1$ which are not known to the investigator in this simulation study. A second explanatory variable W is seen by the investigator as a possible predictor of Y. The investigator measures both of these predictors with a classical measurement error model:

$$
\begin{aligned}
Z^* &= Z + V_1 \\
X^* &= X + V_2
\end{aligned}
$$

where V_1 and V_2 are independent of each other and of Z and X. The predictors $Z \sim N(0,1)$ and $X \sim N(0,1)$ have correlation ρ which induces the collinearity. The variances of the measurement errors U and V are σ_{V1}^2 and σ_{V2}^2, respectively. The investigator thinking both of these predictors are relevant fits a Poisson regression model with mean dependent on both Z^* and X^*, i.e $E(Y) = \exp\{\beta_0 + \beta_1 Z^* + \beta_2 W^*\}$. It turns out that the significance of the estimated coefficient $\hat{\beta}_2$ grows as the measurement error in Z and the collinearity between W and Z increases. When $\rho = 0.9$, the effect is dramatic and $\hat{\beta}_2$ dominates $\hat{\beta}_1$ when σ_{v1} reaches 0.5 which is 50% of the α_1. It is now W that appears to have the greatest effect, even though Z is the true cause of the outcome.

12.3   Ecological bias

Often health data are only available as aggregated daily counts, meaning that an eco-logical regression model is required. Such models answer fundamentally different epidemiological questions from individual level models and the results should not be stated in terms of a causal link between air pollution and health. For a review see Plummer and Clayton (1996). The use of ecological studies in this context is con-tentious, and has been discussed by Richardson, Stücker, and Hémon (1987); Green-land and Morgenstern (1989); Wakefield and Salway (2001). A causal relationship between air pollution and health can only be estimated from individual level data, but personal mortality or morbidity events may be unavailable for confidentiality reasons, while pollution exposures are expensive and impractical to obtain for more than a few individuals.

As such data are largely unavailable, a small number of researchers have at-tempted to estimate a link between air pollution and health by aggregating individ-ual exposure–response models to the population level, see for example Zeger et al. (2000); Sheppard and Damian (2000); Wackernagel (2003); Wakefield and Shaddick (2006). However such individual level models are typically nonlinear, as they are based on Bernoulli observations, meaning that the aggregation cannot be done ex-actly. The resulting error is known as ecological bias. The description focuses on the context of a spatial ecological study, but the problem is equally applicable in time series studies (Shaddick, Lee, & Wakefield, 2013).

12.3.1   Individual level model

Consider a spatial study where the study region A is split into N subareas A_l $(A_1, \ldots, A_{N_l})$, each of which contains n_l individuals. Then for a prespecified time interval, let $Y_{il}^{(1)}$ denote a Bernoulli disease indicator variable, which is one if in-dividual i in area l has the disease and zero if not. In addition consider p expo-sures $\mathbf{Z}_{il} = (Z_{il1}, \ldots, Z_{ilp})$ measured for these individuals and q area level covari-ates $\mathbf{X}_l = (X_{l1}, \ldots, X_{lq})$, which are constant for all individuals within area l. For this description, we assume that the true exposures are available and later consider the case using measurements of the true exposure, $Y_{il}^{(2)}$, which may contain error. If $(Y_{il}^{(1)}, \mathbf{Z}_{il}, \mathbf{X}_l)$ are all available then an individual level model is given by

$$Y_{il}^{(1)} \quad \sim \quad \text{Bernoulli}(p_{il}) \tag{12.1}$$

$$\ln\left(\frac{p_{il}}{1 - p_{il}}\right) \quad = \quad \beta_0 + \mathbf{Z}_{il}'\beta_z + \mathbf{X}_l'\beta_x$$

where $\beta = (\beta_0, \beta_z, \beta)$ are the regression parameters to be estimated. If the disease in question is rare then p_{il} will be small, so $\log(p_{il}/(1 - p_{il})) \approx \log(p_{il})$ and the logit link can be replaced by a log link.

12.3.2 Aggregation if individual exposures are known

In many cases the disease indicators $Y_{il}^{(1)}$ are not known, and only the aggregate disease count for each area, $Y_l = \sum_{i=1}^{n_l} Y_{il}^{(1)}$, is available. The aim is now to aggregate the individual level model shown in Equation 12.1 up to a model for Y_l which is known. If we initially assume that the exposures for all individuals are constant within each area, that is $\mathbf{Z}_{il} = \mathbf{Z}_l$, then $p_{il} = p_l$ meaning that the risk of disease is constant for all individuals across each area. Conditional on the exposures and covariates in this simplified case the health model for $Y_l^{(1)}$ is given by:

$$
Y_l^{(1)} \sim \text{Binomial}(n_l, p_l)
$$
$$
\ln\left(\frac{p_l}{1-p_l}\right) = \beta_0 + \mathbf{Z}_l'\beta_z + \mathbf{X}_l'\beta_x
$$

and the individual level parameters in the model given in Equation 12.1 are the same as those in the ecological model above. In contrast, if individual exposures are not constant within an area then the aggregate counts $y_k^{(1)}$ are not Binomial as the individual Bernoulli disease probabilities are not equal. In this case the distribution of $Y_l^{(1)}$ inherited from $Y_{il}^{(1)}$ is non-standard, and instead we focus on the mean and variance structure of $Y_l^{(1)}$. If we can assume that the individual level exposures $\mathbf{Z}_{il}$ are known and that the disease indicators $Y_{il}^{(1)}$ are independent, then

$$
\mathbb{E}[Y_l^{(1)}|\mathbf{Z}_{il}] = \mathbb{E}[\sum_{i=1}^{n_l} Y_{il}^{(1)}|\mathbf{Z}_{il}]
$$
$$
= \sum_{i=1}^{n_l} \mathbb{E}[Y_{il}^{(1)}|\mathbf{Z}_{il}]
$$
$$
= \sum_{i=1}^{n_l} p_{il}
$$

$$
\text{Var}[Y_l^{(1)}|\mathbf{Z}_{il}] = \text{Var}[\sum_{i=1}^{n_l} Y_{il}^{(1)}|\mathbf{Z}_{il}]
$$
$$
= \sum_{i=1}^{n_l} \text{Var}[Y_{il}^{(1)}|\mathbf{Z}_{il}] \qquad \text{(assuming independence)}
$$
$$
= \sum_{i=1}^{n_l} p_{il} - p_{il}^2
$$

where again

$$
g(p_{il}) = \beta_0 + \mathbf{Z}_{il}'\beta_z + \mathbf{X}_l'\beta_x
$$

Here g is typically a log or logit link function. In this case one approach would be to specify a Poisson distribution for $Y_l^{(1)}$ as they are counts with a log link, giving the model (again conditional on $\mathbf{Z}$):

$$Y_l^{(1)} \sim \text{Poisson}(\mu_l)$$

$$\mu_l = \exp(\beta_0 + \mathbf{X}_l'\beta_x) \sum_{i=1}^{n_l} \exp(\mathbf{Z}_{il}'\beta_z)$$

In the above model the individual level parameters β_z can still be estimated. However in view of the variance function, this model should be altered to allow for under dispersion as $Var(Y_l^{(1)}|\mathbf{Z}_{il}) = \mathbb{E}[Y_l^{(1)}|\mathbf{Z}_{il}] - \sum_{i=1}^{n_l} p_{il}^2$.

12.3.3 Aggregation if the individual exposures are not known

In ecological studies the individual exposures $\mathbf{Z}_{il}$ are unknown, and only a summary measure of exposure for each area $\tilde{\mathbf{Z}}_l$ is available. Such an exposure is likely to be the population mean for the area, or be a surrogate obtained from a centrally sited monitor. In this situation the mean and variances of $Y_l^{(1)}|\tilde{\mathbf{Z}}_l$ can be derived from the equations $\mathbb{E}[U] = \mathbb{E}[\mathbb{E}[U|V]]$ and $Var\ U = \mathbb{E}[\ Var\ U|V]] + Var\ [\mathbb{E}[U|V]]$ as:

$$
\begin{aligned}
\mathbb{E}[Y_l^{(1)}|\tilde{\mathbf{Z}}_l] &= \mathbb{E}[\sum_{i=1}^{n_l} Y_{il}^{(1)}|\tilde{\mathbf{Z}}_l] \\
&= \sum_{i=1}^{n_l} \mathbb{E}[Y_{il}^{(1)}|\tilde{\mathbf{Z}}_l] \\
&= \sum_{i=1}^{n_l} \mathbb{E}[\mathbb{E}[Y_{il}^{(1)}|\mathbf{Z}_{il}]|\tilde{\mathbf{Z}}_l] \\
&= \sum_{i=1}^{n_l} \mathbb{E}[p_{il}|\tilde{\mathbf{Z}}_l] \\
&= \sum_{i=1}^{n_l} \mathbb{E}[p_{Z_k}|\tilde{\mathbf{Z}}_l] \\
&= n_l \mathbb{E}[p_{Z_k}|\tilde{\mathbf{Z}}_l]
\end{aligned}
$$

where p_{Z_k} is the disease risk in area l dependent upon the exposure distribution in area l and the observed summary measure $\tilde{\mathbf{Z}}_l$. This exposure distribution is represented by a random variable $\mathbf{Z}_l \sim f(.|\tilde{\mathbf{Z}}_l)$, and each unknown individual exposure $\mathbf{Z}_{il}$ is assumed to be a realisation from this distribution. Therefore the expectation in the above equation is with respect to the distribution $\mathbf{Z}_l|\tilde{\mathbf{Z}}_l$, and assuming a log rather than logit link as the disease risks will be small, the expectation function is given by

$$
\begin{aligned}
\mathbb{E}[Y_l^{(1)}|\tilde{\mathbf{Z}}_l] &= n_l\mathbb{E}[p_{Z_k}|\tilde{\mathbf{Z}}_l] \\
&= n_l\mathbb{E}[\exp(\beta_0 + \mathbf{x}_l'\beta_x + \mathbf{Z}_l'\beta_z)|\tilde{\mathbf{Z}}_l] \\
&= \exp(\beta_0 + \mathbf{x}_l'\beta_x)\mathbb{E}[\exp(\mathbf{Z}_l'\beta_z)|\tilde{\mathbf{Z}}_l]
\end{aligned}
\tag{12.2}
$$

where the expectation is with respect to $\mathbf{Z}_l|\tilde{\mathbf{Z}}_l$. In the last line n_l is a constant and has been absorbed into the intercept term β_0. The associated variance is given by

$$
\begin{aligned}
Var(Y_l^{(1)}|\tilde{\mathbf{Z}}_l) &= Var(\sum_{i=1}^{n_l} Y_{il}^{(1)}|\tilde{\mathbf{Z}}_l) \\
&= n_l\mathbb{E}[p_{Z_k}|\tilde{\mathbf{Z}}_l] - n_l\mathbb{E}[p_{Z_k}|\tilde{\mathbf{Z}}_l]^2
\end{aligned}
\tag{12.3}
$$

Again p_{Z_k} is the disease risk in area l dependent upon the exposure distribution $\mathbf{Z}_l|\tilde{\mathbf{Z}}_l$. Now from the mean function (12.2), an appropriate mean model for disease counts $Y_l^{(1)}$ is given by

$$
\begin{aligned}
Y_l^{(1)} &\sim \text{Poisson}(\mu_l) \\
\mu_l &= \exp(\beta_0 + \mathbf{X}_l'\beta_x)\mathbb{E}[\exp(\mathbf{Z}_l'\beta_z)|\tilde{\mathbf{Z}}_l]
\end{aligned}
\tag{12.4}
$$

where again the expectation is with respect to $\mathbf{Z}_l|\tilde{\mathbf{Z}}_l$. However this model assumes the mean and variance are equal, which from the above derivations is not true as the theoretical variance will be less than the mean. However the naive ecological model used by the majority of studies is worse still, having the general form

$$
\begin{aligned}
Y_l^{(1)} &\sim \text{Poisson}(\mu_l) \\
\mu_l &= \exp(\beta_0 + \mathbf{X}_l'\beta_x)\exp(\tilde{\mathbf{Z}}_l'\beta_z^*)
\end{aligned}
\tag{12.5}
$$

where the difference is because the risk function is exponential, meaning that in general

$$
\exp(\mathbb{E}[\mathbf{Z}_l|\tilde{\mathbf{Z}}_l]'\beta_z^*) \neq \mathbb{E}[\exp(\mathbf{Z}_l'\beta_z)|\tilde{\mathbf{Z}}_l]
$$

where $\mathbb{E}[\mathbf{Z}_l|\tilde{\mathbf{Z}}_l] = \tilde{\mathbf{Z}}_l$. Therefore adopting the naive ecological model means that you are estimating the ecological rather than individual level relationship between exposure and health, so that in general $\beta_z \neq \beta_z^*$. Therefore the naive ecological model (12.5) uses the wrong mean function and incorrectly assumes that the variance is equal to the mean. This is known as pure specification bias. Note that pure specification bias is not a problem if either exposure is constant within an area, or the disease model is linear, as in both cases models (12.4) and (12.5) are equivalent.

12.4 Acknowledging ecological bias

There have been two major approaches for incorporating ecological bias, which were first suggested by Prentice and Sheppard (1995) and Richardson et al. (1987) respectively. These are now described with the focus on estimating the mean function in Equation 12.4.

12.4.1 Aggregate approach

The aggregate approach assumes that personal exposures $\mathbf{Z}_{il}$ are available, but not the individual health indicators $Y_{il}^{(1)}$. A simple, but unrealistic, situation is that exposure for all individuals $\mathbf{Z}_{il}^{n_l}$ are available, which is exactly the case discussed in Section 12.3.2. In this setting the mean and variance are given by

$$\mathbb{E}[Y_l^{(1)}|\mathbf{Z}_{il}^{n_l}] = \mu_l = \exp(\beta_0 + \mathbf{x}_l'\beta_x)\sum_{i=1}^{n_l}\exp(\mathbf{Z}_{il}'\beta_z)$$

$$Var(Y_l|\mathbf{Z}_{il}^{n_l}) = \exp(\beta_0 + \mathbf{X}_l'\beta_x)\sum_{i=1}^{n_l}\exp(\mathbf{Z}_{il}'\beta_z) \qquad (12.6)$$

$$- \exp(\beta_0 + \mathbf{X}_l'\beta_x)^2\sum_{i=1}^{n_l}\exp(2\mathbf{Z}_{il}'\beta_z)$$

Prentice and Sheppard (1995) do not make any parametric assumptions about the distribution for $Y_l^{(1)}$ (such as Poisson) and instead adopt a quasi-likelihood estimation approach based on the mean and variance above.

12.4.2 Parametric approach

The parametric approach was introduced by Richardson et al. (1987), and assumes that the individual exposures come from a parametric distribution. That is we assume that $\mathbf{Z}_{il} \sim f(.|\phi_l)$, where the parameters ϕ_l will change with area l. The desired individual level model given by Equation 12.1 is the conditional model $Y_{il}^{(1)}|\mathbf{Z}_{il}, \beta \sim \text{Bernoulli}(p_{il})$, with link function $g(p_{il}) = \beta_0 + \mathbf{Z}_{il}'\beta_z + \mathbf{X}_l'\beta_x$. However in this situation we only know that $\mathbf{Z}_{il}$ comes from a distribution $f(.|\phi_l)$, so the desired model is for $Y_{il}^{(1)}|\beta, \phi_l$, where the exposure distribution f has been averaged over. In this scenario $Y_{il}^{(1)}|\beta, \phi_l \sim \text{Bernoulli}(p_{il}^*)$ as it is still a Bernoulli indicator, and assuming a log link p_{il}^* is given by

$$p_{il}^* = \mathbb{E}[Y_{il}^{(1)}|\beta, \phi_l]$$
$$= \mathbb{E}[\mathbb{E}[Y_{il}^{(1)}|\mathbf{Z}_{il}, \beta, \phi_l]|\beta, \phi_l]$$
$$= \mathbb{E}[p_{il}|\beta, \phi_l]$$
$$= \mathbb{E}[\exp(\beta_0 + \mathbf{Z}_{il}'\beta_z + \mathbf{x}_l'\beta_x)|\beta, \phi_l] \qquad (12.7)$$
$$= \exp(\beta_0 + \mathbf{X}_l'\beta_x)\mathbb{E}[\exp(\mathbf{Z}_{il}'\beta_z)|\beta, \phi_l] \qquad (12.8)$$

where $\mathbb{E}[\exp(\mathbf{Z}'_{il}\beta_z)|\beta,\phi_l]$ is the moment generating function of the exposure distribution $f(.|\phi_l)$. We now consider two cases separately, firstly we assume that the exposures $\mathbf{Z}_{il}$ are independent across individuals i, and we then look at the more realistic situation that they are correlated.

If the exposures $\mathbf{Z}_{il}$ are independent across individuals i then from the equation above $p^*_{il} = p^*_l$ because it is the same for each individual in area l (the unknown exposures have been averaged over). Therefore as all individuals in an area have the same risk function and the Bernoulli outcomes are independent, the resulting aggregate model for Y_l is given by

$$\begin{aligned} Y^{(1)}_l &\sim \text{Binomial}(n_l, p^*_l) \\ p^*_l &= \exp(\beta_0 + \mathbf{X}'_l\beta_x)\mathbb{E}[\exp(\mathbf{Z}'_{il}\beta_z)|\beta,\phi_l] \end{aligned}$$

which is often approximated by a Poisson distribution as the events are rare giving

$$\begin{aligned} Y^{(1)}_l &\sim \text{Poisson}(\mu_l) \\ \mu_l &= \exp(\beta_0 + \mathbf{X}'_l\beta_x)\mathbb{E}[\exp(\mathbf{Z}'_{il}\beta_z)|\beta,\phi_l] \end{aligned} \qquad (12.9)$$

where $\mu_l = n_l p^*_l$ and n_l has been absorbed into the constant β_0. We now discuss a number of forms for $f(.|\phi_l)$.

If we assume that $\mathbf{Z}_{il} \sim N(\phi_{1k}, \Phi_{2l})$ then the moment generating function is given by $\exp(\phi'_{1l}\beta_z + \beta'_z\Phi_{2l}\beta_z/2)$. Therefore the mean function for the Poisson model (12.9) becomes

$$\mu_l = \exp(\beta_0 + \mathbf{X}'_l\beta_x)\exp(\phi'_{1l}\beta_z + \beta'_z\Phi_{2l}\beta_z/2)$$

with the same results holding for the Binomial model. If (ϕ_{1k}, Φ_{2l}) are estimated from a sample, then this model differs from the naive ecological model in that it incorporates the variance matrix in the mean function as well as the mean. Note that if either the variance is zero or it does not depend on area l (so $\beta'_z\Phi_{2l}\beta_z/2$ is merged into the intercept term), then this model is identical to the naive ecological model and pure specification bias will not be a problem. This mean function holds true for all values of β_z.

12.5 Exposure pathways

12.5.1 Concentration and exposure response functions

Concentration response functions (CRFs) are estimated primarily through epidemiological studies, by relating changes in ambient concentrations of pollution to a specified health outcome such as mortality (see Daniels et al. (2004) for example). In contrast exposure response functions (ERFs) have been estimated through exposure chamber studies, where the physiological reactions of healthy subjects are assessed at safe levels of the pollutant (see EPA (2006) for example). However ERFs cannot be ethically established in this way for the most susceptible populations such as the very old and very young who are thought to be most adversely effected by pollution exposure. This paper presents a method for estimating the ERF based on ambient concentration measures.

We specifically consider the case of particulate air pollution, which has attained great importance in both the health and regulatory contexts. For example they are listed in the USA as one of the so-called criteria pollutants that must be periodically reviewed. Such a review by the US Environmental Protection Agency led to a 2006 revision of the US air quality standards (EPA, 2004). These require that in US urban areas daily ambient concentrations of PM_{10} do not exceed 150 μgm^{-3} more than once a year on average over three years. Concern for human health is a driving force behind these standards, as the US Clean Air Act of 1970 states they must be set and periodically reviewed to protect human health without consideration of cost while allowing for a margin of error.

Example 12.2. *The short-term effects of air pollution on health*

The majority of studies relating air pollution with detrimental effects on health have focused on short-term relationships, using daily values of aggregate level (ecological) data from a fixed geographical region, such as a city. In such studies, the association between ambient pollution concentrations and mortality is of interest for regulatory purposes primarily because it is only ambient pollution concentrations that are routinely measured. However personal exposures are based on indoor as well as outdoor sources, and are likely to be different from ambient concentrations (see for example Dockery & Spengler, 1981 and Lioy, Waldman, Buckley, Butler, & Pietarinen, 1990) because the population spend a large proportion of their time indoors. Therefore to obtain more conclusive evidence of the human health impact of air pollution via an ERF, exposures actually experienced by individuals as well as any subsequent health events are required. Ideally, these would be obtained by individual level studies conducted under strict conditions, such as in randomised controlled trials, but issues of cost and adequate confounder control make them relatively rare (a few examples are given by Neas, Schwartz, & Dockery, 1999, Yu, Sheppard, Lumley, Koenig, & Shapiro, 2000 and Hoek, Brunekreef, Goldbohm, Fischer, & van den Brandt, 2002). An alternative approach is to obtain only individual level pollution exposures, which can be related to routinely available (aggregated) health and confounder data. However such exposures are still prohibitively expensive to obtain for a large sample of the population, and consequently only a small amount of personal exposure data has been collected (see for example Lioy et al., 1990 and Ozkaynak et al., 1996).

Form of an exposure response function (ERF)

Often the association between an environmental hazard and health outcomes is modelled using a log–linear or logistic model as seen in Chapter 2 with, in the case of Poisson regression, the log rate being modelled as a function of the levels of the hazard in question and potential covariates as seen in Equation 2.11 in Chapter 2,

$$\log \mu_l = \beta_0 + g(\beta_z Z_l) + \beta_x X_l \qquad (12.10)$$

where $g(\beta_x Z_l)$ represents the effect of exposure, Z, and β_x is the effect of a covariate, X_l.

Here we specifically consider the nature of the relationship between the exposure and the (log) rate which is encapsulated in the function $g(\beta_z Z)$. This is the exposure response function which is commonly represented by the form given in (2.11) with the simplification $g(x) = x$. However this simplification that $g(x) = x$ may not be appropriate environmental studies, because there must eventually be an upper bound on the effect that the hazard can have on health. An alternative approach is to consider a general function g that satisfies the desirable requirements of: (i) boundedness; (ii) increasing monotonicity; (iii) smoothness (thrice differentiability); and (iv) $g(0) = 0$. Note that these properties are not commonly enforced on CRFs estimated for ambient pollution concentrations using generalised additive models as seen in Chapter 10 (see for example Daniels et al., 2004).

12.6 Personal exposure models

Exposure to an environmental hazard will depend on the temporal trajectories of the population's members which will take individual members of that population through a sequence of micro-environments, such as a car, house or street. For background on the general micro-environmental modelling approach see Berhane et al. (2004).

Information about the current state of the environment may be obtained from routine monitoring, for example in the case of air pollution, or through measurements taken for a specialised purpose. An individual's actual exposure is a complex interaction of behavior and the environment. Exposure to the environmental hazard affects the individual's risk of certain health outcomes, which may also be affected by other factors such as age and smoking behavior. Finally, some individuals will actually contract the health outcome of interest, placing a care and financial burden on society. Thus, for example policy regulation on emission sources can indirectly affect mortality and morbidity and the associated financial costs. However this pathway is complex and subject to uncertainties at every stage.

There will be uncertainty in determining which are the important pathways for the environmental hazard and which one(s) should be used in the assessment. For example, in the case of exposure to mercury, it can be found in air, sea and land. Sources include emissions from power plants since it is found in coal. Once airborne, it can be transported over long distances and may eventually end on land through wet and dry deposition. From there it can be leached into lakes or taken up into the food chain. Levels of ambient mercury are thus determined by a variety of environmental factors of which some like wind direction are random. These ambient levels help determine exposure through such things as food consumption, fish or shellfish (that contain a very toxic form called methylmercury) being a primary source. Alternatively, exposure can be from breathing air containing mercury vapor, particularly in warm or poorly ventilated rooms, that comes from evaporation of liquid mercury which is found in many places like school laboratories. In any case, exposure will depend on the micro-environments that are visited, time spent in them and the internal sources within. Clearly individual factors such as age determine the time–activity patterns that affect the levels of exposure to the hazard, for example, breast milk is a source of mercury in many micro-environments for infants. Careful consideration of the population being studied is therefore very important, especially when the intention is to apply the results more generally.

12.6.1   Micro-environments

The identification of appropriate micro-environments (MEs) and the associated estimation of levels of the environmental hazard are an essential part of linking external measurements with actual individual exposures. Figure 12.1 shows an example of a $2 \times 3 \times 24$ concentration array that highlights the path followed by an imaginary individual through time. Their activities take the individual from home, to outdoor, to indoor (at work), to outdoor, to indoor and back to home at times 5, 6, 14, 16 and 17. The sequence of personal exposures is obtained by reading the array at the highlighted cells from left to right. The choice of these micro-environments by an individual may be a reflection of the ambient conditions.

There are two types of micro-environment: closed and open. A closed micro-environment is one for which the derivation of the local concentration involves a mass balance equation. Such a micro-environment may have local sources that produce amounts of the environmental hazard, and its volume as well as other quantities are in the mass balance equation that is used to derive the resulting local concentration. In this case the concentration to which its occupants are exposed will be increased. On the other hand, as air from within is exchanged with air from outside, there is a tendency for the micro-environment local concentration to adjust itself to the ambient one. In contrast, an open micro-environment is one for which there is no source and for which the local concentration is a simple linear transformation of the ambient one.

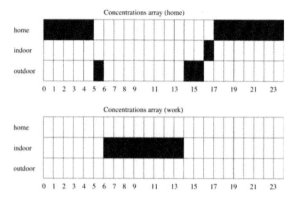

Figure 12.1: The path followed by an randomly selected individual through 24 hours of time. The $2 \times 3 \times 24$ local array categorizes their exposure to pollution at home or work and whether they are exposed to sources either in the home, (other) indoor or outdoor.

To determine local concentrations of the environmental hazard within the different micro-environments, both open and closed, requires some modelling assumptions, which themselves introduce uncertainty. Firstly, we must identify an appropriate set of micro-environments for the environmental hazard under consideration, which itself can introduce uncertainty if this is misspecified or important micro-environments are missing. In a closed micro-environment, the mass balance equation is a differential equation and may involve approximations to solve. It will also require various input data, for example air exchange rates, or rates at which local sources produce the pollutants; these will need to be estimated and so will introduce uncertainty. For an open micro-environment, the relationship between local and ambient levels and the estimation of the parameters controlling that relationship will both be subject to uncertainty.

The state of the environment only partly determines the exposure to the risk factor that is experienced by individuals. Estimating the exposure involves characterising the link between levels of the environmental hazard and actual exposure. There are two elements to this; the first is how being in a specific environment translates into an exposure which could involve complex modelling and the second concerns how individuals move around through different micro-environments and thus determines cumulative exposure. This can be influenced to some extent by policy and other decision makers, albeit indirectly; for example reporting high pollution days on local weather forecasts may encourage susceptible individuals to stay at home where possible.

Example 12.3. *Estimating personal exposures of particulate matter*

There are a number of implementations of the micro-environment frame-work, with early models being largely deterministic in nature (Ott, Thomas, Mage, & Wallace, 1988; MacIntosh, Xue, Ozkaynak, Spengler, & Ryan, 1994). More recently, models have incorporated levels of uncertainty into the estimation procedure (Burke, Zufall, & Ozkaynak, 2001; Law, Zelenka, Hu-ber, & McCurdy, 1997; Zidek et al., 2005; Zidek, Shaddick, Meloche, Chat-field, & White, 2007).

In this example, we describe the use of a probabilistic exposure model known as pCnem (Zidek et al., 2005, 2007). Zidek et al. (2005) predicts expo-sure distributions for particulate matter in London in 1997 for the subpopu-lation of working females. Daily data from eight PM_{10} monitoring sites were uploaded to the pCNEM site together with maximum daily temperatures for London for 1997. Each site represents a pCNEM's exposure district, that is a geographical area surrounding it whose ambient PM_{10} level can realistically be imputed to be that of its central monitoring site. The spatial distribution of PM_{10} in London for this period has been shown to be relatively homogeneous (Shaddick & Wakefield, 2002), meaning that the boundaries of the exposure regions are less critical than might otherwise be the case.

The subpopulation considered here is that of working women who smoke, live in Brent in semi-detached dwellings that use gas as the cooking fuel and work in the Bloomsbury exposure district. Two sub-cases are considered, one covering the spring and the other the summer of 1997. For each of the two cases, the output from a single pCNEM run consists of a sequence of 'expo-sure events' for a randomly selected member of the subpopulation. That indi-vidual is composite; a different time–activity is selected for each succeeding day. This composite individual better represents her subpopulation's activity patterns than any single member would do. Each event in the sequence takes place in a random micro-environment and consists of exposure to a randomly varying concentration of PM_{10} for a random number of minutes.

The output from the pCNEM model, consisting of 30 replicates generated for each of the two different cases, can be analysed in a variety of ways. The results for daily averages appear in Figure 12.2. The working females experience high exposures for a period of about 4 days near the beginning of the year, during which a number of replicates have high levels of average daily exposure, approaching or exceeding 50 μgm^{-3}. This is still well below the standard for daily average level for PM_{10} of 150 μgm^{-3} in both the UK and the US. During August of 1997, similar peaks are seen in daily average exposures and again, there are notable extremes among the replicate values, at about or exceeding 50 μgm^{-3}.

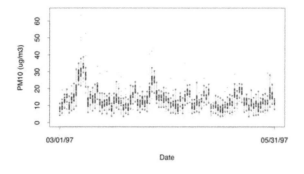

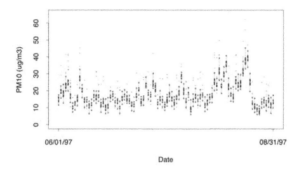

Figure 12.2: Boxplots depict the estimated predictive distributions of particulate matter (PM_{10}) exposure in 1997 for a random member of a selected subpopulation of
Londoners. Here distributions are for average daily exposures. Both panels are for
women who live in semi-detached Brent dwellings, work in Bloomsbury, smoke, and
cook with gas. The top and bottom panels refer to spring and summer respectively.

Finally, Figure 12.3 shows the differential impact of a hypothetical 20%
deflation termed 'rollback' of actual PM_{10} levels in the spring of 1997. The
capacity of pCNEM to enable such 'scenario' analyses proves to be one of
the program's most important features, giving regulators a way to check the
impact of proposed changes on subpopulation groups. The deflated scenario
values are set by the user and computed as the linear reduction, *baseline* +
$p \times (x - baseline)$ for every hourly datum, x, for both Brent, the residential
exposure district and Bloomsbury where they worked. In our example, the
reduction factor and baseline were chosen somewhat arbitrarily to be p=0.8
and baseline = 15 μgm^{-3}. The result of the rollback in both summer and
spring is similar, showing a reduction of a little less than 1 μgm^{-3}, although
from differing baseline levels.

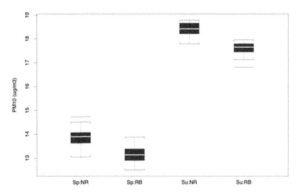

Figure 12.3: Boxplots compare the estimated predictive distributions of particulate matter (PM_{10}) exposure in 1997 for a random member of a selected subpopulation of Londoners. Here distributions are for average daily exposures on spring and summer days for working females (labelled 'Sp' and 'Su' respectively) before ('NR') and after ('RB') a 20% rollback in (hourly) ambient levels. The women are smokers, use gas as a cooking fuel, live in a semi-detached dwelling in Brent and work in Bloomsbury.

12.7 Summary

This chapter contains a discussion of the differences between causality and association. It also covers specific issues that may be encountered in this area when investigating the effects of environmental hazards on health. The reader will have gained an understanding of the following topics:

- Issues with causality in observational studies.
- The Bradford–Hill criteria which are a group of minimal conditions necessary to provide adequate evidence of a causal relationship.
- Ecological bias which may occur when inferences about the nature of individuals are made using aggregated data.
- The role of exposure variability in determining the extent of ecological bias.
- Approaches to acknowledging ecological bias in ecological studies.
- Concentration and exposure response functions.
- Models for estimating personal exposures including micro-environments.

Exercises

Exercise 12.1. For the model seen in 12.1 we are going to investigate the effect of transfer of causality with different levels of measurement error and collinearity.

(i) Write R code to perform the example shown in the text with $\alpha_0 = 0, \alpha_1 = 1$ and $Z \sim N(0,1), Z \sim N(0,1)$. Verify that with $\rho = 0.9$ the 'transfer' occurs when $\sigma_{v1} = 0.5$.

(ii) Investigate how this threshold changes with different levels of collinearity, i.e. different levels of ρ. Produce a graph of threshold against ρ.

(iii) Repeat the investigation performed in part (ii) for lower values of ρ. What do you conclude?

(iv) Investigate the effect that measurement error variance has in the 'transfer of causality'. What do you conclude?

Exercise 12.2. Using the model shown in Equation 12.9 if we assume that $Z_{il} \sim N(\phi_{1l}, \Phi_{2l})$ mean function for the Poisson is

$$\mu_l = \exp(\beta_0 + \mathbf{X}_l^T \beta_x) \exp(\phi_{1k}^T \beta_z + \beta_z^T \Phi_{2l} . \beta_z / 2).$$

Exercise 12.3. Again using the model shown in Equation 12.9, find the mean function if we assume the exposures are univariate and have distribution, $z_{it} \sim$ Gamma(ϕ_{1l}, Φ_{2l}), where (ϕ_{1l}, Φ_{2l}) represent the mean and variance of the exposure distribution.

Exercise 12.4. Return to the model shown in Equation 12.9.

(i) Why might it be difficult to use a log–normal distribution for exposures in the same way?

(ii) Show how a Taylor series expansion for $\mathbb{E}[\exp(\mathbf{Z}_{il}'\beta_z)|\beta, \phi_l]$ could be used in order to find the mean function where exposures are log–normally distributed.

Exercise 12.5. The exposures for individuals i in an area l may be correlated, for example if the exposure is spatially correlated and the exact locations of each individual are known. In this case $(Y_{il}^{(1)}, y_{jl}^{(1)})$ are still Bernoulli variables but are not independent, as the correlation in $(\mathbf{Z}_{il}, \mathbf{Z}_{jk})$ induces correlation into the health data. Therefore the sum $Y_l^{(1)} = \sum_{i=1}^{n_l} Y_{il}^{(1)}$ is not a Binomial random variable.

(i) Find the mean and variance of $Y_l^{(1)}$.

(ii) Find the covariance of $Y_l^{(1)}$.

(iii) Under what circumstances will the covariance you found in part (iii) be zero?

Exercise 12.6. The output from the pCNEM model for PM$_{10}$ exposures in working women for the London example are shown graphically in Figure 12.2. The output from pCNEM for these women together with a sample of seniors are included in the online resources. Using these samples,

(i) Perform similar aggregation to that seen in Example 12.3, i.e. produce summaries of the distribution of exposures by day.

(ii) Using ggplot2, produce a plot which enables a clear comparison between the exposures experienced by the working women and the seniors.

(ii) Are there any seasonal differences between the exposures experienced by the two groups?

Exercise 12.7. How would you choose a suitable time series model to the output used in Exercise 12.6?

(i) Choose a suitable ARMA model for the daily average values over all individuals in the output from pCNEM.

(ii) Choose a suitable ARMA model for the daily mean values for a single individual in the output from pCNEM.

(iii) Repeat part (ii) for a selection of individuals. What do you conclude about the suitability of fitting a single time series model to the mean exposures over all individuals?

Chapter 13

Better exposure measurements through better design

13.1 Overview

The famous London fog of 1952 left no doubt about the human health risks associated with air pollution and in particular airborne particulates. An estimated 4000 excess deaths were attributed to the fog (Ministry of Health, 1954), which was due to a considerable extent to coal combustion. The result was the general recognition that levels of air pollution should be regulated. This led to the Clean Air Acts of the United Kingdom (1956) and the United States (1970). These Acts set air quality standards whose enforcement required networks of monitors. By the early 1970s the UK had over 1200 monitoring sites measuring black smoke (BS) and sulphur dioxide (SO$_2$).

Mitigation measures, which included restrictions on the burning of coal, were successful and the levels of these pollutants declined over time. In turn that led to the perception that the number of monitoring sites could be reduced. The process by which the sites were selected for elimination is undocumented but empirical analysis of the results over time show that they tended be the ones at which the levels of BS were lowest (Shaddick & Zidek, 2014). The result was that the annual average level of BS in the UK was substantially overestimated (Zidek et al., 2014).

The analysis cited above for BS in the UK has not been repeated for other environmental monitoring networks but it is plausible that many of them were also adaptively changed over time with sites retained at locations with higher levels of the environmental hazard field (Guttorp & Sampson, 2010).

These considerations highlight the importance of good data in statistical analysis, something that is commonly ignored by statistical analysts who use data without due regard to their quality. A primary determinant of that quality in environmental epidemiology is the representativeness of the monitoring site data as the example described above makes clear. This chapter introduces the topic of designing, or redesigning, an environmental monitoring network. A recent handbook article provides a comprehensive overview of the topic (Zidek & Zimmerman, 2010).

Within this chapter we explore issues arising and methods needed when designing environmental monitoring networks with particular focus on ensuring that they can provide the information that is required to assess the health impacts of an environmental hazard.

13.2   Design objectives?

Introductory courses on design of experiments emphasise the need to clearly articulate the objectives of the experiment as the first step in design. An example where this was done through a preliminary workshop convened by the contractor is seen in Example 11.1. There the goal of the USA's National Oceanic and Atmospheric Agency (NOAA) was the detection of change in the concentration of trace metals in the seabed before and after the startup of exploratory drilling for oil on the north slopes of Alaska (Schumacher & Zidek, 1993). In other words, it was to test the hypothesis of no change. However not all, possibly even very few, of the existing networks were designed in accordance with the ideal of meeting a specific objective. For those that were, the objective may have been so vague that it could not lead to a scientific approach to choosing specific locations for sampling sites. This section will give examples of important environmental monitoring networks and describe how they were established.

We begin by listing the myriad of purposes for which environmental monitoring networks have been established:

- Monitoring a process or medium such as drinking water to ensure quality or safety.
- Determining the environmental impact of an event, such as a policy-induced intervention or the closure of an emissions source.
- Detecting non-compliance with regulatory standards.
- Enabling health risk assessments to be made and to provide accurate estimates of relative risk.
- To determine how well sensitive sub-populations are protected including all life, not only human.
- Issuing warnings of impending disaster.
- Measuring process responses at critical points, for example near a new smelter using arsenic or near an emitter of lead.

In other cases, the goals have been more technical in nature, even when the ultimate purpose concerned human health and welfare:

- Monitoring an easy-to-measure surrogate for a process or substance of real concern.
- Monitoring the extremes of a process.
- Enabling predictions of unmeasured responses.
- Enabling forecasts of future responses.

- Providing process parameter estimates for physical model parameters or stochastic model parameters, e.g. covariance parameters.
- Assessing temporal trends, for example assessing climate change.

As the examples that follow will show, a network's purpose may also change over time and its objectives may conflict with one another. For example, non-compliance detection suggests siting the monitors at the places where violations are seen as most likely to occur. However, an environmental epidemiologist would want to divide the sites equally between high- and low-risk areas in order to maximise contrasts and hence maximise the power of their health effects analyses. Even objectives that seem well-defined, e.g. monitoring to detect extreme values of a process over a spatial domain, may lead to a multiplicity of objectives when it comes to implementation (Chang, Fu, Le, & Zidek, 2007).

As noted by Zidek and Zimmerman (2010), often many different variables of varying importance are to be measured at each monitoring site. To minimise cost, the network designer may elect to measure different variables at different sites. Further savings may accrue from making the measurements less frequently, forcing the designer to consider the inter-measurement times. In combination, these many choices lead to a bewildering set of objective functions to optimise simultaneously. That has led to the idea of designs based on multi-attribute theory, ones that optimise an objective function that embraces all the purposes (Zhu & Stein, 2006; Müller & Zimmerman, 1999).

However, such an approach will not be satisfactory for long-term monitoring programs when the network's future uses cannot be foreseen, as in the example described in Zidek, Sun, and Le (2000). Moreover in some situations the 'client' may not even be able to precisely specify the network's purposes (Ainslie, Reuten, Steyn, Le, & Zidek, 2009). As noted above, the high cost of network construction and maintenance will require the designer to select a defensible number of approaches that may provide such a justification.

Example 13.1. *The Metro Vancouver air quality network*

As with most modern urban areas in developed countries, Metro Vancouver, together with the Fraser Valley Regional District, operate an air quality monitoring network (http://www.gvrd.bc. ca/air/monitoring.htm). Like many such networks, it was not planned as an integrated whole (Ainslie et al., 2009) but instead grew without planned structure from a small initial nucleus of stations. Although it was seen from its inception as growing to enable monitoring air quality over the entire city and surrounding areas there was no structure in the ways it would be expanded.

A redesign was undertaken in 2008 to develop a strategy that was consistent with the joint Air Quality Management Plan of Metro Vancouver and the Fraser Valley

The network included 27 ozone monitoring sites (stations) as well as sites for the other criteria air pollutants. Using the methods described in this chapter, the redesign analysis suggested a number of changes. One particular recommendation was that one suburban site was redundant and that it could profitably be relocated between the two most easterly sites in the Fraser Valley (Ainslie et al., 2009). Overall the redesign showed that the quality of the network could be improved at relatively low cost.

Even when an objective is prescribed for a network, meeting it may not be easy for both technical as well as conceptual reasons. For example, detecting noncompliance with regulations may be the objective of an urban network (Guttorp & Sampson, 2010). but interpreting that objective in a concrete quantitative way can pose difficult conceptual challenges (Chang et al., 2007). For example what does "detecting non-compliance" actually mean? And what might be an optimal network design for that purpose during a three-year period might not be optimal for the next, depending on different climate regimes. How should a compromise be made between conflicting objectives?

Some networks were designed for one purpose but developed to serve others. Some networks are a synthesis of networks that were set up at different times.

Example 13.2. *The CAPMoN Network*

During the 1980s, acid precipitation was deemed to be a serious problem worldwide because of its deleterious effects on the terrestrial as well as aquatic regions. Impacts were seen on fish and forest for example due to acids that were formed in the atmosphere during precipitation events and were then deposited on the earth's surface. This in turn had adverse effects on human welfare.

As a result, acid rain monitoring networks were established to monitor trends. The CAPMoN (Canadian Acid and Precipitation Monitoring Network) was one such network, however in reality it consisted of the union of three monitoring networks established at different times for various purposes (Zidek et al., 2000; Ro et al., 1988; Sirois & Fricke, 1992).

CAPMoN's history is particularly instructive (Zidek et al., 2000). It was established in 1978 with just three sites in remote areas but in 1983 its size increased when it was merged with the Air and Precipitation Network (APN). However the merged network ended up with a second purpose; to trace source–receptor relationships in acid rain generation and monitoring sites were added close to urban areas. Later a third purpose was identified, that of discovering relationships between air pollution and human health (Burnett et al., 1994; Zidek et al., 1998)

This example shows us that the purpose of a network may not have been foreseen when it was established and it may therefore not be optimal for its current purpose.

Example 13.3. *The Mercury Deposition Monitoring Network*

When ingested, mercury (Hg) is dangerous to human health. In general it comes in three forms: elemental mercury, such as that found in thermometers and dental amalgams; organic mercury, mainly methylmercury, found in foods such as fish; and nonelemental mercury, found in batteries and some disinfectants. The consumption of fish is a primary pathway for exposure to methylmercury (MeHg), the most bioaccumulative form of mercury in humans and wildlife (Schmeltz et al., 2011).

Mercury is a neurotoxin that can affect many areas of the brain leading to diminished fine motor capacity and language. People exposed to mercury can develop acrodynia commonly known as pink disease whose symptoms include leg cramps, irritability, redness and the peeling of skin of the hands, nose and soles of the feet. Other symptoms include itching, fever, sweating, salivating, rashes including the 'baboon syndrome', which produces rashes in the buttocks, anal and genital regions, and sleeplessness. Mercury also has severe impacts on biota such as fish and birds.

Concern about toxicity of mercury in all its forms led to the establishment in the US Clean Air Act Amendments of 1990 (CAA) of section 112(n)(1)(B) that requires the U.S. EPA to study the impacts of mercury air pollution.

That, and the recognition of the serious human health risk associated with mercury, led to the establishment of MercNet. This is a multi-objective National US network that embraces a number of monitoring programs, notably the Mercury Deposition Network (MDN). The latter, working under the aegis of the National Atmospheric Deposition Network (NADP), consists of the monitoring site locations depicted in Figure 13.1. Data are available from the NADP site (nadp.sws.uiuc.edu/data/mdn/weekly.aspx). Atmospheric Hg is deposited on the earth's surface through precipitation and is a major source of mercury in the environment. As originally conceived, MercNet would provide a comprehensive multi-objective mercury network that would amongst other things enable the development, calibration, and refinement of predictive mercury models to guide effective management (Schmeltz et al., 2011).

13.3 Design paradigms

The spatial domain in which the monitors are to be sited is denoted by $\mathscr{S}$ and it can be treated as either a continuum or a discrete set, the latter commonly being chosen

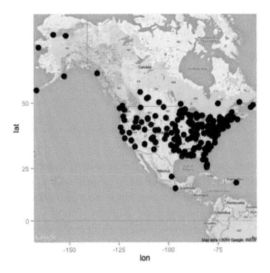

Figure 13.1: Locations of the sites maintained in the United States to monitor atmospheric deposition of mercury. The metadata containing these locations were obtained from National Atmospheric Deposition Program (NRSP–3), 2007, NADP Program Office, Illinois State Water Survey, Nation2204 Griffith Dr., Champaign, IL
61820.

due to practical considerations. The potential sites $s \in D$ could be points on a regular
or irregular grid. They could mark features in a geographical region, such as rivers
that could be selected for chemical assessment. (See Example 13.10 below.) They
could be centroids of grid cells to be sampled. The process $Z = Z_s$ to be monitored
over time and space is parametrized by those points. The network designer is required
to specify a finite set of points at which monitors are to be placed. In the Le–Zidek
approach (Le & Zidek, 2006) in a multivariate setting, each site s is equipped with
a small, finite set of pseudo-site locations where individual monitors can be placed.
The design problem becomes one of deciding at which pseudo-sites gauges (monitors) should be placed and the choice to a considerable extent is subjective.

Example 13.4. *The acid deposition monitoring network*

The National Atmospheric Deposition Network referred to above, stipu-
lates that when siting network monitors:

"The COLLECTOR should be installed over undisturbed land on its
standard 1 meter high aluminum base. Naturally vegetated, level areas
are preferred, but grassed areas and up or down slopes up to 15% will
be tolerated. Sudden changes in slope within 30 meters of the collector

should also be avoided. Ground cover should surround the collector for a distance of approximately 30 meters. In farm areas a vegetated buffer strip must surround the collector for at least 30 meters."

Most design strategies fall into one of three categories or "paradigms" (Le & Zidek, 2006; Müller, 2007; Dobbie, Henderson, & Stevens Jr, 2008), which we describe in the next three sections. The first are the geometry-based designs. These sites are selected from the population $\mathscr{S}$ without reference to the responses indexed by $\mathscr{S}$ that are to be measured. The second category are the probability-based designs, the domain of the survey sampler. Like the geometry–based designs, these are selected by randomly sampling from the list $\mathscr{S}$, again without reference to the process $\mathscr{S}$ or the joint distribution of those processes. Finally we have the model-based designs where now the joint distribution of the $\{Z_s\}$ responses over $\mathscr{S}$ is taken into account. Thus for example these designs will tend not to pick sites that are geographically close to one another. Here the focus may be parameters, as for example the coefficients in a model that relates covariates to the process. Or it may be on prediction as was the case for most designs developed in geostatistics over the past half century.

13.4 Geometry-based designs

This approach seems to be based on the heuristic idea of covering a region and would have included things like corner points of a grid spread over a region as well as space-filling designs (Nychka & Saltzman, 1998). Müller (2007) points to their use when the purpose of the network is exploratory while Cox, Cox, and Ensor (1997) rationalize this choice for multi-objective analysis.

Example 13.5. *Redesigning a freshwater monitoring network*

Olea (1984) performed an interesting simulation study to find an optimal geometric sampling plan in a geostatistical setting where prediction performance through universal kriging was the criterion for assessing network quality. Their result showed sites arranged in hexagonal patterns proved best among the twelve geometric patterns he considered. He used his result to redesign the network for monitoring the freshwater levels in the Equus Beds of Kansas. The original design, which was not the result of careful planning, proved quite unsatisfactory judged by the criteria they used.

Example 13.6. *Space-filling design*

Royle and Nychka (1998) give an algorithm for finding the optimum space-filling design. Given a set $S \subset D$ as a proposed set of monitoring points define the distance from a point $s \in D$ to S to be

$$d_p(s,D) = \left(\sum_{s' \in S} || s - s' ||^p \right)^{1/p}$$

Then the overall coverage criterion is given by

$$C_{p,q}(S) = \left(\sum_{s \in D} d_p(s,D)^q \right)^{1/q}$$

The (sub-) optimal design is found using a point swapping algorithm: given a starting set S, visit each of S's points in successive steps; at any given step replace the point in S with one not in S and keep it if and only if it improves (reduces) the coverage criterion; go on to the next point in S and repeat the search for an improvement using points not in S including those that have been dumped in an earlier round; continue until convergence.

Royle and Nychka (1998) apply their algorithm to an ozone network in the city of Chicago. They impose a grid of 720 candidate sites over the City excluding points in Lake Michigan but located in the convex hull of the existing sites. In their first analysis they reduce the existing network of 21 down to 5, by picking the best space-filling subset using the coverage criterion. Notice that no ozone data are required for this analysis.

The space-filling approach to design is very appealing due to its simplicity. It is also robust, since unlike other approaches it does not rely on data from the environmental fields and associated stochastic modelling.

13.5   Probability-based designs

This approach to design goes back at least half a century and is now widely used for such things as public opinion polling. This is the domain of the 'survey sampler'. The approach has appeal due to its apparent objectivity, since the sample is drawn purely at random from a list of the population elements, the *sampling frame*. Designers need not know anything about the population distribution to complete that process, except possibly for optimising the sample size.

In our application, the sampling frame would consist of the elements s of the finite population $\mathscr{S}$. The responses of interest would be the $\{Z_s\}$ at a specific time (or possibly a sequence of responses over time) depending on the nature of the enquiry. However, these would not be considered random, even if they were unknown. The survey assisted approach to survey sample assumes that $\mathscr{S}$ itself is a random sample from a superpopulation and that therefore the $\{Z_s\}$ have a joint probability distribution whose unknown parameter vector θ would itself be unknown. And the unknown 'estimate' of θ would then be regarded as a population 'parameter'.

Example 13.7. *Superpopulation modelling*

Suppose $D = \{s_1, \ldots, s_N\}$ so that the population is of unknown size N. Assume the $Z_i = Z_{s_i} = (Y_i, X_i)$, $i = 1, \ldots, N$ are drawn independently from the superpopulation each element having a bivariate normal distribution with $\mu_{1\cdot 2} = E(Y_i \mid X_i)\alpha + \beta X_i$ while $\sigma_{1\cdot 2} = Var(Y_i \mid X_i) = \sigma_Y(1 - \rho^2)$. Then it is easily shown that the maximum likelihood estimator for β would be

$$\hat{\beta} = \frac{\sum_i Y_i X_i}{\sum_i X_i X_i} \tag{13.1}$$

The sample $\tilde{d}$ from $\mathscr{S}$ would now yield the data needed to estimate $\hat{\beta}$ and yield $\hat{\hat{\beta}}$, an estimate of the "estimate".

The superpopulation helps define the relevant finite population parameters. They would commonly be linear functions of $\{Z_s\}$, such as the average pollution level for a given year, $\bar{Z} = \sum_{s \in S} Z_s / N_S$ where N_S denotes the number of elements in S. Or they may be combinations of linear functions as in Equation 13.1.

However these so-called *population parameters* are unknown and need to be estimated and hence a sample needs to be selected from $\mathscr{S}$. The probability-based sample design is specified by the probabilities of selecting a random sample say $\tilde{d} \subset D$, i.e. by the list $\{P(\tilde{d} = s) : d \subset D\}$. Of particular importance are the selection probabilities $\pi_s = P(s \in \tilde{d})$. Equiprobability sampling would mean $\pi_s \equiv K$ for some constant K. Note $\sum_{s \in D} \pi_s \neq 1$ in general but they can be used to enable inference in complex sampling designs as seen in the following examples.

Example 13.8. *Simple random sampling*

In simple random sampling (SRS) n elements, $\tilde{d}$ are selected without replacement at random from the total of N elements in the sampling frame $\mathscr{S}$ in such a way that $\pi_s = n/N$. The responses Z_s would then be measured for the elements $s \in \tilde{d}$. In our applications these elements would commonly be site locations.

Example 13.9. *Stratified random samples*

This approach can be very effective when $\mathscr{S}$ can be stratified into subsets $\mathscr{S}_j$, $j = 1, \ldots, J$ with $\mathscr{S} = \cup \mathscr{S}_j$ where the responses are thought to be quite similar within the $\mathscr{S}_j$. In an extreme case, suppose all the elements in $\mathscr{S}_j$ were identical. Then you would need to sample only one element in that strata to gain all of the available information. More generally the expectation is that only a small SRS of size n_j would need to be taken from each $\mathscr{S}_j$, with its subpopulation total N_j, to get a good picture of the population of Z responses for all of $\mathscr{S}$. Thus in SRS sampling $\pi_\mathbf{s} = n_j/N_j$, $\mathbf{s} \in \mathscr{S}_j$, $j = 1, \ldots, J$.

To complete this discussion, we turn to inference and the Horwitz–Thompson (HT) estimator. Consider first the case of a population parameter of the form $\hat{\theta} = \sum_{s \in \mathscr{S}} \omega_s Z_s$ where the ωs are known constants. Then the HT estimator is defined by

$$\hat{\hat{\theta}}_{HT} = \sum_{s \in \tilde{d}} \omega_s Z_s / \pi_\mathbf{s} \tag{13.2}$$

An important property of the HT estimator is its design–unbiasedness:

$$E[\hat{\hat{\theta}}_{HT}] = \hat{\theta} \tag{13.3}$$

where the expectation in Equation 13.3 is taken over all possible $\tilde{d}$ (see Exercise 13.9).

However the population parameter may not be a linear function of the process responses Z. If it is a function of such linear combinations as in Example 13.7 then the HT estimator can be used to estimate each of the linear combinations; these will then be unbiased estimators of their population level counterparts, even if the function of them, which defines the parameter, is not. In other words in Example 13.7 we would get

$$\hat{\beta} = \frac{\sum_i Y_i X_i / \pi_i}{\sum_i X_i X_i / \pi_i} \tag{13.4}$$

that is not a design unbiased estimator unlike the numerator and denominator that are.

Probability-based designs have been used in designing environmental sampling plans. The US EPA's Environmental Monitoring and Assessment Program, which ran from 1990 to 2006, was concerned with ecological risk and environmental risk. Example 13.10 illustrates the use of probability-based design in a situation where other approaches would be difficult to use.

Example 13.10. *The National Stream Survey*

The National Stream Survey was carried out as part of the US EPA's National Surface Water Survey. The survey sought to characterise the water chemistry of streams in the United States, 26 quantitative physical and chemical variables being of interest (Mitch, 1990). The elements of target population $\mathscr{S}$ were 'stream reaches' defined as (Stehman & Overton, 1994) the segment of a stream between two confluences or, in the case of a head-waters reach, the segment between the origin of a stream and the first conflu-ence (with some additional technical restrictions). A confluence is a point at which two or more streams flow together. Sampling units were selected using a square dot grid with a density of 1 grid point per 64 square miles, which was imposed (via a transparent acetate sheet) on a 1:250,000 scale topographical map. A reach was included if a grid point fell into its watershed, meaning that the selection probability for a reach was proportional to the area of that watershed. In other words if we denote by X_s the area of the watershed s and $X.$ the population total of these X's (Stehman & Overton, 1994)

$$\pi_s = n\frac{X_s}{X.} \tag{13.5}$$

Much more complex examples may be used as in multi-stage sampling where primary units, for example urban areas, are selected at stage one, with secondary units (e.g. dwelling units) at stage two and tertiary units (e.g. an individual within the dwelling unit) at stage three. The selection probabilities may still be found in such situations.

We have now seen some of the advantages of the probability-based sampling approach. However these can also be seen as some of the disadvantages in that some implicit modelling will be required. For example, choosing the strata effectively will require having enough information to be able to pick homogeneous strata. Modelling can be seen as a good thing in that it enables the analyst to bring a body of back-ground knowledge to the problem; knowledge whose inclusion probability sampling would restrict or exclude altogether. Finally, while stratified sampling has the ad-vantage of forcing the sampling sites to be spread out over an area, you could still end up in the awkward situation of having two sites right next to each other across a strata boundary. These shortcomings have led to the development of model-based alternatives.

13.6 Model-based designs

Broadly speaking, model-based designs optimise some form of inference about the process or its model parameters. This section will present a number of those ap-proaches. There are two main categories of design theory of this type:

- Parameter estimation approaches.
- Random fields approaches.

We begin by looking at parameter estimation approaches.

13.6.1   Regression parameter estimation

This approach was developed outside the framework of spatial design and so we focus more generally on optimising the fitting of regression models. This topic has a long history and was developed for continuous sampling domains (Smith, 1918; Elfving et al., 1952; Kiefer, 1959). The result was an elaborate theory for design that differed greatly from the established theory of the time. That new theory was pioneered by Sir Ronald Fisher (Silvey, 1980; Fedorov & Hackl, 1997; Müller, 2007).

Example 13.11. *An optimal design for regression*

Consider a dataset that consists of a fixed number n of vectors (x_i, y_i), $i = 1, \ldots, n$ that are assumed to be independent realizations of $(x, Y) \in [a, b] \times (-\infty, \infty)$ where the x's are chosen by the experimenter while Y is generated by the model

$$Y = \alpha + \beta x + \varepsilon \tag{13.6}$$

where α and β are unknown parameters and ε has zero mean and variance σ_ε^2. The least squares estimate of β is given by

$$\hat{\beta} = \frac{\sum_i (x_i - \bar{x})((y_i - \bar{y}))}{\sum_i (x_i - \bar{x})^2} \tag{13.7}$$

How should the experimenter choose the x's?

The answer is to ensure that the accuracy of $\hat{\beta}$ is maximised i.e. its standard error (se) is minimised. Conditional on the unknown parameters, that se is the square root of

$$Var(\hat{\beta}) = \frac{\sigma_\varepsilon^2}{\sum_i (x_i - \bar{x})^2} \tag{13.8}$$

Obviously the optimal strategy is to put half of the x's at either end of their range $[a, b]$, a result that can be substantially generalised in spatial sampling (Schumacher & Zidek, 1993).

13.7   An entropy-based approach

So far we have described a number of approaches to designing networks where a specific objective, such as parameter estimation can be prescribed. However, in our experience most major networks are not planned for a single purpose. Instead the purpose may be too ill-defined to yield a specific quantitative objective function to be maximised. There may be a multiplicity of purposes or the purpose, or purposes, may be unforeseen. The conundrum here is that despite those challenges, the designer may need to provide and justify specific monitoring site locations.

One way forward recognises that most networks have a fairly fundamental purpose of reducing uncertainty about some aspect of the random environmental process of interest, unknown parameters or unmeasured responses for example. This uncertainty can be reduced or eliminated by measuring the quantities of interest. An entropy-based approach provides a unified general approach to optimising a network's design. Chapter 3 gives the foundation needed to develop that theory, which has a long history including general theory (Good, 1952; Lindley, 1956) and applications to network design (Shewry & Wynn, 1987; Caselton & Zidek, 1984; Caselton et al., 1992; Sebastiani & Wynn, 2002; Zidek et al., 2000).

The entropy approach is implemented within a Bayesian framework. To explain the approach in a specific case, assume the designer's interest lies in the process response vector Z one time step into the future where T denotes the present time, $T+1$ for all sites in $\mathscr{S}$ i.e. monitored and unmonitored (i.e. gauged and ungauged in alternate terminology).

For simplicity let $\mathscr{D}$ denote the set of all available data upon which to condition and get posterior distributions. Entropy also requires that we specify h_1 & h_2 as baseline reference densities against which to measure uncertainty. With H denoting the entropy and θ the process parameters, define with expectation being conditional on $\mathscr{D}$ and taken over both the unknown Z and θ:

$$H(\mathbf{Z} \mid \theta) = E[-\log(p(\mathbf{Z} \mid \theta, \mathscr{D})/h_1(\mathbf{Z}) \mid \mathscr{D}]$$
$$H(\theta) = E[-\log(p(\theta \mid \mathscr{D})/h_2(\tilde{\theta})) \mid \mathscr{D}]$$

Finally we may restate the fundamental identity in Equation 3.6 as

$$H(\mathbf{Z}, \theta) = H(\mathbf{Z} \mid \theta) + H(\theta) \tag{13.9}$$

Example 13.12. *Elements of the entropy decomposition in the Gaussian case*

Assume that the process in Equation 13.9 has been non-dimensionalised along with its unknown distribution parameters in θ. Then we may take $h_1 = h_2 \equiv 1$ (although there would be more judicious choices perhaps in some applications). Furthermore assume the $\mathscr{S}$ consists of a single point $\mathscr{S} = \{1\}$. We assume the response at time t at this site, Z_{1t} is measured without error so that $Z_{1t} = Y_{1t}$. We now adopt the assumptions made in Example 4.1. Moreover conditional on the unknown mean θ and known variance σ^2, the $Y_{1t}\sigma N(\theta, \sigma^2)$, $t = 1, \ldots, T$ are independently observed to yield the dataset $\mathscr{D} = \{y_{11}, \ldots, y_{1T}\}$ with mean $\hat{y}$. A conjugate prior is assumed so that

$$\theta \sim N(\theta_0, \sigma_0^2) \tag{13.10}$$

The posterior distribution for θ is found to be

$$\theta \sim N(\hat{\theta}_{Bayes}, \sigma_1^2) \tag{13.11}$$

where $\hat{\theta} = w\hat{y} + (1-w)\theta_0$, $\sigma_1^2 = [w\sigma^2 n^{-1}]$ and $w = \sigma_0^2(\sigma_0^2 + \sigma^2/n)$. At the same time $-\log p(Z \mid \theta, \mathscr{D}) = -\log p(Y \mid \theta) = -n/[2\sigma^2](\theta - \hat{y})^2 + \ldots$ where we have ignored for now some known additive constants.

We can now compute the first term in Equation 13.9, call it T_1 for short, as

$$T_1 = \frac{n}{2}\left[\frac{\sigma_1^2}{\sigma^2} + \frac{(\hat{\theta}_{Bayes} - \hat{y})^2}{\sigma^2} + \right.$$
$$\left. \log 2\pi\sigma^2 + \frac{s^2}{\sigma^2}\right] \tag{13.12}$$

where s^2 denotes the sample variance (calculated by dividing by n rather than $(n-1)$. The second term in Equation 13.9, call it T_2 is given by

$$T_2 = \frac{1}{2}[1 + \log 2\pi\sigma_1^2] \tag{13.13}$$

Thus Equation 13.9 reduces to

$$T = T_1 + T_2 \tag{13.14}$$

where we denote the total uncertainty $H(Z, \theta)$ by T. A number of sources of uncertainty contribute to the total uncertainty (while recalling that we have rescaled the responses/parameters a priori to eliminate the units of measurement). First, when we look at T_1 we see that if we ignore additive constants, it is essentially the expected squared prediction error incurred were θ to be used to predict the uncertain future, Y, the log likelihood is essentially $(Y - \theta)^2$, whose expectation is being computed. However we don't know θ, which adds to the uncertainty in T_1. The result is that our uncertainty grows if the following apply:

- The known sampling variance σ^2 is large, although that effect is relatively small since its logarithm is being taken.
- The sample variance proves much larger than the supposedly known sampling variance σ^2, almost as though that prior knowledge is being assessed.
- The posterior mean and sample mean differ a lot relative to the natural variance of the sample mean σ^2/n.
- A large component of w comes from the prior variance σ_0^2 meaning the designer is not sure of his prior opinions about the value of the predictor θ of Y, this same uncertainty will inflate T_2 as well.

In summary, even in this simple example, the uncertainty as discussed in Chapter 3 is complex, even without taking model uncertainty into consideration.

13.7.1 The design of a network

We now consider the design of a network. Assume for simplicity $Y = Z$ is measurable without error and take the network's design goal to be that of adding or subtracting sites. We focus on adding new sites to an existing network to reduce uncertainty in an optimal way at time $T + 1$. We now have all the data $\mathscr{D}$ up to the present time T and we have the sites that provided it. We will be adding new sites and these as well as current sites will provide measurements at time $T + 1$, that will help us predict the measurable but unmeasured responses at the sites that will not be gauged at that time. The question is how do we choose the sites to add.

The total future uncertainty will be the total entropy TOT by $H[\mathbf{Y}_{(T+1)}, \theta \mid \mathscr{D}]$. We can partition the response vector as follows

$$\mathbf{Y}_{(T+1)}\mathbf{Y}_{(T+1)} = (\mathbf{Y}^{(1)}_{(T+1)}, \mathbf{Y}^{(2)}_{(T+1)}) \tag{13.15}$$

where $\mathbf{Y}^{(1)}_{(T+1)}$ and $\mathbf{Y}^{(2)}_{(T+1)}$ are respectively, the responses corresponding to the sites that are not and are currently monitored. The sites without monitors need to split into two groups, corresponding to the ones which will be gauged and those that will not. After relabeling them, this will correspond to a partition of the vector of responses $\mathbf{Y}^{(1)}_{(T+1)} = (\mathbf{Y}^{(rem)}_{(T+1)}, \mathbf{Y}^{(add)}_{(T+1)})$, $\mathbf{Y}^{(rem)}_{(T+1)}$ and $\mathbf{Y}^{(add)}_{(T+1)}$ being respectively the responses at sites that will remain ungauged and those that will be added to the existing network and yield measurements at time $T + 1$.

Let's simplify notation by letting

$$\mathbf{U} = \mathbf{Y}^{(rem)}_{(T+1)}, \ \mathbf{G} = (\mathbf{Y}^{(add)}_{(T+1)}, \mathbf{Y}^{(2)}_{(T+1)}), \ \mathbf{Y}_{(T+1)} = [\mathbf{U}, \mathbf{G}] \tag{13.16}$$

Then we get a revised version of the fundamental identity:

$$\boxed{\text{TOT} = \text{PRED} + \text{MODEL} + \text{MEAS}}$$

where

$$PRED = E[-\log(f(\mathbf{U} \mid \mathbf{G}, \theta, \mathscr{D})/h_{11}(\mathbf{U})) \mid \mathscr{D}] \tag{13.17}$$
$$MODEL = E[-\log(f(\theta \mid \mathbf{G}, \mathscr{D})/h_2(\theta)) \mid \mathscr{D}] \tag{13.18}$$

and

$$MEAS = E[-\log(f(\mathbf{G} \mid \mathscr{D})/\mathbf{h_{12}}(\mathbf{G})) \mid \mathscr{D}] \tag{13.19}$$

Here TOT denotes the total amount of uncertainty. The identity states that it has three components. The first PRED is the residual uncertainty in the field after it has been predicted. The predictor involves the uncertain process parameter vector θ, so that also is reflected in PRED. MOD represents model uncertainty due to its unknown parameters. An extended version of this identity could include model uncertainty, and given the model, e.g. Gaussian, the parameter uncertainty for the model. Finally MEAS is the amount of uncertainty that would be removed from TOT simply by measuring the random field at the monitoring sites. We immediately see the remarkable result that maximising MEAS through judicious selection of the sites will simultaneously minimise the combination of uncertainties about the unmonitored sites in the field as well as the model. Since the total uncertainty TOT is fixed, this decomposition must hold, no matter which sites we decide to monitor.

Finding entropy–optimal designs

The following material is included for completeness to briefly describe one way of creating an optimal design. The foundations of the approach are those underlying the the BSP method (Chapter 11, Section 11.5) and the EnviroStat package. The latter includes a function for finding entropy-based optimal designs whose use will be described in Example 13.14.

The BSP assumes that in an initial step the process is transformed and pre–filtered by removing regional level spatio–temporal components. These would include things like trends and periodicities along with regional autocorrelation. These are commonly the largest source of variation. Furthermore, since a single parameter set is fitted for all sites, the standard errors of estimation for these parameters will be essentially negligible. Yet most of the trend and autocorrelation may be taken out of the process spatio–temporal mean field, leaving the site-specific parameters much less work to do. The entropy however, will be unaffected as it is determined by the stochastic variation in the residuals.

For completeness we briefly describe the BSP model for the residuals process, which we denote by Y rather than Z. This is because the previous steps remove the same mean model across all sites, making the residual process observable by subtracting it from the measurements at the monitored sites.

The important result is the posterior predictive distribution for the unmeasured responses as this is needed to find PRED above (Le & Zidek, 2006). We use once again the notation $\mathscr{D}$ to represent the totality of data. The notation $\mathscr{H}$ is used to represent the set of all hyperparameters of which there are many. The latter are estimated by maximising the marginal likelihood as an empirical Bayes step in fitting the model in order to reduce computation times. The superscript u will refer below to an ungauged site, while the superscript g will refer to an observation at a gauged site. The general version of the BSP, which is included in EnviroStat allows a (monotone) staircase pattern in the data matrix. There the superscript g_m is attached to responses at gauged sites that go unmeasured and these too are part of the predictive distribution. This staircase pattern is commonly seen due to the variable start or termination dates of the sites in a network (Shaddick & Zidek, 2014; Zidek et al., 2014). They are sometimes important due to the need to reconstruct exposure histories of subjects involved in the study of the chronic effects of exposure to environmental hazards.

An example is the study of the relationship between air pollution and cancer. The latter has a long latency period so Le and Zidek (2010) used the BSP to infer those long-term exposures of patients in a cancer register. The result did not show a significant result although the measurement error in such a study would be large, greatly reducing the power to detect an association.

However for instructional purposes, we are going to present the simplest version of the model—no staircase and just one response per site—not the multivariate version of the BSP. This models the random response field over the time period $1, \ldots, (T+1)$ the last of these time points being one time step into the future. Here is the model (Zidek et al., 2000) where as follows:

$$\mathbf{Y}^{(T+1) \times p} \mid \beta, \Sigma \sim N_p(\mathbf{X}^{(T+1) \times k} \beta^{k \times p}, I_{(T+1)} \otimes \Sigma)$$

$$\beta \mid \Sigma, \beta_0, F \sim N(\beta_0, \mathbf{F}^{-1} \otimes \Sigma) \tag{13.20}$$

$$\Sigma \sim IW(\Phi, \delta) \; \# \text{ Inverted Wishart distribution}$$

where k denotes the number of covariates at each time point. Here the notation of Dawid (1981) is used for the inverted Wishart (IW) distribution which is replaced by the generalised inverted Wishart (GIW) distribution in the general theory (Brown, Le, & Zidek, 1994) The distribution theory that is required for this can be found in Appendix 14.5. Note that the β matrix represents the site-specific effects that the covariates have, after removing the regional version which is embedded in β_0. The model is more flexible than it appears at first glance. For example monthly effects can be represented by eleven indicator variables in X as covariates to capture seasonality. In that case a column of the B matrix for a site at location s can represent the impact of different months on different sites. Those near the mountain would respond differently in winter to those near the sea. The regional model would not capture that difference.

The posterior predictive distributions we need for $\mathbf{U}^{1 \times u}$ and $\mathbf{G}^{1 \times g}$ as defined above in Equation 13.16, given the data to time T $\mathscr{D}$ are (Le & Zidek, 1992, 2006)

$$\mathbf{G} \mid \mathscr{D} \quad \sim \quad t_g \left(\mu^g, \frac{c}{l} \hat{\Phi}, l) \right)$$

$$\mathbf{U} \mid \mathbf{G} = \mathbf{g}, \mathscr{D} \quad \sim \quad t_u \left(\mu^u, \frac{d}{q} \Phi_{u|g}, q). \right) \tag{13.21}$$

The product of the distributions in Equation 13.21 yields the joint distribution of all responses at time $T+1$. We see in that equation, a lot of undefined symbols.

It turns out we do not need to know them. For a start it can be shown that for any multivariate normal random variable with covariance $\Sigma \sim IW(\Phi, \delta)$ must have entropy of a remarkably simple form, that of the multivariate–t distribution (Caselton et al., 1992)

$$\frac{1}{2} \log |\Phi| + c \tag{13.22}$$

with a known constant c whose value is irrelevant. Our design goal is to add sites to the existing sites to maximize the entropy of the resulting network's response $\mathbf{G}$—that will make maximizing the uncertainty we remove by measurement at time $T + 1$. After permuting coordinates appropriately we may partition $\mathbf{G}$ as $(\mathbf{G}^{add}, \mathbf{G}^{original})$. Elementary probability tells us that we write the joint distribution $[\mathbf{G}]$ as the product of the joint distribution $[\mathbf{G}^{original}]$ and the conditional distribution $[\mathbf{G}^{add} \mid \mathbf{G}^{original}]$. However Equation 13.21 shows that each of these distributions will have a multivariate–t distribution. Moreover taking logarithms of the product of the densities leads to the sum of their logarithms. In other words

$$H(\mathbf{G} \mid \mathscr{D}) = H(\mathbf{G}^{add} \mid \mathbf{G}^{original}, \mathscr{D}) + H(\mathbf{G}^{original} \mid \mathscr{D}) \qquad (13.23)$$

The second term in Equation 13.23 is fixed—those sites are not to be removed. Thus to maximize MEAS, in other words $H(\mathbf{G} \mid \mathscr{D})$ means maximizing $H(\mathbf{G}^{add} \mid \mathbf{G}^{original}, \mathscr{D})$. In other words the responses for the add sites should be the ones that are least predictable from $\mathbf{G}^{original}$ and hence maximally uncertain. As noted above that conditional distribution must be a multivariate–t and hence its entropy must be of the form given in Equation 13.22, in other words,

$$H(\mathbf{G}^{add} \mid \mathbf{G}^{original}, \mathscr{D}) = \frac{1}{2} \mid \Phi_{add \mid original} \mid + c* \qquad (13.24)$$

where in an obvious notation

$$\Phi_{add \mid original} = \Sigma_{add,add} - \Sigma_{add,original} \Sigma_{original,original}^{-1} \Sigma_{original,add}$$

the Φ being that which appears in the inverted Wishart prior as a matrix of hyperparameters.

This matrix can be specified in a variety of ways but in line with our empirical Bayes approach, EnviroStat estimates it for the data at the original sites using a maximum likelihood approach . For a staircase data pattern the EM algorithm needs to be employed due to the missing data (Le & Zidek, 2006). This estimate can then be extended to the entire network of sites $\mathscr{S}$ by fitting a spatial semivariogram and using this to determine the missing elements of Φ. That second step may require use of the Sampson–Guttorp method to address field non-stationarity. We will see this approach in Example 13.13.

Example 13.13. *Redesign of Metro-Vancouver's PM$_{10}$ network*

Le and Zidek (2006) present a case study which anticipates the redesign described in Example 13.1 although they had no forewarning of it. It is concerned with Metro Vancouver's air quality monitoring sites as seen in Figure 13.2. Purely as a hypothetical case study 10 PM$_{10}$ sites scattered through the region were chosen as a network to which an additional 6 sites

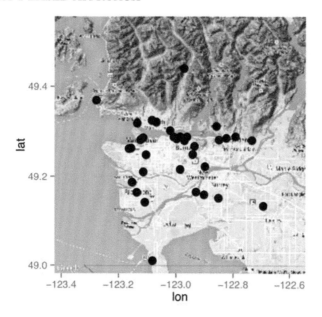

Figure 13.2: Locations of the sites maintained in the Lower Mainland of British Columbia, Canada to monitor the air quality in the region. Notice the topography of the region, which in summer combines with the prevailing wind from the west to push the pollution cloud eastward into the Fraser valley. Ten of the sites were selected in a hypothetical redesign of the PM_{10} network as described in the text.

were to be added from 20 potential sites located in census tracts with relatively large populations. The entropy-based approach was used with an interesting result.

Their study used the approach described above and described in the vignette accompanying the EnviroStat package for network design. After transforming and prefiltering the data, a posterior predictive distributive distribution was created as described above. The Sampson–Guttorp approach (Sampson & Guttorp, 1992) had to be applied and the estimated hypercovariance for the field of all 30 sites estimated with its help (an exponential semi-variogram was used).

They began by ranking the 20 potential sites by the size of the posterior variance of their future PM_{10} concentrations and labelling the sites from [1] to [20], site [20] being the one with the largest variance. They then applied the greedy algorithm to find one-by-one, the best 6 sites as far as uncertainty reduction was concerned and labelled these from (1) to (6), where (1) being the single site that reduces uncertainty the most.

The results were quite interesting. The site (1) chosen for first place was not a surprise, its posterior variance was large placing it second from the top, giving it a label of [19] according to the variance ranking and hence its measurements were quite uncertain. A lot could be gained by making measurements there. The next five best choices had variance ranks [18], [20], [10], [16], and [17]. Things went more or less as expected as sites (1) to (3) were picked, but then we see an abrupt divergence, site [10] gets selected for inclusion in the "add" set. Why is that? The answer is that the entropy recognises that once (1)–(3) have been included, more information could be gained from the variance ranked site [10] than from any of the sites [17]–[11]. It turned out that [10] was on the edge of the network and so able to provide information not available from other sites. Curiously, this is close to the site that was chosen in the redesign described in Example 13.1. These results shows the subtlety of the entropy criterion in the way it picks its optimal sites.

13.7.2   Redesigning networks

In this subsection we will see step-by-step how to redesign an existing network. The approach in the example follows that of the vignette for the EnviroStat package, used here since it includes a function for warping to achieve non-stationarity as well as one for entropy-based design.

Example 13.14. *Adding temperature monitoring sites in California*

This example follows Example 9.6 and concerns what may be increasingly seen as an environmental hazard as discussed in that example. We have simplified the modelling of the field for this illustrative example. In fact finding good spatio–temperature models is quite challenging and the subject of current research (Kleiber et al., 2013).

We begin by plotting the locations of the 18 temperature monitoring sites we selected for this example. The ultimate goal is that of adding 5 additional sites to this network we have artificially created. First we need to load the necessary libraries along with the necessary data. Then we plot the 18 sites that constitute the 'current' network.

```
### The libraries
library(EnviroStat)
library(sp)
library(rgdal)
library(ggmap)

### The data
dat <- read.csv("MaxCaliforniaTemp.csv", header=T)
crs.org <-  read.table("metadataCA.txt", header=T)

### Now we plot the sites
Cal.sp = crs.org
```

```
coordinates(Cal.sp) = ~Long+Lat
proj4string(Cal.sp) = CRS("+proj=longlat
        +ellps=WGS84")
latLongBox = bbox(Cal.sp)
location = c(-126, 32, -114, 42)
CaliforniaMap = get_map(location = location, source
        = "google", maptype = "roadmap")
CaliforniaMap = ggmap(CaliforniaMap)
CaliforniaMap = CaliforniaMap +
        geom_point(data = metadata,
        aes(x = Long, y = Lat), size = 3, col = 2)
        + xlab("Longitude") + ylab("Latitude")
```

We see the result in Figure 13.3, a set of sites selected to cover the State. The datasets consists of two files. The first (metadataCA.txt) is the metadata for the 18 sites: elevations above sea level (feet); geographic coordinates (latitude, longitude); (in the two right hand most columns for each site) a reference point's coordinates on the west coast of California that are closest to the site to enable a calculation of its distance from the ocean. The second (MaxTempCalifornia.csv) gives the maximum daily temperatures in 1/10th of a degree Fahrenheit for those sites from Jan 1, 2012 to Dec 30, 2012. For convenience we divide 2012 into 13 "months" of 28 days each by dropping days at the end of the year. This is not necessary but it makes life easier.

The analysis begins with standard data analysis. The assumption of a conditional Gaussian distribution underlying EnviroStat seems valid, as a result of a site-by-site analysis using standard methods such as the qq-plot. In any case, EnviroStat is fairly robust against departures from that

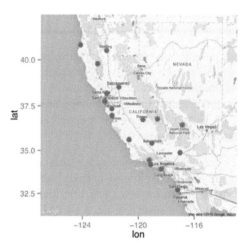

Figure 13.3: The current network of eighteen temperature monitoring sites in California to which five sites must be added.

assumption provided that the distribution is symmetric since the marginal process model conditional on the hyperparameters is the large family of multivariate (or matric) t distributions, conditional on hyperparameters. The ultimate test of all such assumptions is the performance of the resulting spatial process model (Fu, Le, & Zidek, 2003), e.g. by cross validation although that was not done for this example.

Next we compute the monthly averages of data at each site and then their grand mean for each month. This gives thirteen 28-day averages at each site and then thirteen regional averages. The latter are estimated from a lot of data so effectively can be treated as constants. Subtract the regional averages from the site-specific monthly averages to get the site-specific deviations for each month. The regional means will capture a very large amount of temporal variation due to the movement of the sun around earth, which produces the winter to summer variation. The residuals are also important since they represent the deviations due to such things as distance from the ocean. A plot of these 13 residuals for San Francisco is V-shaped with a minimum in the summer, which is pretty cool compared with the overall state average. That for Death Valley is the exact opposite, the desert reaches its maximum in the summer. This analysis can be captured in the EnviroStat model by setting up the design matrix which removes these major time trends along with the big site-specific deviations as follows:

```
month <- 1:13day <- 1:28
dgrid <- expand.grid(d=day, m=month)
dat <- dat[-365, ]
ZZ <-  model.matrix(~as.factor(dgrid\$m))
```

We can now fit the hyperparameters in the model using EnviroStat's <staircase.EM> function designed to handle a staircase pattern in the data matrix. In particular this gives us estimates of the covariance parameters.

Next we project latitude and longitude using the Lambert projection to get a flat surface on which intersite distances are measured in kilometres. Recall that the lines of longitude are not parallel like those of latitude so for an area as large as California, Euclidean distance cannot be calculated using geographic coordinates. That job is done using EnviroStat's Flamb2 function. The code is included in the online resources.

The next check takes the fitted covariance matrix and assesses it for nonstationarity using the Sampson–Guttorp warping method. Two versions of the warping approach are shown here. The first finds the best possible fit when there is no smoothing involved, i.e. $\lambda = 0.0$. The result is seen in Figure 13.4, where the right panel shows a near perfect fit of an exponential variogram while the left hand panel in that row shows with vectors the tremendous degree to which the sites would need to be moved to overcome the non-stationarity reflected in the covariance matrix fitted to the 18 temperature sites.

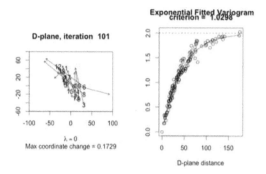

Figure 13.4: The effect of applying the Sampson–Guttorp warping method is seen in this figure where no smoothing has been used $\lambda = 0$. The left hand panel shows how the sites must be moved around on California's surface to achieve stationarity and a near perfect variogram.

```
X11()
par(mfrow=c(1, 2))
### First approach
sg.est = Falternate3(disp, scoord, max.iter=100,
+ alter.lim=100, model=1)

### Second approach
apply(scoord, 2, range)
coords.grid = Fmgrid(range(scoord[,1]), range(scoord[,2]))
par(mfrow=c(1,2))
temp = setplot(scoord, ax=T)
deform  = Ftransdraw(disp=disp, Gcrds=scoord, MDScrds=sg.est
    \$ncoords, gridstr=coords.grid)

Tspline = sinterp(scoord, sg.est\$ncoords, lam = 50 )
```

The second piece of code above requires the analyst to get interactively involved by choosing increasing values of λ starting from $\lambda = 0$ until a satisfactory degree of smoothing is achieved. Thus the first fit in the first row of Figure 13.5 shows a crumpled mess instead of the flat surface of California in the right panel. Moving to $\lambda = 20$ resolves that surface to some extent.

Most of the basic work is now done. We now move onto the new design and here we omit a lot of the relevant code for brevity and refer the reader to the online resources for the code. The first step is that of creating potential new design points, five of which are to be selected. That we do by overlaying a regular grid based on equal latitude and longitude spacing. That grid will have points that lie outside California's boundaries, so we need to intersect the grid with those boundaries to keep the sites inside the state. Finally we have to project the new site with the Lambert projection and extend the original

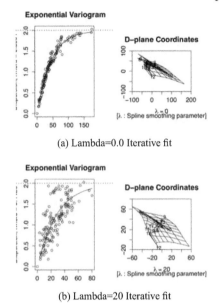

(a) Lambda=0.0 Iterative fit

(b) Lambda=20 Iterative fit

Figure 13.5: The effect of applying the Sampson–Guttorp warping method is seen in these figures beginning with the first row where we see the results for $\lambda = 0$. The analyst must then become involved and experimentally try a sequence of λs until $\lambda = 20$ is reached. The first row shows the same variogram fit, but also in the right hand panel, that the surface of California would be a crumpled mess. To achieve some interpretability and a more recognizable surface we must go to $\lambda = 20$ in the second row.

estimated spatial covariance structure to cover all the points in the grid. The principal functions and steps are as follows, the last two being for the selection of the five new sites as shown in Figure 13.6.

```
Tspline.var = sinterp(allcrds[(u+1):(u+18),],matrix(diag(cov
    .est),ncol=1),lam=50)
###
varfit = seval(allcrds,Tspline.var)\$y
temp = matrix(varfit,length(varfit),length(varfit))
covfit = corr.est\$cor * sqrt(temp * t(temp))
###
hyper.est = staircase.hyper.est(emfit= em.fit,covfit=covfit,
    u =u, p=1)
###
nsel = 5
yy = ldet.eval((hyper.est\$Lambda.0+ t(hyper.est\$Lambda.0))
    /2,nsel,all =F)
```

Figure 13.6: A redesigned network in California for monitoring temperatures. Here the entropy design approach has been used to add 5 new sites to a hypothetical network of 18 sites, the latter picked from sites already existing in the State.

In Example 9.6 applying the Sampson–Guttorp approach showed the temperature field residuals after removing a simple mean model to be highly nonstationary. This points in the first instance to the need to try much more sophisticated mean models as functions of the geographic coordinates and time (e.g. months). Nevertheless the approach does do a great job of removing the nonstationarity albeit at the price of losing some interpretability due to a heavily warped geographic space.

13.8   Implementation challenges

Finding the optimal design entails finding a best subset of sites from the entire set of possibilities which may be infinite in the case of a continuous domain $\mathscr{S}$. In the rest of this chapter we will continue to assume that for practical reasons $\mathscr{S}$ is finite. (The continuous case is discussed in Section 13.6.1.) Thus the problem is one of combinatorial optimisation whose solution is famously difficult. First of all it is an NP–hard problem: finding an exact solution when a large number of sites are involved is essentially impossible. Le and Zidek (2006) provide an exact branch-and-bound algorithm developed given in Ko, Lee, and Queyranne (1995). It can find optimum designs up to about the level of selecting 40 out of 80 sites when the objective function like ours, is a monotone function of the determinant of a matrix. Many designs will involve far more sites than that. The greedy algorithm is often used when the goal is to expand the network; first add the best site and then find the second best and so on. This process would be reversed should the goal be the contraction of a network.

Deciding how many sites to include is also a practical issue that can prove difficult to solve, although the issue is mute since budget is the deciding factor in our experience. However, more formal analyses are possible. One approach would treat this as a two criterion optimisation problem: maximize the entropy but minimize the cost. That cost will entail the fixed startup costs—the equipment will be expensive and so is acquiring a site—the running costs are more modest when the equipment is automated and online, but that equipment has to be monitored and fixed or replaced when it breaks down during its long operating life. Basic multi-attribute optimisation theory tells us that the optimal subset of sites $\tilde{S}^o$ given by

$$\tilde{S}^o = argmax_{\tilde{S}} entropy(\tilde{S}) - \lambda cost(\tilde{S}) \qquad (13.25)$$

in a simplified notation, where λ is the information to setup and operating cost trade-off coefficient. We have not found any systematic way of finding that coefficient. A more practical approach was suggested in personal communication by Dr Larry Phillips, maximise the "information bang per sampling buck" criterion.

$$\tilde{S}^o = argmax_{\tilde{S}} \ entropy(\tilde{S})/cost(\tilde{S}) \qquad (13.26)$$

The curve of $entropy(\tilde{S}^o)/cost(\tilde{S}^o)$ as a function of the number $N_{\tilde{S}}$ of sites in $\tilde{S}$ will increase to a maximum and then decrease when one reaches the point of diminishing information returns from adding another site. We have found this works quite well.

Convincing a user of the importance of a good design can be a problem meaning that inevitably a suboptimal plan may be chosen. For example, a survey of fresh water bodies involved sending float equipped helicopters into pristine areas to collect water samples to determine the levels of toxic substances. "Were these samples representative?" the second author asked. "Oh yes" came the reply, "We took two separate samples in different spots in each lake and then combined them to ensure the result was representative before we sent it to the lab for analysis." In another case the second author learned that the mayor of a local area had applied unsuccessfully for years to have an air quality monitor installed in her area. She was unsuccessful every time. The reason: "Your air pollution is too low so you don't need a monitor." We hope by this stage of the book, the reader will recognise the fallacies in both of these cases.

Of course the optimal design methods are indeed too simplistic since they overlook practical problems on the ground. Site locations thought to be optimal may be inaccessible. Costs will generally be hard to pin down since a lot of these will be soft costs e.g. the administration time involved in managing a monitoring network. Then again despite the best of intentions the process of setting up the sites will involve committees, politics, negotiations, setting up the requisite infrastructure and so on.

Even an initially optimal design can decline over time in value for various reasons, changing societal concerns being an important one where the costs are to be borne by that society. For example the dominance of acid rain as an issue gave way to the health effect impacts of air pollution and that in turn to climate change. The purpose of the design may then change (Example 13.2) along with its suitability for its new purposes, pointing anew to the need for regular reviews of a network in terms of its current purposes.

The redesign of the Metro Vancouver air quality monitoring network (Example 13.1) is encouraging. Firstly administrators recognised that the network had grown somewhat haphazardly since it started and that it might therefore not be satisfactory for current uses. Secondly, they contracted out the planned redesign of the network to experts. Also, they were willing to listen to the recommendations made by the experts and act on them. This is not a unique example, but in our experience it is very rare!

13.9 Summary

In the plethora of objectives we have seen in this chapter, we see the emergence of a central purpose; to explore or reduce uncertainty about aspects of the environmental processes of interest. One form of uncertainty, aleatory, cannot be reduced by definition whereas with the other, epistemic, where uncertainty can be reduced (see Chapter 3). However that reduction does not stop the original network from becoming sub-optimal over time, pointing to the need to regularly reassess its performance. From that perspective we see that the design criteria must allow for the possibility of 'gauging' (adding monitors to) sites that

- maximally reduce uncertainty at their space–time points (measuring their responses eliminates their uncertainty);
- best minimise uncertainty at other locations;
- best inform about process parameters;
- best detect non-compliers.

In this Chapter the reader will have gained an understanding of many of the challenges that the network designer may face. These involve the following topics:

- A multiplicity of valid design objectives.
- Unforeseen and changing objectives.
- A multiplicity of responses at each site, i.e. which should be monitored.
- A need to use prior knowledge and to characterise prior uncertainty.
- A need to formulate realistic process models.
- A requirement to take advantage of, and integrate with, existing networks.
- The need to be realistic, meaning to contend with economic as well as administrative demands and constraints.

Exercises

Exercise 13.1. Redo the plot in Example 13.3 but this time for just the region around the Great Lakes where a lot of industrial activity is concentrated.

Exercise 13.2. Exposure to lead is, like mercury, a serious human health risk and so is now regulated.

 (i) What are the specific health risks associated with lead?

 (ii) Describe pathways of human exposure to lead.

(iii) Atmospheric transportation of lead in various forms has led to the establishment of monitoring programs for it. Describe some specific features of those programs.

(iv) Plot monitoring sites for lead as found on the US EPA page
       www.epa.gov/airdata/ad_maps.html .

Exercise 13.3. Starting with the California temperature data, use the space–filling design-based approach as in Example 13.6 to create a new temperature monitoring network with a total of twenty-three temperature monitoring sites. This will entail some reading (Royle & Nychka, 1998) as well as downloading the `fields` package. That package has the cover.design function with an example you should be able to follow to complete this exercise.

Exercise 13.4. Returning to Example 13.7, determine an estimator based on the finite population sample $\tilde{d}$, an estimator of the superpopulation parameter σ_{12} and one for the population parameter $\hat{\sigma}_{12}$.

Exercise 13.5. Redo Exercise 13.3 but this time by using the entropy-based approach, to add five sites to the existing network. This will entail downloading the `EnviroStat` package and an investigation of the stationarity of the temperature field.

Exercise 13.6. Prove the result asserted in Equation 13.3. Also find an expression for the variance of the HT estimator.

Exercise 13.7. Carbon monoxide (CO) is called the 'silent killer' because its victims experience few symptoms (dizziness and tiredness) before they become unconscious and die, essentially by asphyxiation. The reason is that CO attaches itself to hemoglobin in the blood to form carbyoxyhemoglobin and blocks oxygen from being carried to the brain. (Tobacco smoking also creates carbyoxyhemoglobin.) For these reasons, CO has long been a criteria pollutant because of its risk to human health.

   In general incomplete burning leads to the production of CO rather than its benign cousin CO_2. Thus indoor sources can be found in gas stoves and heaters when they go out of adjustment. So it is likely to be found in the cold days of winter in the northern climates. A probability-based survey of homes is planned to determine the average level of CO in homes on cold days using portable detectors. Briefly describe how probability-based sampling could be used to survey homes on such days and how the resulting selection probabilities could be calculated.

Exercise 13.8. Return to Example 13.11.

 (i) Prove the claim that it is best to make half the observations at either end of the interval $[a, b]$.

 (ii) We may extend the result to the case where the vector $\varepsilon = (\varepsilon_1, \ldots, \varepsilon_n)) \sim N_n(\mathbf{0}, \Sigma_\varepsilon)$ where $\Sigma_\varepsilon = \sigma_\varepsilon \tau$ and τ is a known positive definite matrix. Determine the maximum likelihood estimator of β and optimal design for estimating it.

Exercise 13.9. Prove the result asserted in Equation 13.3. Also find an expression for the variance of the HT estimator.

Exercise 13.10. Suppose for elements s of the spatial sampling frame $\mathscr{S}$ are listed and their sizes X_s are known. How would you go about selecting a sample of size n from this list with selection probability proportional to size, i.e. so that Equation 13.5 holds.

Exercise 13.11. Chang et al. (2007) carry out simulation studies when the random field distribution follows the assumptions underlying the BSP assumptions so that the field has a marginal multivariate–t distribution. They see a smaller loss of intersite dependence when the t distribution has heavy tails compared to when it has light tails (a large number of degrees of freedom).

Try the simulation experiment yourself and confirm that heavier tails lead to increased intersite correlations.

Exercise 13.12. Return to Example 9.6 and the design of 18 temperature monitoring sites.

(i) Redo the analysis in the Example, but this time removing say 3 sites. Plot the resulting new network.

(ii) In that example 'month' was used to capture the large variation over time. Find a more sophisticated mean temperature model that depends on the spatial coordinates as well, remembering to first use regional models for the whole State and thereby keep the number of parameters used in the model to a small number.

Exercise 13.13. DISCUSSION QUESTION. How might design criteria be arrived at in practice? Who should be responsible for setting them?

Exercise 13.14. RESEARCH QUESTION. Monitor placement should recognize such things as the geographical distribution of impacted populations (e.g. trees or fish). How can an optimal design be determined in such a context?

Exercise 13.15. RESEARCH QUESTION. Develop a design theory in a non-Gaussian context.

Chapter 14

New Frontiers

14.1 Overview

The field of spatio–temporal epidemiology has expanded rapidly in the past 10 years due to the development of statistical techniques that can accommodate variation over both space and time and the increasing availability of high–resolution data measuring a wide variety of environmental processes. Bayesian hierarchical modelling has steadily expanded as has the ability to handle a large number (n) of measurement vectors which may be of high dimension (p). Conventional methods for performing Bayesian analysis may be infeasible due to their high computational demands, paving the way for approximate methods for Bayesian inference such as INLA (see Chapter 5, Section 5.6).

There are a great number of specific areas that are under active development. They include the following:

- Uncertainty Quantification (UQ), which is defined by Wikipedia as

    "the science of quantitative characterisation and reduction of uncertainties in applications. It tries to determine how likely certain outcomes are if some aspects of the system are not exactly known."

    This topic embraces the content of Chapter 3 along with more conventional sources of uncertainty such as unknown parameters in a mathematical, physical or numerical computer models.

- Model-based geostatistics.

- Modelling high dimensional response vectors; the 'big p problem'.

- Handling and analysing datasets with a large number of records; the 'big n problem'.

- Multivariate extreme value theory for high dimensional data.

- Preferential sampling and network design, a topic that is discussed in Chapter 13.

- Non-stationary spatio–temporal covariance structures.

- Physical–statistical modelling

Limitations of space in this book rule out treatment of all these topics. In this chapter, we discuss three that are of particular importance.

14.2   Non-stationary fields

The topic of non-stationarity was discused in Chapter 9 in a spatial setting. Here we expand on this topic, describing two approaches:

Spatial deformation: The Sampson–Guttorp approach (Sampson & Guttorp, 1992) warps the geographic space into dispersion space meaning that strongly correlated sites are moved closer together with uncorrelated ones being moved further apart.

Dimension expansion: The geographic space is kept the same but additional dimensions are added.

14.2.1   Spatial deformation

In the methodology described so far, two major assumptions have been required, stationarity and isotropy, which are unlikely to hold in many environmental problems. One approach to dealing with nonstationary and anisotropic process was suggested by Sampson and Guttorp (1992). The approach begins with the idea of warping the geographical plane, the G–plane, into another latent space, the D–plane, in such a way that two points whose process values are uncorrelated are pushed apart while two that are correlated are pushed together. The result is a stationary and isotropic process on the D–plane. In practice this can be challenging. To find the required 1:1 transformation, g, Sampson and Guttorp (1992) rely on dispersion rather than correlation. Dispersion is similar in concept to the distance between points. For any two sites, this is defined as $s, s' \in \mathscr{S}$ by $D(Z_s, Z_{s'}) = 2(1 - Corr(Z_s, Z_{s'}))$ for a process Z, which we assume here is measured without error. Here, D is assumed to be a variogram model with a suitable parametric form (Meiring, Sampson, & Guttorp, 1998) so that

$$D(Z_s, Z_{s'}) = \Gamma_\xi [\|g(s) - g(s')\|] \tag{14.1}$$

The transformation, g, is selected in order to ensure a smooth transformation and the parameter ξ is estimated in the process.

We can then compute the dispersion, and hence correlation, between any two points by inserting their G–plane coordinates into Equation 14.1. Using the parametric dispersion model we can get a covariance matrix for a process over the entire spatial domain $\mathscr{S}$.

Damian, Sampson, and Guttorp (2000) and Schmidt and O'Hagan (2000) have independently proposed Bayesian approaches for spatial deformation which incorporate the uncertainty in the choice of transformation function and the effect this has on subsequent estimation. Schmidt and O'Hagan (2000) use a Gaussian process prior for the deformation to D–plane whilst Damian et al. (2000) uses a prior based on thin plate splines. Both approaches use MCMC simulation to sample from the posterior.

In considering spatial predictors we adopt the notation used in Chapter 11, Section 11.5 which links the following description to the EnviroStat package which can be used to implement the Sampson–Guttorp method. We will use the terminology used in Section 11.5 and let $U = \{u_1, \ldots, u_{N_u}\}$ and $S = G = \{s_1, \ldots, s_{N_s}\}$ denote the set of ungauged and gauged sites in $\mathscr{S}$, respectively. Thus $U \cup G = \mathscr{S}$.

The process vectors indexed by the sets of sites U and G are $\mathbf{Z}_u^{N_g \times 1}$ and $\mathbf{Z}_g^{N_s \times 1}$ respectively. The process and its distribution over $\mathscr{S}$ are represented by

$$\begin{pmatrix} \mathbf{Z}_u \\ \mathbf{Z}_g \end{pmatrix} \sim N_{N_u + N_s} \left(\mu, \begin{bmatrix} \Sigma_{uu} & \Sigma_{ug} \\ \Sigma_{gu} & \Sigma_{gg} \end{bmatrix} \right) \tag{14.2}$$

where Σ_{gg} is the covariance between gauged sites, Σ_{uu} between ungauged sites and Σ_{gu} between gauged and ungauged sites.

The predictive distribution is then given by

$$(\mathbf{Z}_u | \mathbf{Z}_g = \mathbf{z}_g) \sim N(\Sigma_{ug} \Sigma_{gg}^{-1} (\mathbf{z}_g - \mu), \Sigma_{uu} \Sigma_{ug}' \Sigma_{gg}^{-1} \Sigma_{ug}) \tag{14.3}$$

Example 14.1. *An application of the spatial deformation approach*

We now give some further details of the application described in 13.14. In that example, the geographic surface was spatially deformed to achieve a nonstationary daily maximum temperature field. The design goal there was to add 5 sites from a grid of possible new monitoring sites. The existing grid consisted of 18 of the current monitoring sites in California. The spatial field of residuals after subtracting an estimate of the mean proved highly nonstationary so spatial deformation was used. Although a D–plane was found on which the field of residuals seemed stationary, the deformation needed was severe. Current research of the second author revealed that this was due to a failure to include in the estimated mean field the complex three-way interaction between latitude, longitude and time. In summer unlike winter the maximum temperature field was fairly flat from one longitude to the next but only in the southern latitudes. Much less warping is needed when that interaction is accounted for.

In Chapter 13 we saw examples of how a monitoring network might be expanded; one hypothetical (Example 13.14) and one real (Example 13.1). The second of these related to the need for changes in the Metro Vancouver monitoring network. The population has grown substantially and with it the levels of air pollution. So a study of the Metro Vancouver air pollution monitoring network was undertaken (Ainslie et al., 2009). Deficiencies in the network were found and subsequently the network was modified. Key to the analysis was deformation of the spatial fields of each of the various air pollutants being studied. In this case, the deformation was largely in response to topography, in particular the mountains on the northern edge of the urban area.

The analysis in Example 13.13, which describes the Metro Vancouver's PM_{10} monitoring network, also required spatial deformation. This hypothetical example demonstrated how 6 sites were selected from 20 candidate locations in highly populated areas to augment the existing 10 site PM_{10} network.

Example 14.2. *Extending a composite air pollution network*

The subject of this example was the analysis reported in Zidek et al. (2000) and described in Example 13.2. Briefly, this was a network that grew out of a combination of acid rain monitoring networks that came to be used for monitoring air pollution. Spatial deformation had a critical role to play. The application in Zidek et al. (2000) statistically integrated the disparate networks into a combined network which was then extended in an optimal way.

At the time the combined network consisted of 31 air pollution monitoring sites in southern Ontario. Each of these sites monitored one or more pollutants including ionic sulphate (SO_4), sulphite (SO_2), nitrite (NO_2) and ozone (O_3).

This redesign presented two major challenges: (i) the spatio–temporal process was multivariate in nature; (ii) the data being collected were misaligned in that not all sites measured the same pollutants. The redesign could therefore involve either adding monitors to an existing site and/or creating new sites. A 'quasi-site' referred to existing sites to which monitors could be added. To illustrate, a location, s, with two pollution monitors would be designated as having two quasi-sites indicating that two additional monitors could be attached. This option would be less expensive than creating a new monitoring location elsewhere due to startup costs.

Zidek et al. (2000) analysed monthly data for ozone and sulphate as these had previously been found to be strongly associated with hospital admissions for respiratory morbidity (Burnett et al., 1994; Zidek et al., 1998). There were $2 \times 31 = 62$ quasi-sites of which 10 measured SO_4 and 21 O_3. This left 31 ungauged sites. Monitors were to be located at an additional 15 sites (30 quasi-sites) The spatial domain $\mathscr{S}$ was taken to contain the original 31 sites plus the additional 15 to give a total of 46 sites.

The approach generally followed that described in Chapter 11, Section 11.5) but with some novel elements. The process model assumed temporally uncorrelated sequence of two-dimensional vectors $\tilde{s}_{st}$, $s \in \mathscr{S}$ with coordinates representing the O_3 and SO_4 concentrations. These responses could then be combined into $\tilde{s}_t$ a 46×2 process matrix. It was assumed that the within site correlation matrices were identical and equal to Ω. The hyper–covariance matrix for $\tilde{s}_t$, after adopting an inverted Wishart distribution as the prior for the spatial covariance matrix, was $\Phi = \Lambda \otimes \Omega$ where Λ represents the spatial covariance between sites. Measurement error was ignored and the available data were used to estimate Λ using the EM algorithm which was used to impute the missing data. That provided an estimate $\hat{\Lambda}_{gg}$ for Λ_{gg} from which the intersite dispersions could be estimated. The substantial non-stationarity seen in those estimates led to use of the spatial deformation method to eliminate folds in the geographic surface of southern Ontario. This allowed an approximately stationary covariance to be found.

An exponential variogram was fit based on the spatial deformation analysis and was extrapolated over the entire domain $\mathscr{S}$ of gauged and ungauged sites to obtain a hyper-covariance for the process.

Zidek et al. (2000) then consider the problem of design using the entropy approach where keeping the quasi-site structure was essential. Given that the posterior distribution for the process had already been characterised, this turned out to be relatively straightforward. They transformed the original process vectors as $\tilde{s}_t^* = \tilde{R}^T \tilde{s}_t$ where

$$\tilde{R} = diag\{I_{2u}, R\}$$

and R, defined in Le et al. (1997), simply permutes the coordinates of $\tilde{s}_t$ corresponding to the gauged sites, i.e. $\tilde{s}_{gt}$ so that responses corresponding to the quasi-sites constituted the bottom rows of the $\tilde{s}_{gt}^*$ matrix.

With this transformation and the resulting change in the hypercovariance matrix the entropy approach could be used with the transformed version of the approximately stationary spatial covariance matrix.

In this case, the optimisation needed to account for costs. This was incorporated by using a linear combination of the log determinant from the entropy approach and these costs as the optimisation criteria. More precisely,

$$O(\tilde{S}^{add}) = E(\tilde{S}^{add}) - ED \cdot C(\tilde{S}^{add}) \tag{14.4}$$

where instead of S^{add}, the sites to be added, we use $\tilde{S}^{add}$ to represent the quasi-sites to be added. In Equation 14.4 $E(\tilde{S}^{add})$, ED and $C(\tilde{S}^{add})$ denote the entropy, the entropy to dollar conversion ratio and cost respectively.

The costs were based on estimates from air pollution monitoring authorities. Over a 60-month period, the monthly costs for buying, installing and operating a single gauge at a formerly ungauged site were estimated to be $1,334. For an already gauged site, that monthly cost of adding a gauge would be considerably less at $292. Not surprisingly, the optimal choice of quasi-sites to add depended on the ED ratio; for all ED in the interval considered, $0.05 \le ED \le 0.015$, all additions were pseudo-sites on existing sites and no new sites were added.

In developing methods for coping with non-stationary in the spatio–temporal modelling there are a number of issues that will arise. They include the following:

- Handling multivariate data.
- Dealing with high degrees of non-stationary that even the spatial deformation approach may not be able to handle well.
- The need to provide a degree of robustness against the misspecification of the spatial mean field.
- The need to contend with monotone and misaligned data patterns in the data.

14.2.2   Dimension expansion

Dimension expansion has similarities with spatial deformation but it differs in that the locations in the geographic space are retained, with added flexibility obtained through the extra dimensions. It addresses one of the major issues with the image warping approach, the folding of the space (Bornn, Shaddick, & Zidek, 2012).

This idea goes back a long way; it was described by Edwin A. Abbott in 1884 although it has been reprinted many times since (Abbott, 2009).

> "Place a penny on one of your tables in space; and leaning over look down upon it. It will appear as a circle. But now, drawing back to the edge of the table, gradually lower your eye....and you will find the penny becoming more and more oval...until you have placed your eye exactly on at the edge of the table [when] ...it will become a straight line."   Edwin A. Abbott (1884).

Example 14.3. *Gaussian spatial process on half-ellipsoid*

In this example, Abbott's flatlander lives on a two-dimensional disk as seen in the right hand panel of Figure 14.1. He sees a distorted version of the field in which points on opposite sides of the disk have the same concentrations unlike those in the centre which have much heavier concentrations. However a person living in three dimensions sees a much clearer pattern with the higher values being associated with the magnitude of the third dimension. This is often the case in measuring air pollutants or solar radiation where elevation, along with x and y coordinates will be an important factor. Ignoring such a third dimension would result in a non-stationary field whereas including it in a model may help in achieving stationarity.

This observation is confirmed by the empirical variogram plot in Figure 14.2 based on a random sample of sites, where we see an obvious improvement when distance is now calculated using the three-dimensional coordinate (left hand panel) compared to the original variogram seen on the right.

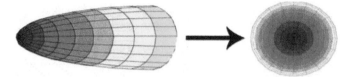

Figure 14.1: The left hand figure shows a Gaussian process on a half-ellipsoid with the right hand figure showing the effect that will be seen when it is compressed to two dimensions which results in a disk centred at the origin. The resulting field is an example of one in which Abbott's flatlander might live.

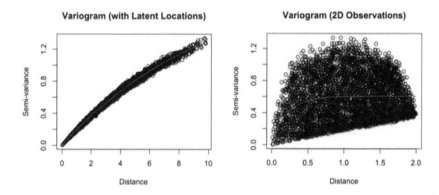

Figure 14.2: Variograms before and after dimension expansion showing again the same effect seen in Figure 14.1. Before expansion there are obvious signs of non-stationarity which have been addressed by the dimension expansion.

The dimension expansion method starts in two-dimensional space, the disk in Figure 14.1, and seeks additional dimensions in a new domain that will resolve non-stationarity. In this example it finds the original three-dimensional surface quite precisely without any prior specification being passed to the algorithm. In contrast, spatial deformation does not resolve the problems of non-stationarity in this example (Bornn et al., 2012).

Adding new dimensions

As seen in Example 14.3, embedding the original field in a space of higher dimension can resolve non-stationarity. This begins with the original site coordinate vectors $s_1, \ldots, s_g \in S$ each of dimension d. These are augmented to get new site coordinate vectors $[s_1, \tilde{s}_1], \ldots, [s_g, \tilde{s}_g]$ each of dimension $d + p$. The ultimate goal is a stationary process $Z_{[s, \tilde{s}]t}$, $[s, \tilde{s}] \in S \times \tilde{S}$, $t \in T$, with variogram

$$\Gamma_\phi(d_{ij}) \tag{14.5}$$

where $d_{ij} = [s_i, \tilde{s}_i] - [s_j, \tilde{s}_j]$.

To define the search algorithm let

$$\mathbf{S} : (g \times d) = \begin{bmatrix} s_1 \\ \vdots \\ s_g \end{bmatrix}, \tilde{\mathbf{S}} : (g \times p) = \begin{bmatrix} \tilde{s}_1 \\ \vdots \\ \tilde{s}_g \end{bmatrix} \tag{14.6}$$

The problem then becomes that of choosing an augmented, coordinate system denoted in shorthand in matrix notation by $[\mathbf{S}, \tilde{\mathbf{S}}]$.

There is theory in support of this approach (Perrin & Schlather, 2007), which shows that (subject to moment conditions) for any Gaussian process Z on $\mathcal{R}^d$ there exists a stationary Gaussian field Z^* on $\mathcal{R}^{d+p}, p \geq 2$ such that Z on $\mathcal{R}^d$ is a realisation of Z^* . However the theory only shows existence and not how to construct Z^*.

Finding the new coordinates

One approach we could take would find the $\tilde{s}_1, \ldots, \tilde{s}_g$ such that

$$\hat{\phi}, \tilde{\mathbf{S}} = \phi, \tilde{\mathbf{S}}' argmin \sum_{i<j} (v_{ij}^* - \Gamma_\phi(d_{ij}([\mathbf{S}, \tilde{\mathbf{S}}'])))^2$$

where d_{ij} is defined in Equation 14.5 except that it now includes the variable coordinate system in its argument $[\mathbf{S}, \tilde{\mathbf{S}}']$. Here v_{ij}^* is an estimate of variogram (spatial dispersion between sites s_j and s_j) e.g. with data Y,

$$v_{ij}^* = \frac{1}{|\tau|} \sum_\tau |Y_{s_i} - Y_{s_j}|^2$$

with $\tau > 1$ indexing some relevant observation pairs.

Given the matrix $\tilde{\mathbf{S}} \in \mathcal{R}^g \times \mathcal{R}^p$ we need to construct an f with $f(\mathbf{S}) \approx \tilde{\mathbf{S}}$. Note that we could follow Sampson and Guttorp (1992) and determine thin plate splines f with a smoothing parameter λ_2. This would in turn give us f^{-1} to carry us from the manifold in $\mathcal{R}^{g+p}$ defined by $\{[\mathbf{S}, f(\mathbf{S})], \mathbf{S} \in \mathcal{R}^d\}$ back to the original space. In other words, $f^{-1}(\tilde{\mathbf{S}}) = \mathbf{S}$ and so no issues arise around the bijectivity of f.

However, this approach would not tell us the number of new coordinates that are required. This could be found using cross-validation or model selection to determine the dimension of $\tilde{S}$. However for parsimony and to regularise in the optimisation step we instead solve

$$\hat{\phi},\tilde{\mathbf{S}} \quad = \quad \phi,\tilde{\mathbf{S}}'argmin \sum_{i<j}(v^*_{s_i,\tilde{s}_j} - \Gamma_\phi(d_{s_i,\tilde{s}_j}([\mathbf{S},\tilde{\mathbf{S}}'])))^2 + \lambda_1 \sum_{k=1}^{p} ||\mathbf{S}'_{.,k}||_1$$

where λ_1 regularizes estimation of $\tilde{S}$ and may be estimated through cross-validation. But other model fit diagnostics or prior information could be used.

To solve the optimisation problem, we could proceed as in traditional multi-dimensional scaling. However this would not work since the objective function would not have unique maximum. Our optimisation is more regularised, due to its penalty function with the result that the optimisation is unique (up to sign and indices of zero/non-zero dimensions). Finally, the gradient projection method (Kim, Kim, & Kim, 2006) is used to carry out the optimisation.

Example 14.4. *Black smoke in the United Kingdom*

In this example we consider a measure of particulate matter, black smoke, which has been measured in the United Kingdom over many decades (Elliott et al., 2007). This field has been shown to be highly non-stationary (Bornn et al., 2012).

The left hand panel of Figure 14.3 shows covariance as a function of the intersite distance and gives a clear indication of non-stationarity. The spatial

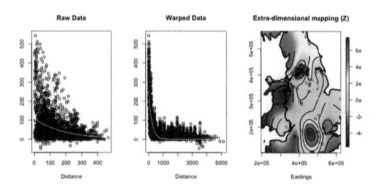

Figure 14.3: Black smoke concentrations in the United Kingdom. The covariance (against distance) of the original data appears in the left hand panel and shows clear signs of non-stationarity. The centre panel shows the result of warping geographic space that shows a field that is much closer to being stationary. The right hand panel shows the modelled surface after dimension expansion.

deformation method transforms geographic coordinates to dispersion coordinates and the result is seen in the middle panel which shows a substantial improvement. The result of dimension expansion appears in the right hand panel.

14.3   Physical–statistical modelling

A modern frontier in spatio–temporal modelling combines deterministic and statistical models. The former embrace scientific knowledge and are deterministic since the future is determined by the past with certainty. Although their parameters may need to be estimated, uncertainty associated with these estimates is not reflected in the outcomes they generate. Climate models give future atmospheric temperatures (that have potential health impacts, see Example 9.5) under various greenhouse gas emission scenarios without any expression of the uncertainty about these calculated temperatures. Statistical analysis of their outcomes has shown them to be biased (Jun, Knutti, & Nychka, 2008). Similarly, chemical transport models (CMTs) predict concentrations of air pollutants over various spatial locations without quantification of their inherent uncertainty. The outputs from such models is called 'simulated data'.

These models have a number of limitations. The numerical computer models that represent these deterministic models can take a long time to run. As a result, they are commonly simplified by avoiding microscale phenomena such as evaporation, condensation and turbulence by running them at larger spatio–temporal scales, e.g. the mesoscale. Climate models may generate output for grid cells of fifty kilometres squared (Kalnay, 2003). Their output may be perfectly valid at that scale of resolution but they need to be downscaled where possible, or subgrid scale processes modelled, when small spatio–temporal resolution is required. Another difficulty faced when using non-linear deterministic models e.g. for weather, is their sensitivity to their initial conditions. This can lead to the so-called 'butterfly effect' where small changes in their inputs lead to large changes in their outputs. Data (observations) may therefore be required in order to periodically restart or adjust the models (data assimilation) with new inputs although their inclusion is often somewhat ad-hoc. Also, as noted above they are not able by their very nature to provide estimates of the uncertainty in their simulated data.

In contrast, statistical models are designed to provide just such estimates of uncertainty. However they do not usually incorporate an extensive base of scientific knowledge even when Bayesian methods are used. The idea of combining these two approaches seems appealing and this is the topic of much current research. It is possible by combining the two approaches in a hierarchical Bayesian model, to estimate the fraction of ground level ozone in Vancouver, Canada, that is transported there at a specified time from outside sources (Kalenderski & Steyn, 2011). Cressie and Wikle (2011) provides a review on the use of deterministic dynamic models in a spatio–temporal setting.

14.3.1 Dynamic processes

In this section deterministic processes are denoted using lower case and random processes using upper case. To aid clarity, we assume that the underlying processes Z are observable and are therefore represented by Y.

We begin with dynamic temporal processes which are indexed by continuous time and are characterised by

$$\frac{\partial y_t}{\partial t} = \dot{y}_t = H[y_t], \ t \geq 0 \tag{14.7}$$

In discrete time they will be characterised by

$$\nabla y_t = y_t - y_{t-1} = H[y_t], \ t \geq 0 \tag{14.8}$$

Note the formal resemblance of these equations to the state space models seen in Chapter 10, Section 10.7. As H and the y_0 are specified, these equations make y a deterministic process. A non-linear H can lead to chaotic behaviour in deterministic systems. In that case small changes in y_0 can lead to huge variations in y_t, i.e. the butterfly effect.

A very simple linear system would be $\nabla Y_t = \theta y_t$. If Y_0 were random we would get a stochastic system. Even a deterministic dynamic system can be subject to random environmental disturbances so that

$$\frac{dY_t}{dt} = \dot{Y}_t = H[Y_t] + v_t$$

where v is random and independent of Y. A Bayesian approach makes parameters in a deterministic system described by H random. In all these cases and others we get a random process even though we start with a conceptually deterministic one.

Example 14.5. *Dynamic growth model*

We use a simple linear dynamic model to illustrate some key ideas. After rescaling y_t so that $y_0 = 1$, it is given by

$$\frac{dy_t}{dt} = \dot{y}_t = \lambda y_t, \ t \geq 1 \tag{14.9}$$

Solving this gives the result $\ln y_t = \lambda(t-1)$ i.e. $y_t = \exp\{\lambda(t-1)\}$.

This will become a stochastic differential equation if the value of λ is uncertain, and hence random in a Bayesian context, i.e. the deterministic model becomes a random process model. This model could be subject to random perturbations and so would seem suitable for embedding in a statistical framework. One approach would be to express Equation 14.9 as a state space model by turning the differential equation into a difference equation:

$$\dot{y}_t \quad \approx \quad \nabla y_t \tag{14.10}$$

$$= \quad y_t - y_{t-1} = \lambda y_{t-1} \tag{14.11}$$

Adding the random perturbation term we get $Y_t = (\lambda + 1)Y_{t-1} + v_t$, $t \geq 1$ a state space model (see Chapter 10, Section 10.7).

Example 14.6. *Infectious disease as a dynamic process*

In this example, the approach illustrated in Example 14.5, can be used for any system of partial differential equations, as long as there are not too many of them.

The left hand panel of Figure 14.4 shows the number of cases of mumps per month in New York City during 1950–1971 (Source: Hipel and McLeod, 1994, in Times Series Data Library, file: epi/mumpsmo). A closer look at the 1970–1972 period is seen in the right hand panel. Notice the apparent seasonal pattern. A vaccine was licensed in 1967 and led to a dramatic (99%) decline in the number of cases in the United States (van Loon et al., 1995).

One could attempt a classical temporal process modelling approach for the number of cases of mumps, however the cyclical nature of an epidemic for an infectious disease is well understood and this understanding should ideally be incorporated in a model. Attempts in that direction began at least as far back as 1927 (Kermark & Mckendrick, 1927) with the publication of the susceptibles–infected–recovered (SIR) model. Let u_t, v_t and w_t denote the

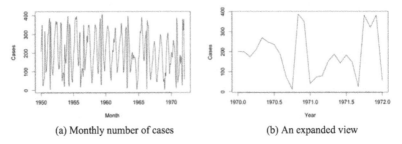

(a) Monthly number of cases   (b) An expanded view

Figure 14.4: The number of cases of mumps in New York City during the period 1950–1971. Panel (a) shows the monthly number of cases from 1950–1971 with a clear seasonal pattern in the number of cases. Panel (b) shows an expanded view for 1970–1972.

numbers in the three categories of the SIR model. The sum of the numbers in the categories will be N, the population size.

The following is a dynamic process model, which describes how cases enter and exit the infective population:

$$
\begin{aligned}
\dot{u}_t &= -\beta_N u_t v_t \\
\dot{v}_t &= \beta_N u_t v_t - \gamma_N v_t \\
\dot{w}_t &= \gamma_N v_t
\end{aligned}
$$

This process can be turned into a stochastic model with random elements U_t, V_t, and W_t replacing their deterministic counterparts (Leduc, 2011). Since U is determined by V and W only these are stochastically modelled. They in turn can be modelled by using classical stochastic process models with their deterministic counterparts serving as model parameters as follows: U_t is a non-homogeneous pure death process with rate $\beta_N^* u_t v_t$ and W_t is a non-homogeneous Poisson (birth) process with rate γv_t.

The approaches seen above will not work in the case of complex models such as CMAQ, the Community multi-scale air quality model chemical transport model for air pollutants that builds in emissions, meteorology, atmospheric chemistry and atmospheric transport. There may be as many as one hundred differential equations just to represent the chemistry alone and hence no hope of turning the deterministic model into a stochastic one by methods like those described above. Instead, these numerical models must be treated as 'black box' simulators and their outputs used as inputs into a statistical modelling framework (Berrocal, Gelfand, & Holland, 2010; Zidek, Le, & Liu, 2011).

14.4   The problem of extreme values

A major component of risk analyses is the modelling of extreme values. Reliability theory classifies events by their return periods. A 1000 year return period for the event of a failure of a nuclear power generating facility, such as Chernobyl, indicates the period in which one failure is expected.

Although the statistical study of extremes spans at least a century (Fisher & Tippett, 1928), new frontiers present themselves due to advances in technology. One developing field is that of multivariate extreme value theory with process response vectors whose dimensions can be in the hundreds of thousands. Conceptual issues may arise where there are differences in the size of the dimensions of the responses. One approach is suggested by Heffernan and Tawn (2004), who propose a conditional approach based on the assumption that the asymptotic form of the joint distribution has an extreme component.

We now describe some specific ways in which extremes arise in environmental epidemiology along with issues arising in network design, especially that of preferentially selecting locations in order to detect non-attainment of regulatory standards.

The inevitable lack of data at extreme values makes modelling the tails of a distribution difficult or even impossible. This lead to a search for distribution theory that could be justified by weak but plausible assumptions. Fisher and Tippett (1928) developed a trinity of distributions that make up the entire family of possibilities for an extreme value distribution for a single data record, e.g. the sequence of monthly maxima of a pollution series at a single site.

The Fisher–Tippett trinity is reached by assuming a sequence of independently and identically distributed process values observed without error Y_t, $t = 1, \ldots, T$ with a common CDF F_Y. Then $M_T = \max\{Y_1, Y_2, \ldots, Y_T\}$ must converge to one of three distributions as $T \to \infty$, the Gumbel distribution, the Fréchet distribution or the Weibull distribution, Later it was recognised that all three could be combined in a single parametric family of distributions now called the generalised extreme value distribution (GEV): More precisely for a sequence of normalising constants a_T and b_T that keep the following limit from being degenerate,

$$P(\frac{M_T - b_T}{a_T} \leq m) \to G(m) \qquad (14.12)$$

as $T \to \infty$ where G must be of the form

$$G(m) = \begin{cases} exp\left[-\left\{1 + \xi\left(\frac{(m-\mu)}{\sigma}\right)\right\}^{-1/\xi}\right], & 1 + \xi(m-\mu)/\sigma > 0, \ \xi \neq 0 \\ exp\left\{-exp\left[-\frac{(x-\mu)}{\sigma}\right]\right\} & \xi = 0 \end{cases} .$$

$$(14.13)$$

For diagnostic analysis, a qqplot is useful. This can be constructed by letting $q_{tT} = (t - 1/2)/T$ and $e_{tT} = \hat{G}^{-1}(q_{tT})$ and plotting q against e. Notice that the q's are just the quantiles of the empirical distribution function.

Further developments have led to a rich theory for the case of a single series (Gumbel, 2012; Leadbetter, 1983; Coles, Bawa, Trenner, & Dorazio, 2001; Embrechts, Klüppelberg, & Mikosch, 1997). An alternatives approach is the class of peak over threshold (POT) models that look at exceedances over high thresholds. The number of such exceedances can then be modelled using a non-homogeneous Poisson process. The response can also be modelled conditional on its exceeding a specified threshold. Its conditional distribution will be the generalised Pareto distribution (Pickands III, 1975; Davison & Smith, 1990) with right-hand tail,

$$G(m) = 1 - \lambda \left\{1 + \frac{\xi(m-u)}{\sigma}\right\}_+^{-1/\xi}, \quad m > u \qquad (14.14)$$

for parameters $\lambda > 0$, $\sigma > 0$, and $\xi \in (-\infty, \infty)$. Note that as $\xi \to 0$ in Equation 14.14, $G(m) \to 1 - \exp[-(m-u)/\sigma]$, $m > u$, in words to an exponential distribution, a surprisingly simple result. If you delete the residuals above u in the series, the sequences of heights of the series above u will have an approximately exponential distribution.

Example 14.7. *Hourly ozone extreme values*

Smith (1989) presented what was one of the first analyses of the extreme values of air pollutants from the perspective of regulatory policy where both their extreme values as well as their exceedances of a specified threshold are important. Amongst other things, the paper investigates the fit of the GEV model to 61 day maxima of hourly O_3 concentrations based on 199,905 observations and found the GEV provides a good fit.

Our illustrative example is much more modest in scope in that we fit the GEV distribution to the daily, one week and two week block maxima for an hourly series of O_3 measurements recorded at a site in Illinois for 2004 to 2014. The data was obtained from the EPA's AQS database (https://ofmext.epa.gov/AQDMRS/aqdmrs.html).

Here we show a selection of the code that was used. The full code is included in the online resources. We compute block maxima for four periods; daily, weekly, two weekly and four weekly. Figure 14.5 shows the series of daily maxima which shows the high degree of variability (volatility) of an extreme value compared to say the sample average.

```
library(fExtremes)
illinoiso04_14.dat <-read.table("illinois_aqs03data2004_
    2014.txt", sep="",header=FALSE)

### blockMaxima computes the block maxima for blocks of a
    specified number
### of hours Daily, Weekly Two Weekly and Four weekly maxima

bm24<-blockMaxima(illinoiso04_14.dat\$V1, block = 24,doplot=
    FALSE)
bm168<-blockMaxima(illinoiso04_14.dat\$V1, block = 168,
    doplot=FALSE)
bm336<-blockMaxima(illinoiso04_14.dat\$V1, block = 336,
    doplot=FALSE)
bm672<-blockMaxima(illinoiso04_14.dat\$V1, block = 672,
    doplot=FALSE)
```

Next we fit a GEV distribution and plot the fit using a histogram. This can be seen in Figure 14.6 and indicates a good fit.

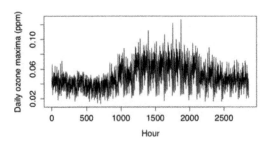

Figure 14.5: Daily ozone maxima (ppm) at a site in Illinois over the period 2004 to 2014. Notice the volatility of this series. Data source: US EPA Air Data Site

```
fit24<-gevFit(illinoiso04_14.dat\$V1, block = 24, type = "
    mle")

######## Results: Estimated Parameters:
######## xi            mu           beta
######## -0.10574035  0.04434469  0.01323267

hist(bm24, nclass = NULL, freq = FALSE,, xlim = c(0,0.12),
    ylim=c(0,30),xlab = "O3 daily maxima (ppm)",  ylab="
    density", main = "")
x = seq(0,0.12, by = 0.001)
lines(x, dgev(x, xi = -0.09164308, mu = 0.04218148, beta =
    0.01400470), col = "black")
```

Remember the upper tail is particularly important and so further diagnostics are important and to assess this we construct a qqplot. This can be seen in Figure 14.7.

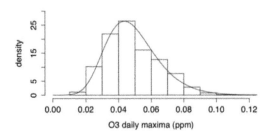

Figure 14.6: A comparison of the fitted generalized extreme value distribution with the raw data for daily ozone maxima (ppm) at a site in Illinois over the period 2004 to 2014. The fit seems reasonably good but note that it's the right tail that counts most. Data source: US EPA Air Data Site

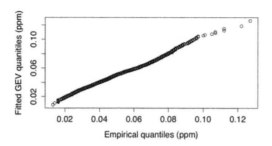

Figure 14.7: A comparison of the fitted generalized extreme value distribution with the raw data through a diagnostic qqplot. Again the fit looks good except for the all important right tail where it droops a little. Data source: US EPA Air Data Site

```
e<-(1:length(bm24)-1/2)/length(bm24)
empq<-sort(bm24)
fitq<-qgev(e,xi = -0.09164308, mu = 0.04218148, beta =
    0.01400470)
plot(empq,fitq,xlab='empirical quantiles (ppm)',ylab='Fitted
    GEV quantiles (ppm)')
```

Again the result looks good except for the all important right tail. On the right hand side, we see that the fitted quantiles are too high relative to their empirical quantiles as seen by the empirical demonstration shown in Figure 14.7.

```
#Interpreting the plot.
length(e)-length(e[e>=0.99])
### Result: 2827
bigq<-cbind(empq[2828:2856],fitq[2828:2856])
bigq
```

The calculations above show that the first empirical quantile above the 0.99 level in empq is 0.091 (ppm) while for the fitted GEV distribution it is $0.095(ppm)$. That result shows the fit to be unsatisfactory as it shows the 99th percentile of the fitted distribution is too high. To emphasise this, suppose if it had been $0.400(ppm)$ instead of 0.095 (ppm). In this case, the model would indicate that the population living near that monitor would experience O_3 concentrations in excess of 400 (ppb) about 3 or 4 days a year.

Example 14.8. *Air quality criteria and nonattainment*

This example shows how problems of extreme values arise in relation to health impact analysis. The National Ambient Air Quality Standards are exemplary; the US Clean Air Act (1970) requires that they be set and regularly reviewed to protect human health and welfare while allowing for a margin of safety. Specifically Section 109(b)i states:

"National primary ambient air quality standards, prescribed, under subsection (a) shall be ambient air quality standards the attainment and maintenance of which in the judgment of the Administrator, based on such criteria and allowing an adequate margin of safety, are requisite to protect the public health. Such primary standards may be revised in the same manner as promulgated."

Regions found to be in non-attainment of the NAAQS are required to submit a plan for getting back into attainment. Specifically Section 172(a)(2) states:

"(2) ATTAINMENT DATES FOR NONATTAINMENT AREAS (A) The attainment date for an area designated nonattainment with respect to a national primary ambient air quality standard shall be the date by which attainment can be achieved as expeditiously as practicable, but no later than 5 years from the date such area was designated nonattainment under section 107(d), except that the Administrator may extend the attainment date to the extent the Administrator determines appropriate, for a period no greater than 10 years from the date of designation as nonattainment, considering the severity of nonattainment and the availability and feasibility of pollution control measures."

In other words, areas in nonattainment must get back into attainment within ten years at most. Severe financial penalties are imposed for nonattainment, so the impact of a change in the standards can cost states with areas not in attainment a small fortune.

The criteria vary for different air pollutants. The eight-hour standards made by an US EPA Clean Air Scientific Advisory Committee for Ozone (on which the second author served) are based on daily maximum 8-hour averages over the entire year. The standard states that the "Annual fourth-highest daily maximum 8-hour concentration, averaged over 3 years shall not exceed 0.075ppm". It is understood that this will be computed for each site in an urban area, with network design standards stipulating a minimum number of sites for such an area based on population size. The area will be in nonattainment if any site is in nonattainment.

This complicated standard was reached after much risk analysis using concentration response functions (CRFs) that had been published by environmental epidemiologists. Other averaging times were considered along with other quantiles. The standards had to make a compromise between competing targets. They had to be relatively stable over time, for example weather conditions influence the production of ozone and three-year averaging was deemed to be sufficiently long as to make the metric sensitive to a structural change in ozone production, while insensitive to changes in weather. On the other hand, the primary goal of the standards was to protect human health and too much smoothing of the series might result in important changes, which may be important with regard to human health, being overlooked.

Of course, underlying all of this analysis were the raw data and these will to a large extent depend on the siting of the monitors. Preferential siting was conclusively demonstrated in the UK as the black smoke monitoring network changed over time (Shaddick & Zidek, 2014; Zidek et al., 2014) although to the knowledge of the authors no such analysis has been performed for the networks in the US.

Example 14.8 shows that where environmental risk is concerned, it's the extreme values, however measured, that matter. Networks should therefore be designed with this in mind. It shows that the extremes can be calculated from complex metrics whose distributions would not have simple analytic forms. This points to the need for spatial predictive distributions that can be used to realistically simulate these metrics over a spatial region of interest. Such predictive distributions would allow calculation of the metric at locations where there are no monitoring sites However, the spatial prediction of extremes is much more difficult than predicting central values. Finally the example points to the need for an optimal design theory and with it clearly stated objectives. Should the network be designed to optimally detect nonattainment? Or to protect human health by monitoring areas with susceptible populations? How many sites are needed to do that effectively? Recall that each site in an area must be in attainment of the air quality standards that are deemed to protect human health, but what would happen if a location that was currently unmonitored was predicted to be in nonattainment?

Chang et al. (2007) explore the problem of finding designs for regulating and controlling air pollution in urban regions. Their findings using a combination of empirical analysis and simulation studies include the following:

1. Current networks may provide data that can characterise the environmental hazard field relatively well but are less able to characterise extremes. That is in part because extremes, for example using the metric in Example 14.8, tend to lose spatial correlation.

2. Even if the objective is to find areas in nonattainment, the interpretation of that goal may be difficult. Presumably it would be based on the probability of detecting nonattainment but should locations be chosen to give the highest probability of detecting nonattainment, i.e. where levels might be expected to be highest, or where nonattainment is least likely to occur? These two approaches give different networks!

3. In the absence of a clear criterion, the entropy-based criterion has been shown to give reasonable results as long as the multivariate–marginal distribution for the environmental field is appropriate.

14.5 Summary

In this chapter the reader will have encountered a selection of new frontiers in spatio–temporal epidemiology including the following:

- A selection of areas that are currently under active development.
- Two modern approaches to addressing the problem of non-stationarity in random spatio–temporal fields; warping and dimension expansion.
- How dimension expansion can be used to dramatically reduce non-stationarity and suggest its possible causes.
- A powerful approach combining both physical and statistical modelling within a single framework.

Exercises

Exercise 14.1. Return to Example 14.7 and

(i) plot the times series for weekly, two weekly and four weekly blocks. Comment on the differences you see between them and the 24-hour version in Example 14.7.

(ii) fit the generalized extreme value distribution for each of the cases in part (i). Comment on the quality of the fits.

Exercise 14.2. Return to the Example 14.7 but this time use the peak over threshold approach with various thresholds including the 0.075ppm which is currently the US Air Quality Standard.

(i) Plot a histogram of the exceedances of your thresholds and comment on its shape vis a vis its resemblance to what might be expected for an exponential distribution.

(ii) Plot a generalized Pareto distribution's PDF over your histogram in Part (i).

Exercise 14.3. Repeat Exercise 14.1 for $PM_{2.5}$.

Appendix 1: Distribution theory

A.1 Overview

This appendix provides some normal and related probability distribution theory. Much more detail about this theory and its application can be found in Le and Zidek (2006) and Brown (2002)

A.2 The multivariate and matric normal distributions

The random vector $\mathbf{Z}^{p \times 1}$ is said to have a multivariate normal distribution, $\mathbf{Z} : p \times 1 \sim N_p(\mu, \Sigma)$, if for any $\mathbf{a} : p \times 1$,

$$\mathbf{a}^T \mathbf{Z} \sim N(a'\mu, \mathbf{a}^T \Sigma \mathbf{a}) \tag{A.1}$$

where $N(\mu, \sigma^2)$ denotes the univariate normal distribution with expectation and variance, respectively, μ and σ^2. If $\Sigma : p \times p > 0$ (i.e. is positive definite and hence invertible) Equation A.1 implies that $\mathbf{Z}$ has the pdf

$$f_{\mathbf{Z}}(\mathbf{z}) = (2\pi)^{-p/2} |\Sigma|^{-1/2} \exp\left\{ -(\mathbf{z} - \mu)^T \Sigma^{-1} (\mathbf{z} - \mu)/2 \right\}$$

Moreover it implies the following properties:

- $E(\mathbf{Z}) = \mu$
- $Cov(\mathbf{Z}) = \Sigma = (\Sigma_{ij})$, $\Sigma_{ij} = E[(Z_i - \mu_i)(Z_j - \mu_j)]$.

For a Gaussian spatial field $\Sigma_{ij} = Cov(Z_i, Z_j)$ is the covariance between process responses at sites i and j.

The multivariate normal distribution is a special case of the matric normal distribution $\mathbf{Z} : n \times m \sim N(\mu, \mathbf{A} \otimes B)$ meaning

$$f_{\mathbf{Z}}(\mathbf{z}) = \frac{1}{(2\pi)^{nm/2}} |\mathbf{A}|^{-m/2} |\mathbf{B}|^{-n/2} \text{etr} \left\{ -\frac{1}{2} [\mathbf{A}^{-1}(\mathbf{z} - \mu)][(\mathbf{z} - \mu)\mathbf{B}^{-1}]^T \right\} \tag{A.2}$$

where $\mathbf{A} = (a_{ij}) : n \times n > 0$ and $\mathbf{B} = (b_{ij}) : m \times m > 0$, and $etr = \exp tr()$ and $tr(\mathbf{A}) = trace(\mathbf{A}) = \sum a_{ii}$ for any matrix $\mathbf{A}$.

Let *vec* denote an operator that converts a matrix to a vector by stacking its transposed rows into a tall column vector. Then the matric-normal distribution defined by Equation A.2 has the following properties:

- $E(\mathbf{Z}) = \mu$
- $\text{var}[vec(\mathbf{Z})] = \mathbf{A} \otimes \mathbf{B}$ and $\text{var}[vec(\mathbf{Z}^T)] = \mathbf{B} \otimes \mathbf{A}$.

- $\mathbf{Z} \sim N(\mu, \mathbf{A} \otimes \mathbf{B})$ if and only if $\mathbf{Z}^T \sim N(\mu^T, \mathbf{B} \otimes \mathbf{A})$.
- $\text{cov}(\mathbf{Z}_i, \mathbf{Z}_j) = a_{ij}\mathbf{B}, \text{cov}(\mathbf{Z}^{(i)}, \mathbf{Z}^{(j)}) = b_{ij}\mathbf{A}$.

  For any matrices $\mathbf{C} : c \times n, \mathbf{D} : m \times d, \mathbf{F} : m \times m$ and $\mathbf{G} : n \times n$:

- $\mathbf{CZD} \sim N(\mathbf{C}\mu\mathbf{D}, \mathbf{CAC}^T \otimes \mathbf{D}^T\mathbf{BD})$
- $E[\mathbf{ZFZ}^T] = \mu\mathbf{E}\mu^T + A\text{tr}(\mathbf{FB})$ and
- $E[\mathbf{Z}^T\mathbf{GZ}] = \mu\mathbf{D}\mu^T + \text{tr}(\mathbf{AG})\mathbf{B}$.

Thus

$$\mathbf{E}[\mathbf{ZB}^{-1}\mathbf{Z}^T] = \mu\mathbf{B}^{-1}\mu^T + m\mathbf{A}$$

and

$$E[\mathbf{Z}^T\mathbf{A}^{-1}\mathbf{Z}] = \mu^T\mathbf{A}^{-1}\mu + n\mathbf{B}$$

A.2.1 Multivariate and matric t-distribution

The multivariate t-distribution is related to the multivariate normal distribution in much the same way as in the univariate case. We designate it by $\mathbf{Z} : p \times 1 \sim t_p(\mu, \mathbf{A}, \nu)$ which means

$$f_\mathbf{Z}(\mathbf{z}) = \frac{\Gamma\left(\frac{p+\nu}{2}\right)\sqrt{|\mathbf{A}|}}{\Gamma(\nu/2)\sqrt{2\pi p}} \times \left[1 + \frac{1}{\nu}(\mathbf{z} - \mu)^T\mathbf{A}(\mathbf{z} - \mu)\right]^{-(p+\mu)/2}$$

where $\mathbf{A} > 0$ represents the precision matrix, $E(\mathbf{Z}) = \mu$ and $Cov(\mathbf{Z}) = \frac{\nu}{\nu-2}\mathbf{A}^{-1}$.

Likewise for the matric t-distribution $Z : n \times m \sim t_{n \times m}(\mu, \mathbf{A} \otimes \mathbf{B}, \delta)$, δ being the degrees of freedom means

$$f_\mathbf{Z}(\mathbf{z}) \propto |\mathbf{A}|^{-m/2}|\mathbf{B}|^{-n/2}|I_n + \delta^{-1}[\mathbf{A}^{-1}(\mathbf{z} - \mu)][(\mathbf{z} - \mu)\mathbf{B}^{-1}]^T|^{-\frac{\delta+n+m-1}{2}}$$

for matrices $\mathbf{A} : n \times n > 0, \mathbf{B} : m \times m > 0$ and $\mu : n \times m$. The normalizing constant is

$$K = (\delta\pi^2)^{-\frac{nm}{2}}\frac{\Gamma_{n+m}[(\delta+n+m-1)/2]}{\Gamma_n[(\delta+n-1)/2]\Gamma_m[(\delta+m-1)/2]}$$

$$\Gamma_p(t) = \pi^{\frac{p(p-1)}{4}}\prod_{i=1}^{p}\Gamma[t - (i-1)/2]$$

We now summarize the properties of the matric t-distribution:

- $\mathbf{Z} \sim t_{n \times m}(\mu, \mathbf{A} \otimes \mathbf{B}, \delta)$, if and only if $\mathbf{Z}^T \sim t_{m \times n}(\mu^T, \mathbf{B} \otimes \mathbf{A}, \delta)$.
- If $n = 1$ and $\mathbf{A} = 1$, $\mathbf{Z}$ has an m-variate t-distribution, i.e.,

$$\mathbf{Z} \sim t_m(\mu, \mathbf{B}, \delta)$$

- If $m = 1$ and $\mathbf{B} = 1$, $\mathbf{Z}$ has an n-variate t-distribution, i.e.,

$$\mathbf{Z} \sim t_n(\mu, \mathbf{A}, \delta)$$

- If $\mathbf{Z} \sim t_{n \times m}(\mu, \mathbf{A} \otimes \mathbf{B}, \delta)$, and $\mathbf{C}_{c \times n}$ and $\mathbf{D}_{m \times d}$ are of full rank (ie. rank c and d respectively), then

$$\mathbf{Y} = \mathbf{CZD} \sim t_{c \times d}(\mathbf{C}\mu\mathbf{D}, \mathbf{CAC}^T \otimes \mathbf{D}^T\mathbf{BD}, \delta)$$

- $E(\mathbf{Z}) = \mu$.
- When $\delta > 2$,

$$\text{var}[vec(\mathbf{Z})] = \delta(\delta - 2)^{-1}\mathbf{A} \otimes \mathbf{B}$$

and

$$\text{cov}(\mathbf{Z}_i, \mathbf{Z}_j) = \delta(\delta - 2)^{-1}a_{ij}B, \ \ \text{cov}(\mathbf{Z}^{(i)}, \mathbf{Z}^{(j)}) = \delta(\delta - 2)^{-1}b_{ij}\mathbf{A}$$

A.2.2   The Wishart distribution

Recall that the chi-squared random variable with p degrees of freedom can be defined by $\chi_p^2 = \sum_{i=1}^{p} Z_i^2$ where the $Z_i \sim N(0,1)$, $i = 1, \ldots p$ are independently distributed. The Wishart random variable generalizes the chi-squared—the one with p degrees of freedom replaces the $\{Z_i\}$ by independent vectors $\mathbf{Z}_i : m \times 1 \sim N_m(\mathbf{0}, \mathbf{I}_m)$, $i = 1, \ldots p$ to get

$$\mathbf{S}_p = \sum_{i=1}^{p} \mathbf{Z}_i\mathbf{Z}_i^T \tag{A.3}$$

This may be extended by replacing $\mathbf{Z}_i$ by $\mathbf{A}^{1/2}\mathbf{Z}$ in Equation A.3 where $\mathbf{A} > 0$. We denote the resulting random variable's distribution by $\mathbf{S}_p : p \times p \sim W_p(\mathbf{A}, m)$. It can be shown that the pdf of $\mathbf{S}_p$ is

$$f_{\mathbf{S}_p}(\mathbf{s}) = \left[2^{mp/2}\Gamma_p(m/2)\right]^{-1}|\mathbf{A}|^{-m/2}|\mathbf{s}|^{(m-p-1)/2}\exp^{-\frac{1}{2}tr(\mathbf{A}^{-1}\mathbf{S})} \tag{A.4}$$

for any $\mathbf{A} > 0$.

Many of the properties of the Wishart are easiest to derive with the help of the representation A.3. For example its expectation is given by $E(\mathbf{S}_p) = n\mathbf{A}$.

A.2.3   Inverted Wishart distribution

This important distribution is simply obtained by inverting the Wishart random variable. In other words, $\mathbf{Z}_p \sim W_p^{-1}(\Psi, \delta)$ if and only if $\mathbf{S}_p = \mathbf{Z}^{-1} \sim W_p(\Psi^{-1}, \delta)$. It may be shown that

$$f_{\mathbf{Z}_p}(\mathbf{z}) = 2^{mp/2}\Gamma_p(m/2)|\Psi|^{-\frac{1}{2}(\delta+p+1)}\exp\{-\frac{1}{2}\mathbf{z}^{-1}\Psi\} \tag{A.5}$$

for some $\Psi > 0$. This distribution generalizes the inverse gamma distribution.

A.2.4 Properties

- If $\mathbf{Z} \sim W_p(\Sigma, \delta)$ then $E(\mathbf{Z}) = \delta\Sigma$ and $E(\mathbf{Z}^{-1}) = \Sigma^{-1}/(\delta - p - 1)$ provided $\delta - p - 1 > 0$.
- If $\mathbf{Y} \sim W_p^{-1}(\Psi, \delta)$, then $E(\mathbf{Y}) = \Psi/(\delta - p - 1)$ and $E(\mathbf{Y}^{-1}) = \delta\Psi^{-1}$.
- If $\mathbf{Y} \sim W_p^{-1}(\Psi, \delta)$, then

$$E \log|\mathbf{Y}| = -p\log 2 - \sum_{i=1}^{p} \eta\left[\frac{1}{2}(\delta - i + 1)\right] + \log|\Psi|$$

where $\eta = $ digamma function $= d[\log\Gamma(x)]/dx$.

A.2.5 Bartlett decomposition

Let

$$\Sigma = \begin{pmatrix} \Sigma_{11} & \Sigma_{12} \\ \Sigma_{21} & \Sigma_{22} \end{pmatrix}$$

$$\Delta = \begin{pmatrix} \Sigma_{1|2} & 0 \\ 0 & \Sigma_{22} \end{pmatrix} \text{ and}$$

$$T = \begin{pmatrix} I & \tau \\ 0 & I \end{pmatrix}$$

where $\Sigma_{1|2} \equiv \Sigma_{11} - \Sigma_{12}\Sigma_{22}^{-1}\Sigma_{21}, \tau \equiv \Sigma_{12}\Sigma_{22}^{-1}$. Then

$$\Sigma = T\Delta T^T \tag{A.6}$$

Hence

$$\Sigma = \begin{pmatrix} \Sigma_{1|2} + \tau\Sigma_{22}\tau^T & \tau\Sigma_{22} \\ \Sigma_{22}\tau^T & \Sigma_{22} \end{pmatrix} \tag{A.7}$$

A problem with the Wishart distribution is that only a single degree of freedom is available to characterize uncertainty about all the parameters in the distribution. This deficiency leads us to the distribution in Subsection A.2.6.

A.2.6 Generalized Inverted Wishart

The two block version of the inverted Wishart above can be easily be extended to k blocks as follows. First apply the Bartlett decomposition to

$$\Sigma = \begin{pmatrix} \Sigma_{11} & \Sigma_{12} \\ \Sigma_{21} & \Sigma_{22} \end{pmatrix}$$

Then $\mathbf{Z} \sim GIW(\Psi, \delta)$ would mean that

$$\Sigma_{22} \sim IW(\Psi_{22}, \delta_2)$$
$$\Sigma_{1|2} \sim IW(\Psi_{1|2}, \delta_1), \text{ and}$$
$$\tau \mid \Sigma_{1|2} \sim N(\tau_{01}, H_1 \otimes \Sigma_{1|2})$$

where

$$\Psi_{1|2} = \Psi_{11} - \Psi_{12}(\Psi_{22})^{-1}\Psi_{21}$$

In the next step we would partition ($\Psi_{1|2}$ and so on leaving a sequence of degrees of freedom to represent the uncertainties that go with the different blocks in the decomposition.

Appendix 2: Entropy decomposition

A.3 Overview

This appendix provides details of the entropy decomposition for the multivariate t-distribution.

Assume for the $p \times 1$ vector $\mathbf{Z}$

$$\begin{aligned}
\mathbf{T} \mid \Sigma &\sim N_p(0, \Sigma) \\
\Sigma \mid \Psi, \delta &\sim W_p^{-1}(\Psi, \delta)
\end{aligned}$$

Then

$$\mathbf{T} \mid \Psi, \delta \sim t_p(0, \delta^{*-1}\Psi, \delta^*)$$

with $\delta^* = \delta - p + 1$.

It can be shown that

$$\Sigma \mid \mathbf{T}, \Psi, \delta \sim W_p^{-1}(\Psi + \mathbf{TT}', \delta + 1)$$

Conditional on Ψ & δ

$$\begin{aligned}
H(\mathbf{T}, \Sigma) &= H(\mathbf{T} \mid \Sigma) + H(\Sigma) \\
H(\mathbf{T}, \Sigma) &= H(\Sigma \mid \mathbf{T}) + H(\mathbf{T})
\end{aligned}$$

Hence

$$H(\mathbf{T}) = H(\mathbf{T} \mid \Sigma) + H(\Sigma) - H(\Sigma \mid \mathbf{T})$$

with

$$\begin{aligned}
H(\mathbf{T} \mid \Sigma) &= \frac{1}{2}E(\log|\Sigma| \mid \Psi) + \frac{p}{2}(\log(2\pi) + 1) \\
&= \frac{1}{2}E(\log|\Psi|) + \frac{1}{2}E(\log|\Sigma\Psi^{-1}|) + \frac{p}{2}(\log(2\pi) + 1) \\
&= \frac{1}{2}E(\log|\Psi|) + c(p, \delta)
\end{aligned}$$

the constant $c(g, \delta)$ depending on g and δ and noting that $\Psi\Sigma^{-1} \sim W_p(I_p, \delta)$.

Using $h(\mathbf{T}, \Sigma) = h(\mathbf{T})h(\Sigma) = |\Sigma|^{-(g+1)/2}$ we obtain

$$
\begin{aligned}
H(\Sigma) &= E[\log f(\Sigma)/h(\Sigma)] \\
&= \frac{1}{2}\delta \log|\Psi| - \frac{1}{2}\delta E(\log|\Sigma|) \\
&\quad - \frac{1}{2}E(tr\Psi\Sigma^{-1}) + c_1(p, \delta) \\
&= -\frac{1}{2}\delta E(\log|\Sigma\Psi^{-1}|) - \frac{1}{2}E(tr\Psi\Sigma^{-1}) + c_1(p, \delta) \\
&= \frac{1}{2}\delta E(\log|\Sigma^{-1}\Psi|) - \frac{1}{2}E(tr\Psi\Sigma^{-1}) + c_1(p, \delta) \\
&= c_2(p, \delta)
\end{aligned}
$$

Similarly

$$
\begin{aligned}
H(\Sigma \mid \mathbf{T}) &= \frac{1}{2}(\delta + 1)\log|\Psi| - \frac{1}{2}(\delta + 1)E(\log|\Psi + \mathbf{T}\mathbf{T}^T)|) \\
&\quad + c_3(p, \delta) \\
&= -\frac{1}{2}(\delta + 1)E(\log|1 + \mathbf{T}'\Psi^{-1}\mathbf{Z})|) + c_3(p, \delta) \\
&= c_4(p, \delta)
\end{aligned}
$$

Note $|\Psi|(1 + \mathbf{Z}^T\Psi^{-1}\mathbf{Z}) = |\Psi + \mathbf{Z}\mathbf{Z}^T)|$ and $\mathbf{Z}^T\Psi^{-1}\mathbf{Z} \sim F$ with degrees of freedom depending only on g, δ and thus $H(\mathbf{Z}) = \frac{1}{2}\log|\Psi| + c_5(p, \delta)$.

References

Abbott, E. (2009). *Flatland*. Broadview Press.

Ainslie, B., Reuten, C., Steyn, D. G., Le, N. D., & Zidek, J. V. (2009). Application of an entropy-based Bayesian optimization technique to the redesign of an existing monitoring network for single air pollutants. *Journal of Environmental Management*, *90*(8), 2715–2729.

Ashagrie, A., De Laat, P., De Wit, M., Tu, M., & Uhlenbrook, S. (2006). Detecting the influence of land use changes on floods in the Meuse River basin? the predictive power of a ninety-year rainfall-runoff relation? *Hydrology and Earth System Sciences Discussions*, *3*(2), 529–559.

Aven, T. (2013). On Funtowicz and Ravetz's "Decision stake–system uncertainties" structure and recently developed risk perspectives. *Risk Analysis*, *33*(2), 270–280.

Bandeen-Roche, K., Hall, C. B., Stewart, W. F., & Zeger, S. L. (1999). Modelling disease progression in terms of exposure history. *Statistics in Medicine*, *18*, 2899-2916.

Banerjee, S., Carlin, B. P., & Gelfand, A. E. (2015). *Hierarchical Modeling and Analysis for Spatial data. Second edition*. CRC Press.

Banerjee, S., Gelfand, A. E., & Carlin, B. P. (2004). *Hierarchical Modeling and Analysis for Spatial Data*. CRC Press.

Basu, D. (1975). Statistical information and likelihood [with discussion]. *Sankhyā: The Indian Journal of Statistics, Series A*, 1–71.

Baxter, P. J., Ing, R., Falk, H., French, J., Stein, G., Bernstein, R., ... Allard, J. (1981). Mount St Helens eruptions, May 18 to June 12, 1980: an overview of the acute health impact. *Journal of the American Medical Association*, *246*(22), 2585–2589.

Bayarri, M., & Berger, J. O. (1999). Quantifying surprise in the data and model verification. *Bayesian statistics*, *6*, 53–82.

Bayarri, M., & Berger, J. O. (2000). P values for composite null models. *Journal of the American Statistical Association*, *95*(452), 1127–1142.

Bergdahl, M., Ehling, M., Elvers, E., Földesi, E., Körner, T., Kron, A., ... Nimmergut, A. *et al.*. (2007). Handbook on data quality assessment methods and tools. *Handbook on Data Quality Assessment Methods and Tools*, 9–10.

Berger, J. (2012). Reproducibility of science: P-values and multiplicity. *SBSS Webinar, Oct, 4*.

Berger, J., De Oliveira, V., & Sansó, B. (2001). Objective Bayesian analysis of spatially correlated data. *Journal of the American Statistical Association*, *96*,

1361-1374.

Berhane, K., Gauderman, W. J., Stram, D. O., & Thomas, D. C. (2004). Statistical issues in studies of the long-term effects of air pollution: The Southern California Children's Health Study. *Statistical Science*, *19*(3), 414–449.

Bernardo, J. M., & Smith, A. F. M. (2009). *Bayesian Theory* (Vol. 405). John Wiley & Sons.

Berrocal, V. J., Gelfand, A. E., & Holland, D. M. (2010). A bivariate space-time downscaler under space and time misalignment. *The Annals of Applied Statistics*, *4*(4), 1942.

Berry, G., Gilson, J., Holmes, S., Lewinshon, H., & Roach, S. (1979). Asbestosis: A study of dose-response relationship in an asbestos textile factory. *British Journal of Industrial Medicine*, *36*, 98-112.

Berry, S. M., Carroll, R. J., & Ruppert, D. (2002). Bayesian smoothing and regression splines for measurement error problems. *Journal of the American Statistical Association*, *97*(457), 160–169.

Besag, J. (1974). Spatial interaction and the statistical analysis of lattice systems. *Journal of the Royal Statistical Socety, Series B*, *36*, 192-236.

Bishop, Y., Feinburg, S., & Holland, P. (1975). *Discrete Multivariate Analysis: Theory and Practice*. MIT Press.

Bivand, R. S., Pebesma, E. J., & Gómez-Rubio, V. (2008). *Applied Spatial Data Analysis with R* (Vol. 747248717). Springer.

Bodnar, O., & Schmid, W. (2010). Nonlinear locally weighted kriging prediction for spatio-temporal environmental processes. *Environmetrics*, *21*, 365-381.

Bornn, L., Shaddick, G., & Zidek, J. V. (2012). Modeling nonstationary processes through dimension expansion. *Journal of the American Statistical Association*, *107*(497), 281–289.

Bowman, A. W., Giannitrapani, M., & Scott, E. M. (2009). Spatiotemporal smoothing and sulphur dioxide trends over Europe. *Journal of the Royal Statistical Society: Series C (Applied Statistics)*, *58*(5), 737–752.

Box, G. E., & Draper, N. R. (1987). *Empirical Model–Building and Response Surfaces*. John Wiley & Sons.

Breslow, N., & Clayton, D. (1993). Approximate inference in Generalized Linear Mixed Models. *Journal of the American Statistical Association*, *88*(421), 9-25.

Breslow, N., & Day, N. (1980). *Statistical Methods in Cancer Research, Volume 2 — The Analysis of Cohort Studies*. Scientific Publications No. 82. Lyon: International Agency for Research on Cancer.

Breslow, N., Lubin, J. H., Marek, P., & Langholz, B. (1983). Multiplicative models and cohort analysis. *Journal of the American Statistical Association*, *78*(381), 1-12.

Briggs, D. J., Sabel, C. E., & Lee, K. (2009). Uncertainty in epidemiology and health risk and impact assessment. *Environmental Geochemistry and Health*, *31*(2), 189–203.

Brook, D. (1964). On the distinction between the conditional probability and the joint probability approaches in the specification of nearest-neighbour systems.

Biometrika, 481–483.

Brown, P. E., Diggle, P. J., Lord, M. E., & Young, P. C. (2001). Space-time calibration of radar-rainfall data. *Journal of the Royal Statistical Society: Series C (Applied Statistics)*, *50*(2), 221 - 241.

Brown, P. E., Karesen, K. F., Roberts, G. O., & Tonellato, S. (2000). Blur-generated non-separable space-time models. *Journal of the Royal Statistical Society: Series B (Statistical Methodology)*, *62*, 847-860.

Brown, P. J. (2002). *Measurement, regression, and calibration*. Oxford University Press.

Brown, P. J., Le, N. D., & Zidek, J. V. (1994). Inference for a covariance matrix. In A. F. M. Smith & P. R. Freeman (Eds.), *Aspects of Uncertainty: a Tribute to DV Lindley*. Wiley.

Burke, J. M., Zufall, M. J., & Ozkaynak, H. (2001). A population exposure model for particulate matter: case study results for PM2. 5 in Philadelphia, PA. *Journal of Exposure Analysis and Environmental Epidemiology*, *11*(6), 470–489.

Burnett, R. T., Dales, R. E., Raizenne, M. E., Krewski, D., Summers, P. W., Roberts, G. R., ... Brook, J. (1994). Effects of low ambient levels of ozone and sulfates on the frequency of respiratory admissions to ontario hospitals. *Environmental Research*, *65*(2), 172–194.

Burton, P., Gurrin, L., & Sly, P. (1998). Extending the simple linear regression model to account for correlated responses: an introduction to generalized estimating equations and multi-level modelling. *Statistics in Medicine*, *11*, 1825-1839.

Calder, C. A., & Cressie, N. (2007). Some topics in convolution-based spatial modeling. *Proceedings of the 56th Session of the International Statistics Institute*, 22–29.

Cameletti, M., Lindgren, F., Simpson, D., & Rue, H. (2011). Spatio-temporal modeling of particulate matter concentration through the spde approach. *AStA Advances in Statistical Analysis*, 1–23.

Carlin, B. P., & Louis, T. A. (2000). *Bayes and Empirical Bayes methods for data analysis*. Chapman and Hall/CRC.

Carlin, B. P., Xia, H., Devine, O., Tolbert, P., & Mulholland, J. (1999). Spatio-temporal hierarchical models for analyzing Atlanta pediatric asthma ER visit rates. In *Case studies in Bayesian Statistics* (pp. 303–320). Springer.

Carroll, R. J., Chen, R., Li, T. H., Newton, H. J., Schmiediche, H., Wang, H., & George, E. I. (1997). Ozone exposure and population density in Harris County, Texas. *Journal of the American Statistical Association*, *92*, 392-413.

Carroll, R. J., Ruppert, D., & Stefanski, L. A. (1995). *Measurement Error in Nonlinear Models*. Chapman and Hall, London.

Carstairs, V., & Morris, R. (1989). Deprivation: explaining differences between mortality between Scotland and England. *British Medical Journal*, *299*, 886-889.

Carter, C., & Kohn, R. (1994). On Gibbs sampling for state space models. *Biometrika*, *81*, 541-53.

Caselton, W. F., Kan, L., & Zidek, J. V. (1992). Quality data networks that minimize entropy. *Statistics in the Environmental and Earth Sciences*, 10–38.

Caselton, W. F., & Zidek, J. V. (1984). Optimal monitoring network designs. *Statistics & Probability Letters*, *2*(4), 223–227.

Chang, H., Fu, A. Q., Le, N. D., & Zidek, J. V. (2007). Designing environmental monitoring networks to measure extremes. *Environmental and Ecological Statistics*, *14*(3), 301–321.

Chang, H. H., Peng, R. D., & Dominici, F. (2011). Estimating the acute health effects of coarse particulate matter accounting for exposure measurement error. *Biostatistics*, *12*(4), 637–652.

Chatfield, C. (1995). Model uncertainty, data mining and statistical inference. *Journal of the Royal Statistical Society, Series A*, *158*, 419-466.

Chatfield, C. (2000). *Time–Series Forecasting*. Chapman and Hall/CRC.

Chatfield, C. (2013). *The Analysis of Time Series: An Introduction*. CRC press.

Chatfield, C., & Collins, A. (1980). *Introduction to Multivariate Analysis*. Chapman and Hall.

Chen, J. (2011). A partial order on uncertainty and information. *Journal of Theoretical Probability*, 1–11.

Chen, J., van Eeden, C., & Zidek, J. V. (2010). Uncertainty and the conditional variance. *Statistics & Probability Letters*.

Clayton, D., & Hills, M. (1993). *Statistical models in epidemiology*. Oxford Scientific Publications.

Coles, S., Bawa, J., Trenner, L., & Dorazio, P. (2001). *An Introduction to Statistical Modeling of Extreme Values*. Springer.

Cox, D. D., Cox, L. H., & Ensor, K. B. (1997). Spatial sampling and the environment: some issues and directions. *Environmental and Ecological Statistics*, *4*(3), 219–233.

Craigmile, P. F., Guttorp, P., & Percival, D. B. (2005). Wavelet-based parameter estimation for polynomial contaminated fractionally differenced processes. *Signal Processing, IEEE Transactions on*, *53*(8), 3151–3161.

Cressie, N. (1985). Fitting variogram models by weighted least squares. *Journal of the International Association for Mathematical Geology*, *17*(5), 563–586.

Cressie, N. (1986). Kriging nonstationary data. *Journal of the American Statistical Association*, *81*(396), 625-634.

Cressie, N. (1990). The origins of kriging. *Mathematical Geology*, *22*(3), 239–252.

Cressie, N. (1993). *Statistics for Spatial Data, revised edition*. John Wiley, New York.

Cressie, N. (1997). Discussion of Carroll, R.J. et al. 'Modeling ozone exposure in Harris county, Texas'. *Journal of the American Statistical Association*, *92*, 392-413.

Cressie, N., & Hawkins, D. M. (1980). Robust estimation of the variogram: I. *Mathematical Geology*, *12*(2), 115–125.

Cressie, N., & Wikle, C. K. (1998). Discussion of Mardia *at al.*, the kriged Kalman filter. *Test*, *7*, 257-263.

Cressie, N., & Wikle, C. K. (2011). *Statistics for Spatio-Temporal Data*. Wiley.

Cressie, N. A. C. (1993). *Statistics for Spatial Data*. Wiley.

Cullen, A. C., & Frey, H. C. (1999). *Probabilistic Techniques in Exposure Assessment: a Handbook for Dealing with Variability and Uncertainty in Models and Inputs.* Springer.

Damian, D., Sampson, P. D., & Guttorp, P. (2000). Bayesian estimation of semiparametric non-stationary spatial covariance structures. *Environmetrics, 12*(2), 161-178.

Daniels, M. J., Dominici, F., Zeger, S. L., & Samet, J. M. (2004). The National Morbidity, Mortality, and Air Pollution Study Part III: Concentration-Response Curves and Thresholds for the 20 Largest US Cities. *HEI Project, 96-97,* 1-21.

Davison, A. C., & Smith, R. L. (1990). Models for exceedances over high thresholds. *Journal of the Royal Statistical Society. Series B (Methodology),* 393–442.

Dawid, A. P. (1981). Some matrix-variate distribution theory: notational considerations and a bayesian application. *Biometrika, 68*(1), 265–274.

DeGroot, M. H. (1986). *Probability and Statistics.* Addison–Wesley.

Dellaportas, P., & Smith, A. F. M. (1993). Bayesian inference for generalized linear and proportional hazards models via Gibbs sampling. *Applied Statistics,* 443–459.

Denby, B., Costa, A., Monteiro, A., Dudek, A., & Erik, S. (2007). Uncertainty mapping for air quality modelling and data assimilation. *Proceedings of the 11th International Conference on Harmonisation within Atmospheric Dispersion Purposes.*

Denison, D. G. T., Mallick, B. K., & Smith, A. F. M. (1998). Automatic Bayesian curve fitting. *Journal of the Royal Statistical Society, Series B, 60*(2), 333-350.

De Oliveira, V. (2012). Bayesian analysis of conditional autoregressive models. *Annals of the Institute of Statistical Mathematics, 64*(1), 107–133.

DETR. (1998). *Review and Assessment: Monitoring Air Quality.* Department of the Environment, Transport and the Regions. LAQM.TGI.

Dewanji, A., Goddard, M. J., Krewski, D., & Moolgavkar, S. H. (1999). Two stage model for carcinogenesis: number and size distributions of premalignant clones in longitudinal studies. *Mathematical Bioscience, 155,* 1-12.

Diggle, P. J. (1991). *Time Series, a Biostatistical Introduction.* Oxford University Press.

Diggle, P. J. (1993). Point process modelling in environmental epidemiology. *Statistics for the Environment, 1.*

Diggle, P. J. (2013). *Statistical Analysis of Spatial and Spatio–Temporal Point Patterns.* CRC Press.

Diggle, P. J., Heagerty, P., Liang, K.-Y., & Zeger, S. L. (2002). *Analysis of Longitudinal Data.* Oxford University Press.

Diggle, P. J., Menezes, R., & Su, T. (2010). Geostatistical inference under preferential sampling. *Journal of the Royal Statistical Society: Series C (Applied Statistics), 59*(2), 191–232.

Diggle, P. J., & Ribeiro, P. J. (2007). *Model–Based Geostatistics.* Springer.

Diggle, P. J., Tawn, J. A., & Moyeed, R. A. (1998). Model-based geostatistics (with discussion). *Journal of the Royal Statistical Society (C), 47,* 299-350.

Director, H., & Bornn, L. (2015). Connecting point-level and gridded moments in the analysis of climate data. *Journal of Climate*.

Dobbie, M., Henderson, B., & Stevens Jr, D. (2008). Sparse sampling: spatial design for aquatic monitoring. *Statistics Surveys, 2*, 113–153.

Dockery, D., & Spengler, J. (1981). Personal exposure to respirable particulates and sulfates. *Journal of Air Pollution Control Association, 31*, 153-159.

Dominici, F., Samet, J. M., & Zeger, S. L. (2000a). Combining evidence on air pollution and daily mortality from the 20 largest cities: a hierarchical modelling strategy. *Journal of the Royal Statistical Society, Series A, 163*(3), 263-302.

Dominici, F., Samet, J. M., & Zeger, S. L. (2000b). Combining evidence on air pollution and daily mortality from the 20 largest US cities: a hierarchical modelling strategy. *Journal of the Royal Statistical Society: Series A (Statistics in Society), 163*(3), 263–302.

Dominici, F., & Zeger, S. L. (2000). A measurement error model for time series studies of air pollution and mortality. *Biostatistics, 1*, 157-175.

Dou, Y. (2007). *Dynamic Linear and Multivariate Bayesian Models for Modelling Environmental Space–Time Fields*. Unpublished doctoral dissertation, University of British Columbia, Department of Statistics.

Dou, Y., Le, N. D., & Zidek, J. V. (2007). *A Dynamic Linear Model for Hourly Ozone Concentrations* (Tech. Rep. No. TR228). UBC.

Dou, Y., Le, N. D., & Zidek, J. V. (2012). Temporal forecasting with a bayesian spatial predictor: Application to ozone. *Advances in Meteorology, 2012*.

Dou, Y., Le, N. D., Zidek, J. V., et al. (2010). Modeling hourly ozone concentration fields. *The Annals of Applied Statistics, 4*(3), 1183–1213.

Draper, N. R., Guttman, I., & Lapczak, L. (1979). Actual rejection levels in a certain stepwise test. *Communications in Statistics, A8*, 99-105.

EC. (1980). *Council directive 80/779/EEC of 15 July 1980 on Air Quality Limit Values and Guide Values for Sulphur Dioxide and Suspended Particulates*. European Commision.

Elfving, G., et al. (1952). Optimum allocation in linear regression theory. *The Annals of Mathematical Statistics, 23*(2), 255–262.

Elliott, P., Shaddick, G., Kleinschmidt, I., Jolley, D., Walls, P., Beresford, J., & Grundy, C. (1996). Cancer incidence near municipal solid waste incinerators in Great Britain. *British Journal of Cancer, 73*(5), 702.

Elliott, P., Shaddick, G., Wakefield, J. C., de Hoogh, C., & Briggs, D. J. (2007). Long-term associations of outdoor air pollution with mortality in great britain. *Thorax, 62*(12), 1088–1094.

Elliott, P., Wakefield, J. C., Best, N. G., & Briggs, D. J. (2000). *Spatial Epidemiology: Methods and Applications*. Oxford University Press.

Elliott, P., & Wartenberg, D. (2004). Spatial epidemiology: current approaches and future challenges. *Environmental Health Perspectives*, 998–1006.

Embrechts, P., Klüppelberg, C., & Mikosch, T. (1997). *Modelling Extremal Events: for Insurance and Finance*. Springer.

EPA. (2004). *Air Quality Criteria for Particulate Matter*. U.S. Environmental Protection Agency.

EPA. (2006). *Air Quality Criteria for Ozone and Related Photochemical Oxidants.* U.S. Environmental Protection Agency.

Eynon, B., & Switzer, P. (1983). The variability of rainfall acidity. *Canadian Journal of Statistics*, *11*(1), 11–23.

Fahrmeir, L., & Tutz, G. (1994). *Multivariate statistical modelling based on generalized linear models.* Springer.

Fanshawe, T. R., Diggle, P. J., Rushton, S., Sanderson, R., Lurz, P. W. W., Glinianaia, S. V., . . . Pless-Mulloli, T. (2008). Modelling spatio-temporal variation in exposure to particulate matter: a two-stage approach. *Environmetrics*, *19*(6), 549–566.

Fedorov, V. V., & Hackl, P. (1997). *Model-oriented design of experiments* (Vol. 125). Springer Science & Business Media.

Finazzi, F., Scott, E. M., & Fassò, A. (2013). A model-based framework for air quality indices and population risk evaluation, with an application to the analysis of scottish air quality data. *Journal of the Royal Statistical Society: Series C (Applied Statistics)*, *62*(2), 287–308.

Finley, A., Banerjee, S., & Carlin, B. (2007). spBayes: an R package for univariate and multivariate hierarchical point-referenced spatial models. *Journal of Statistical Software*, *19*(4), 1–24.

Fisher, R. A., & Tippett, L. H. C. (1928). Limiting forms of the frequency distribution of the largest or smallest member of a sample. In *Mathematical Proceedings of the Cambridge Philosophical society* (Vol. 24, pp. 180–190).

Frey, H. C., & Rhodes, D. S. (1996). Characterizing, simulating, and analyzing variability and uncertainty: an illustration of methods using an air toxics emissions example. *Human and Ecological Risk Assessment*, *2*(4), 762–797.

Fu, A. Q., Le, N. D., & Zidek, J. V. (2003). *A Statistical Characterization of a Simulated Canadian Annual Maximum Rainfall Field* (Tech. Rep.). TR 2003-17, Statistical and Mathematical Sciences Institute, North Carolina.

Fuentes, M. (2002a). Interpolation of nonstationary air pollution processes: a spatial spectral approach. *Statistical Modelling*, *2*(4), 281–298.

Fuentes, M. (2002b). Spectral methods for nonstationary spatial processes. *Biometrika*, *89*(1), 197–210.

Fuentes, M., Chen, L., & Davis, J. M. (2008). A class of nonseparable and nonstationary spatial temporal covariance functions. *Environmetrics*, *19*(5), 487–507.

Fuentes, M., Song, H. R., Ghosh, S. K., Holland, D. M., & Davis, J. M. (2006). Spatial association between speciated fine particles and mortality. *Biometrics*, *62*(3), 855–863.

Fuller, G., & Connolly, E. (2012). *Reorganisation of the UK Black Carbon network.* Department for Environment, Food and Rural Affairs.

Fuller, W. A. (1987). *Measurement Error Models.* Wiley and Sons.

Funtowicz, S. O., & Ravetz, J. R. (1990). *Uncertainty and quality in science for policy* (Vol. 15). Springer.

Funtowicz, S. O., & Ravetz, J. R. (1993). Science for the post-normal age. *Futures*, *25*(7), 739–755.

Gamerman, D., & Lopes, H. (2006). *Markov Chain Monte Carlo: Stochastic Simulation for Bayesian Inference*. Chapman & Hall/CRC.

Gamerman, D., & Migon, H. (1993). Dynamic hierarchical models. *Journal of the Royal Statistical Society: Series B (Statistical Methodology)*, *55*, 629-642.

Gamerman, D., & Smith, A. F. M. (1996). Bayesian analysis of longitudinal data studies. In J. Bernardo, J. Berger, A. Dawid, & A. Smith (Eds.), *Fifth Valencia International Meeting on Bayesian Statistics*. Oxford University Press.

Gatrell, A. C., Bailey, T. C., Diggle, P. J., & Rowlingson, B. S. (1996). Spatial point pattern analysis and its application in geographical epidemiology. *Transactions of the Institute of British Geographers*, *256–274*.

Gelfand, A. E., Banerjee, S., & Gamerman, D. (2005). Spatial process modelling for univariate and multivariate dynamic spatial data. *Environmetrics*, *16*(5), 465–479.

Gelfand, A. E., Kim, H. J., Sirmans, C. F., & Banerjee, S. (2003). Spatial modeling with spatially varying coefficient processes. *Journal of the American Statistical Association*, *98*(462), 387–396.

Gelfand, A. E., Sahu, S. K., & Holland, D. M. (2012). On the effect of preferential sampling in spatial prediction. *Environmetrics*, *23*(7), 565–578.

Gelfand, A. E., Zhu, L., & Carlin, B. P. (2001). On the change of support problem for spatio-temporal data. *Biostatistics*, *2*(1), 31.

Gelman, A., Carlin, J. B., Stern, H. S., Dunson, D. B., Vehtari, A., & Rubin, D. B. (2013). *Bayesian Data Analysis* (3rd ed.). Chapman & Hall / CRC.

Gelman, A., & Rubin, D. B. (1992). Inference from iterative simulation using multiple sequences. *Statistical science*, *7*(4), 457–472.

Geman, S., & Geman, D. (1984). Stochastic relaxation, Gibbs distributions, and the Bayesian restoration of images. *Pattern Analysis and Machine Intelligence, IEEE Transactions on*(6), 721–741.

Gerland, P., Raftery, A. E., Ševčíková, H., Li, N., Gu, D., Spoorenberg, T., ... Lalic, N. *et al.*. (2014). World population stabilization unlikely this century. *Science*, 234-237.

Giannitrapani, M., Bowman, A., Scott, E. M., & Smith, R. (2007). Temporal analysis of spatial covariance of SO2 in Europe from 1990 to 2001. *Environmetrics*, *1*, 1–12.

Gilks, W. R., Richardson, S., & Spiegelhalter, D. J. (1996). *Markov Chain Monte Carlo in Practice*. Chapman and Hall.

Gneiting, T., Genton, M. G., & Guttorp, P. (2006). Geostatistical space-time models, stationarity, separability, and full symmetry. *Monographs On Statistics and Applied Probability*, *107*, 151.

Goldstein, H. (1987). *Multilevel Models in Educational and Social Research*. Charles Griffin.

Good, I. J. (1952). Rational decisions. *Journal of the Royal Statistical Society. Series B (Methodological)*, 107–114.

Gotway, C. A., & Young, L. J. (2002). Combining incompatible spatial data. *Journal of the American Statistical Association*, *97*(458), 632–648.

Green, P. J., & Silverman, B. W. (1994). *Nonparametric Regression and Generalized Linear Models: a Roughness Penalty Approach*. Chapman & Hall/CRC.

Greenland, S., & Morgenstern, H. (1989). Ecological bias, confounding, and effect modification. *International journal of epidemiology*, *18*(1), 269–274.

Gryparis, A., Paciorek, C. J., Zeka, A., Schwartz, J., & Coull, B. A. (2009). Measurement error caused by spatial misalignment in environmental epidemiology. *Biostatistics*, *10*(2), 258–274.

Gu, C. (2002). *Smoothing spline ANOVA models*. Springer.

Gumbel, E. J. (2012). *Statistics of Extremes*. Courier Dover Publications.

Guttorp, P., Meiring, W., & Sampson, P. D. (1994). A space-time analysis of ground-level ozone data. *Environmetrics*, *5*(3), 241–254.

Guttorp, P., & Sampson, P. D. (2010). Discussion of Geostatistical inference under preferential sampling by Diggle, P. J., Menezes, R. and Su, T. *Journal of the Royal Statistical Society: Series C (Applied Statistics)*, *59*(2), 191–232.

Haas, T. C. (1990). Lognormal and moving window methods of estimating acid deposition. *Journal of the American Statistical Association*, *85*(412), 950–963.

Hamilton, J. (1994). *Time Series Analysis*. Princeton University Press.

Handcock, M., & Wallis, J. (1994). An approach to statistical spatial-temporal modelling of meteorological fields (with discussion). *Journal of the American Statistical Association*, *89*, 368-390.

Harris, B. (1982). Entropy. In S. Kotz & N. Johnson (Eds.), (pp. 512–516). Wiley, New York.

Harrison, J. (1999). *Bayesian Forecasting & Dynamic Models*. Springer.

Harrison, P. J., & Stevens, C. F. (1971). A Bayesian approach to short-term forecasting. *Operational Research Quarterly*, 341–362.

Harvey, A. (1993). *Time Series Models* (2nd ed.). MIT Press.

Haslett, J., & Raftery, A. E. (1989). Space-time modelling with long-memory dependence: assessing Ireland's wind power. *Applied Statistics*, *38*, 1-50.

Hastie, T., & Tibshirani, R. (1990). *Generalized Additive Models*. Chapman and Hall, London.

Hastings, W. (1970). Monte Carlo sampling methods using Markov chains and their applications. *Biometrika*, *57*(1), 97.

Hauptmann, M., Berhane, K., Langholz, B., & Lubin, J. H. (2001). Using splines to analyse latency in the Colorado Plateau uranium miners cohort. *Journal of Epidemiology and Biostatistics*, *6*(6), 417–424.

Heffernan, J. E., & Tawn, J. A. (2004). A conditional approach for multivariate extreme values (with discussion). *Journal of the Royal Statistical Society: Series B (Statistical Methodology)*, *66*(3), 497–546.

Helton, J. C. (1997). Uncertainty and sensitivity analysis in the presence of stochastic and subjective uncertainty. *Journal of Statistical Computation and Simulation*, *57*(1-4), 3–76.

Hempel, S. (2014). *The Medical Detective: John Snow, Cholera and the Mystery of the Broad Street Pump*. Granta Books.

Hickling, J., Clements, M., Weinstein, P., & Woodward, A. (1999). Acute health effects of the Mount Ruapehu (New Zealand) volcanic eruption of June 1996. *International Journal of Environmental Health Research*, *9*(2), 97–107.

Higdon, D. (1998). A process-convolution approach to modelling temperatures in the North Atlantic Ocean. *Environmental and Ecological Statistics*, *5*(2), 173–190.

Higdon, D. (2002). Space and space-time modeling using process convolutions. In *Quantitative Methods for Current Environmental Issues* (pp. 37–56). Springer.

Higdon, D., Swall, J., & Kern, J. (1999). Non-stationary spatial modeling. *Bayesian statistics*, *6*(1), 761–768.

Hill, A. B. (1965). The environment and disease: association or causation? *Proceedings of the Royal Society of Medicine*, *58*(5), 295.

Hoek, G., Brunekreef, B., Goldbohm, S., Fischer, P., & van den Brandt, P. A. (2002). Associations between mortality and indicators of traffic-related air pollution in the Netherlands: a cohort study. *Lancet*, *360*, 1203-1209.

Holland, D. M., De Oliveira, V., Cox, L. H., & Smith, R. L. (2000). Estimation of regional trends of sulfur dioxide over the eastern United States. *Environmetrics*, *11*, 373-393.

Hosseini, R., Le, N. D., & Zidek, J. V. (2011). A characterization of categorical Markov chains. *Journal of Statistical Theory and Practice*, *5*(2), 261–284.

Huerta, G., Sansó, B., & Stroud, J. (2004). A spatiotemporal model for Mexico city ozone levels. *Journal of the Royal Statistical Society: Series C (Applied Statistics)*, *53*(2), 231–248.

Hume, D. (2011). *An Enquiry Concerning Human Understanding*. Broadview Press.

Iman, R. L., & Conover, W. (1982). A distribution-free approach to inducing rank correlation among input variables. *Communications in Statistics-Simulation and Computation*, *11*(3), 311–334.

Ioannidis, J. P. (2005). Contradicted and initially stronger effects in highly cited clinical research. *Journal of the American Medical Association*, *294*(2), 218–228.

Jaynes, E. T. (1963). Information theory and statistical mechanics (notes by the lecturer). In *Statistical Physics 3* (Vol. 1, p. 181).

Journel, A., & Huijbregts, C. (1978). Mining geostatistics. *New York*.

Jun, M., Knutti, R., & Nychka, D. W. (2008). Spatial analysis to quantify numerical model bias and dependence: How many climate models are there? *Journal of the American Statistical Association*, *103*(483), 934–947.

Kaiser, M. S., & Cressie, N. (2000). The construction of multivariate distributions from Markov random fields. *Journal of Multivariate Analysis*, *73*(2), 199–220.

Kalenderski, S., & Steyn, D. G. (2011). Mixed deterministic statistical modelling of regional ozone air pollution. *Environmetrics*, *22*(4), 572–586.

Kalnay, E. (2003). *Atmospheric Modeling, Data assimilation, and Predictability*. Cambridge University Press.

Kass, E., & Wasserman, L. (1995). A reference Bayesian test for nested hypothesis and its relationship to the Schwarz criterion. *Journal of the American Statistical Association*, *90*(431), 928-934.

Kass, R. E., & Raftery, A. E. (1995). Bayes factors. *Journal of the American Statistical Association*, *90*, 773-795.

Katsouyanni, K., Schwartz, J., Spix, C., Touloumi, G., Zmirou, D., Zanobetti, A., ... Anderson, H. (1995). Short term effects of air pollution on health: a European approach using epidemiologic time series data: the APHEA protocol. *Journal of Epidemiology and Public Health*, *50 (Suppl 1)*, S12-S18.

Kazianka, H., & Pilz, J. (2012). Objective Bayesian analysis of spatial data with uncertain nugget and range parameters. *Canadian Journal of Statistics*, *40*(2), 304–327.

Kelsall, J. E., Zeger, S. L., & Samet, J. M. (1999). Frequency domain log-linear models; air pollution and mortality. *Journal of the Royal Statistical Society: Series C (Applied Statistics)*, *48*(3), 331–344.

Kermark, M., & Mckendrick, A. (1927). Contributions to the mathematical theory of epidemics. Part I. *Proc. R. Soc. A*, *115*(5), 700–721.

Kiefer, J. (1959). Optimum experimental designs. *Journal of the Royal Statistical Society. Series B (Methodological)*, 272–319.

Kim, Y., Kim, J., & Kim, Y. (2006). Blockwise sparse regression. *Statistica Sinica*, *16*(2), 375.

Kinney, P., & Ozkaynak, H. (1991). Associations of daily mortality and air pollution in Los Angeles County. *Environmental Research*, *54*, 99-120.

Kitanidis, P. (1986). Parameter uncertainty in estimation of spatial functions: Bayesian analysis. *Water Resources Research*, *22*(4), 499–507.

Kleiber, W., Katz, R. W., Rajagopalan, B., et al. (2013). Daily minimum and maximum temperature simulation over complex terrain. *The Annals of Applied Statistics*, *7*(1), 588–612.

Kleinschmidt, I., Hills, M., & Elliott, P. (1995). Smoking behaviour can be predicted by neighbourhood deprivation measures. *Journal of Epidemiology and Community Health*, *49 (Suppl 2)*, S72-7.

Knorr-Held, L. (1999). Conditional Prior Proposals in Dynamic Models. *Scandinavian Journal of Statistics*, *26*, 129-144.

Ko, C.-W., Lee, J., & Queyranne, M. (1995). An exact algorithm for maximum entropy sampling. *Operations Research*, *43*(4), 684–691.

Kolmogorov, A. (1941). Interpolated and extrapolated stationary random sequences. *Izvestia an SSSR, Seriya Mathematicheskaya*, *5*(2), 85–95.

Krige, D. (1951). *A Statistical Approach to Some Mine Valuation and Allied Problems on the Witwatersrand.* Unpublished doctoral dissertation, University of the Witwatersrand.

Laird, N., & Ware, J. (1982). Random-effects models for longitudinal data. *Biometrics*, *38*, 963-974.

Law, P., Zelenka, M., Huber, A., & McCurdy, T. (1997). Evaluation of a probabilistic exposure model applied. *Journal of the Air & Waste Management Association*, *47*(3), 491–500.

Lawson, A. B. (2013). *Statistical Methods in Spatial Epidemiology*. John Wiley & Sons.

Le, N. D., Sun, W., & Zidek, J. V. (1997). Bayesian multivariate spatial interpolation with data missing by design. *Journal of the Royal Statistical Society: Series B (Statistical Methodology)*, *59*(2), 501–510.

Le, N. D., & Zidek, J. V. (1992). Interpolation with uncertain spatial covariances: a Bayesian alternative to Kriging. *Journal of Multivariate Analysis*, *43*(2), 351–374. doi: 10.1016/0047-259X(92)90040-M

Le, N. D., & Zidek, J. V. (1994). Network designs for monitoring multivariate random spatial fields. *Recent Advances in Statistics and Probability*, 191–206.

Le, N. D., & Zidek, J. V. (2006). *Statistical Analysis of Environmental Space–Time Processes*. Springer Verlag.

Le, N. D., & Zidek, J. V. (2010). Air pollution and cancer. *Chronic Diseases in Canada*, *29, Supplement 2*(144–163).

Leadbetter, M. (1983). Extremes and local dependence in stationary sequences. *Probability Theory and Related Fields*, *65*(2), 291–306.

Leduc, H. (2011). *Estimation de Paramètres dans des Modéles d'Épidémies*. Unpublished master's thesis, Département de mathématiques, University of Quebec.

Lee, D. (2013). CARBayes: an R package for bayesian spatial modeling with conditional autoregressive priors. *Journal of Statistical Software*, *55*(13), 1–24.

Lee, D., Ferguson, C., & Scott, E. M. (2011). Constructing representative air quality indicators with measures of uncertainty. *Journal of the Royal Statistical Society: Series A (Statistics in Society)*, *174*(1), 109–126.

Lee, D., & Shaddick, G. (2008). Modelling the effects of air pollution on health using bayesian dynamic generalised linear models. *Environmetrics*, *19*(8), 785–804.

Lee, D., & Shaddick, G. (2010). Spatial modeling of air pollution in studies of its short-term health effects. *Biometrics*, *66*, 1238-1246.

Lee, J.-T., Kim, H., Hong, Y.-C., Kwon, H.-J., Schwartz, J., & Christiani, D. C. (2000). Air pollution and daily mortality in seven major cities of Korea, 1991–1997. *Environmental Research*, *84*(3), 247–254.

Li, B., Sain, S., Mearns, L. O., Anderson, H. A., Kovats, S., Ebi, K. L., ... Patz, J. A. (2012). The impact of extreme heat on morbidity in Milwaukee, Wisconsin. *Climatic Change*, *110*(3-4), 959–976.

Li, K., Le, N. D., Sun, L., & Zidek, J. V. (1999). Spatial-temporal models for ambient hourly PM10 in Vancouver. *Environmetrics*, *10*, 321-338.

Liang, K.-Y., & Zeger, S. L. (1986). Longitudinal data analysis using generalized linear models. *Biometrika*, *73*, 13-22.

Lindgren, F., Rue, H., & Lindström, J. (2011). An explicit link between Gaussian fields and Gaussian markov random fields: the stochastic partial differential equation approach. *Journal of the Royal Statistical Society: Series B (Statistical Methodology)*, *73*(4), 423–498.

Lindley, D. V. (1956). On a measure of the information provided by an experiment. *The Annals of Mathematical Statistics*, 986–1005.

Lioy, P., Waldman, J., Buckley, Butler, J., & Pietarinen, C. (1990). The personal, indoor and outdoor concentrations of pm-10 measured in an industrial community during the winter. *Atmospheric Environment. Part B. Urban Atmosphere*, *24*(1), 57–66.

Little, R. J. A., & Rubin, D. B. (2014). *Statistical Analysis with Missing Data*. John Wiley & Sons.

Livingstone, A. E., Shaddick, G., Grundy, C., & Elliott, P. (1996). Do people living near inner city main roads have more asthma needing treatment? case-control study. *British Medical Journal, 312*(7032), 676–677.

Lubin, J. H., Boice, J. D., Edling, C., Hornung, R. W., Howe, G. R., Kunz, E., ... others (1995). Lung cancer in radon-exposed miners and estimation of risk from indoor exposure. *Journal of the National Cancer Institute, 87*(11), 817–827.

Lumley, T., & Sheppard, L. (2000). Assessing seasonal confounding and model selection bias in air pollution epidemiology using positive and negative control analyses. *Environmetrics, 11*(6), 705–717.

Lunn, D., Jackson, C., Best, N., Thomas, A., & Spiegelhalter, D. (2012). *The BUGS book: a Practical Introduction to Bayesian Analysis*. CRC press.

Lunn, D., Thomas, A., Best, N., & Spiegelhalter, D. (2000). WinBUGS - a Bayesian modelling framework: concepts, structure, and extensibility. *Statistics and computing, 10*(4), 325–337.

MacIntosh, D., Xue, J., Ozkaynak, H., Spengler, J., & Ryan, P. (1994). A population-based exposure model for benzene. *Journal of Exposure Analysis and Environmental Epidemiology, 5*(3), 375–403.

MacLeod, M., Fraser, A. J., & Mackay, D. (2002). Evaluating and expressing the propagation of uncertainty in chemical fate and bioaccumulation models. *Environmental Toxicology and Chemistry, 21*(4), 700–709.

Mar, T. F., Norris, G. A., Koenig, J. Q., & Larson, T. V. (2000). Associations between air pollution and mortality in Phoenix, 1995-1997. *Environmental Health Perspectives, 108*(4), 347.

Mardia, K., Goodall, C., Redfern, E., & Alonso, F. (1998). The kriged Kalman filter. *Test, 7*, 217-276.

Matern, B. (1986). *Spatial variation*. Springer Verlag.

Matheron, G. (1963). Principles of geostatistics. *Economic Geology, 58*, 1246-1266.

Mazumdar, S., Schimmel, H., & Higgins, I. (1982). Relation of daily mortality to air pollution: an analysis of 14 London winters, 1958/59-1971/72. *Archives of Environmental Health, 37*, 213-20.

McCullagh, P., & Nelder, J. (1989). *Generalized Linear Models* (2nd ed.). Chapman and Hall.

McKay, M., Beckman, R., & Conover, W. (2000). A comparison of three methods for selecting values of input variables in the analysis of output from a computer code. *Technometrics, 42*(1), 55–61.

Meinhold, R., & Singpurwalla, N. (1983). Understanding the Kalman filter. *American Statistician, 37*, 123-127.

Meiring, W., Sampson, P. D., & Guttorp, P. (1998). Space-time estimation of grid-cell hourly ozone levels for assessment of a deterministic model. *Environmental and Ecological Statistics, 5*(3), 197–222.

Metropolis, N., Rosenbluth, A. W., Rosenbluth, M. N., Teller, A. H., & Teller, E. (1953). Equation of state calculations by fast computing machines. *The Journal of Chemical Physics, 21*(6), 1087.

Metropolis, N., & Ulam, S. (1949). The Monte Carlo method. *Journal of the American Statistical Association, 44*(247), 335–341.

Ministry of Health. (1954). *Mortality and Morbidity During the London Fog of December 1952.* London: HMSO.

Mitch, M. E. (1990). *National stream survey database guide.* US Environmental Protection Agency, Office of Research and Development.

Moolgavkar, S. H. (2000). Air pollution and daily mortality in three US counties. *Environmental Health Perspectives, 108*(8), 777.

Moolgavkar, S. H., Luebeck, E. G., Hall, T. A., & Anderson, E. L. (1995). Air pollution and daily mortality in Philadelphia. *Epidemiology, 6,* 476-484.

Morgan, M. G., & Small, M. (1992). *Uncertainty: a Guide to Dealing with Uncertainty in Quantitative Risk and Policy Analysis.* Cambridge University Press.

Müller, W. G. (2007). *Collecting spatial data: Optimum design of experiments for random fields* (3rd ed.). Heidelberg: Physica-Verlag.

Müller, W. G., & Zimmerman, D. L. (1999). Optimal design for variogram estimation. *Environmetrics,* 23-27.

Murrell, P. (2005). *R Graphics.* CRC Press.

Neas, L., Schwartz, J., & Dockery, D. (1999). A Case-Crossover Analysis of Air Pollution and Mortality in Philadelphia. *Environmental Health Perspectives, 107,* 629-631.

Nuzzo, R. (2014). Scientific method: statistical errors. *Nature, 506,* 150-2.

Nychka, D., & Saltzman, N. (1998). Design of air–quality monitoring networks. In *Case Studies in Environmental Statistics* (pp. 51–76). Springer.

Oden, N. L., & Benkovitz, C. M. (1990). Statistical implications of the dependence between the parameters used for calculations of large scale emissions inventories. *Atmospheric Environment. Part A. General Topics, 24*(3), 449–456.

Oh, H., & Li, T. (2004). Estimation of global temperature fields from scattered observations by a spherical-wavelet-based spatially adaptive method. *Journal of the Royal Statistical Society: Series B (Statistical Methodology), 66*(1), 221–238.

O'Hagan, A. (1995). Fractional Bayes factors for model comparison. *Journal of the Royal Statistical Society. Series B (Methodology),* 99–138.

Olea, R. A. (1984). Sampling design optimization for spatial functions. *Journal of the International Association for Mathematical Geology, 16*(4), 369–392.

Omar, R., Wright, E., Turner, R., & Thompson, S. (1999). Analysing repeated measurements data: a practical comparison of methods. *Statistics in Medicine, 18,* 1587-1603.

Omre, H. (1984). The variogram and its estimation. *Geostatistics for Natural Resources Characterization, 1,* 107–125.

Ott, W., Thomas, J., Mage, D., & Wallace, L. (1988). Validation of the simulation of human activity and pollutant exposure (shape) model using paired days from the Denver, CO, carbon monoxide field study. *Atmospheric Environment (1967)*, *22*(10), 2101–2113.

Ozkaynak, H., Xue, J., Spengler, J., Wallace, L., Pellizzari, E., & Jenkins, P. (1996). Personal exposure to airborne particles and metals: Results from the particle team study in Riverside California. *Journal of Exposure Analysis and Environmental Epidemiology*, *6*, 57-78.

Paciorek, C. J., & Schervish, M. J. (2006). Spatial modelling using a new class of nonstationary covariance functions. *Environmetrics*, *17*(5), 483–506.

Paciorek, C. J., Yanosky, J. D., Puett, R. C., Laden, F., & Suh, H. (2009). Practical large-scale spatio-temporal modeling of particulate matter concentrations. *Annals of Applied Statistics*, *3*(1), 370–397.

Pati, D., Reich, B. J., & Dunson, D. B. (2011). Bayesian geostatistical modelling with informative sampling locations. *Biometrika*, *98*(1), 35–48.

Pebesma, E. J., & Heuvelink, G. B. (1999). Latin hypercube sampling of Gaussian random fields. *Technometrics*, *41*(4), 303–312.

Peng, R. D., & Bell, M. L. (2010). Spatial misalignment in time series studies of air pollution and health data. *Biostatistics*, *11*, 720 - 740.

Peng, R. D., & Dominici, F. (2008). Statistical methods for environmental epidemiology with R. In *R: a Case Study in Air Pollution and Health.* Springer.

Perrin, O., & Schlather, M. (2007). Can any multivariate Gaussian vector be interpreted as a sample from a stationary random process? *Statist. Prob. Lett.*, *77*, 881–4.

Peters, A., Skorkovsky, J., Kotesovec, F., Brynda, J., Spix, C., Wichmann, H. E., & Heinrich, J. (2000). Associations between mortality and air pollution in central Europe. *Environmental Health Perspectives*, *108*(4), 283.

Pickands III, J. (1975). Statistical inference using extreme order statistics. *The Annals of Statistics*, 119–131.

Plummer, M. (2014). Cuts in Bayesian graphical models. *Statistics and Computing*, 1–7.

Plummer, M., & Clayton, D. (1996). Estimation of population exposure in ecological studies. *Journal of the Royal Statistical Society. Series B (Statistical Methodology)*, 113–126.

Prentice, R., & Sheppard, L. (1995). Aggregate data studies of disease risk factors. *Biometrika*, *82*, 113-125.

Raftery, A. E., & Richardson, S. (1996). Bayesian biostatistics. In (p. 321-353). Marcel-Dekker, New York.

Reagan, P. L., & Silbergeld, E. K. (1990). Establishing a health based standard for lead in residential soils. *Trace Substances in Environmental Health*, *12*.

Reinsel, G., Tiao, G. C., Wang, M. N., Lewis, R., & Nychka, D. (1981). Statistical analysis of stratospheric ozone data for the detection of trends. *Atmospheric Environment*, *15*(9), 1569–1577.

Renyi, A. (1961). On measures of entropy and information. In *Fourth Berkeley Symposium on Mathematical Statistics and Probability* (pp. 547–561).

Richardson, S., & Gilks, W. R. (1993). Conditional independence models for epidemiological studies with covariate measurement error. *Statistics in Medicine*,

12, 1703–1722.

Richardson, S., Leblond, L., Jaussent, I., & Green, P. J. (2002). Mixture models in measurement error problems, with reference to epidemiological studies. *Journal of the Royal Statistical Society: Series A (Statistics in Society)*, *165*(3), 549–566.

Richardson, S., Stücker, I., & Hémon, D. (1987). Comparison of relative risks obtained in ecological and individual studies: some methodological considerations. *International Journal of Epidemiology*, *16*(1), 111–120.

Ripley, B., & Corporation, E. (1987). *Stochastic simulation*. Wiley Online Library.

Ro, C., Tang, A., Chan, W., Kirk, R., Reid, N., & Lusis, M. (1988). Wet and dry deposition of sulfur and nitrogen compounds in Ontario. *Atmospheric Environment (1967)*, *22*(12), 2763–2772.

Roberts, G. O., & Rosenthal, J. S. (2009). Examples of adaptive MCMC. *Journal of Computational and Graphical Statistics*, *18*(2), 349–367.

Robins, J., & Greenland, S. (1989). The probability of causation under a stochastic model for individual risk. *Biometrics*, 1125–1138.

Rothman, K. J., & Greenland, S. (1998). *Modern Epidemiology*. Lippencott-Raven.

Royle, J., & Nychka, D. (1998). An algorithm for the construction of spatial coverage designs with implementation in SPLUS. *Computers & Geosciences*, *24*(5), 479–488.

Rubinstein, R. Y., & Kroese, D. P. (2011). *Simulation and the Monte Carlo Method* (Vol. 707). John Wiley & Sons.

Rue, H., & Held, L. (2005). *Gaussian Markov Random Fields: Theory and Applications*. CRC Press.

Rue, H., Martino, S., & Chopin, N. (2009). Approximate Bayesian inference for latent Gaussian models by using integrated nested Laplace approximations. *Journal of the Royal Statistical Society: Series B (Statistical Methodology)*, *71*(2), 319–392.

Ruppert, D., Wand, M. P., & Carroll, R. J. (2003). *Semiparametric Regression*. Cambridge University Press.

Sahu, S. K., Gelfand, A. E., & Holland, D. M. (2006). Spatio-temporal modeling of fine particulate matter. *Journal of agricultural, biological, and environmental statistics*, *11*(1), 61–86.

Sahu, S. K., Gelfand, A. E., & Holland, D. M. (2007). High-resolution space-time ozone modeling for assessing trends. *Journal of the American Statistical Association*, *102*(480), 1221–1234.

Sahu, S. K., & Mardia, K. (2005). A Bayesian kriged Kalman model for short-term forecasting of air pollution levels. *Journal of the Royal Statistical Society: Series C (Applied Statistics)*, *54*(1), 223–244.

Sampson, P. D., & Guttorp, P. (1992). Nonparametric estimation of nonstationary spatial covariance structure. *Journal of the American Statistical Association*, *87*, 108–119.

Schabenberger, O., & Gotway, C. A. (2000). *Statistical Methods for Spatial Data Analysis*. Chapman and Hall/CRC.

Schafer, J. (1997). *Analysis of Incomplete Multivariate Data*. Chapman & Hall.

Schmeltz, D., Evers, D. C., Driscoll, C. T., Artz, R., Cohen, M., Gay, D., ... Morris, K. *et al.*. (2011). MercNet: a national monitoring network to assess responses to changing mercury emissions in the United States. *Ecotoxicology, 20*(7), 1713–1725.

Schmidt, A. M., & O'Hagan, A. (2000). *Bayesian Inference for Nonstationary Spatial Covariance Structure via Spatial Deformations*. Unpublished doctoral dissertation.

Schumacher, P., & Zidek, J. V. (1993). Using prior information in designing intervention detection experiments. *The Annals of Statistics*, 447–463.

Schwartz, J. (1991). Particulate air pollution and daily mortality in Detroit. *Environmental Research, 56*, 204-213.

Schwartz, J. (1994a). Air pollution and daily mortality: a review and meta analysis. *Environmental Research, 64*, 36-52.

Schwartz, J. (1994b). Nonparamteric smoothing in the analysis of air pollution and respiratory health. *Canadian Journal of Statistics, 22*(4), 471-487.

Schwartz, J. (1994c). Total suspended particulate matter and daily mortality in Cincinnati, Ohio. *Environmental Health Perspectives, 102*, 186-189.

Schwartz, J. (1995). Short term flucuations in air pollution and hospital admissions of the eldery for respiratory disease. *Thorax, 50*, 531-538.

Schwartz, J. (1997). Air pollution and hospital admissions. *Epidemiology, 8*(4), 371-377.

Schwartz, J. (2000). The distributed lag between air pollution and daily deaths. *Epidemiology, 11*, 320-326.

Schwartz, J. (2001). Air pollution and blood markers of cardiovascular risk. *Environmental Health Perspectives, 109*(Suppl 3), 405.

Schwartz, J., & Dockery, D. (1992a). Increased mortality in Philadelphia associated with daily air pollution concentrations. *American Review of Respiratory Disease, 145*, 600-604.

Schwartz, J., & Dockery, D. (1992b). Particulate air pollution and daily mortality in Steubenville, Ohio. *American Journal of Epidemiology, 135*, 12-29.

Schwartz, J., & Marcus, A. (1990). Mortality and air pollution in London: A time series analysis. *American Journal of Epidemiology, 131*, 185-194.

Schwartz, J., Slater, D., Larson, T. V., Pierson, W. E., & Koenig, J. Q. (1993). Particulate air pollution and hospital emergency room visits for asthma in Seattle. *American Review of Respiratory Disease, 147*(4), 826–831.

Scott, A., & Wild, C. (2001). Maximum likelihood for generalised case-control studies. *Journal of Statistical Planning and Inference, 96*(1), 3–27.

Scott, A., & Wild, C. (2011). Fitting binary regression models with response-biased samples. *Canadian Journal of Statistics, 39*(3), 519–536.

Sebastiani, P., & Wynn, H. P. (2002). Maximum entropy sampling and optimal bayesian experimental design. *Journal of the Royal Statistical Society: Series B (Statistical Methodology), 62*(1), 145–157.

Sellke, T., Bayarri, M. J., & Berger, J. O. (2001). Calibration of p-values for precise null hypotheses. *The American Statistician*.

Shaddick, G., Lee, D., & Wakefield, J. (2013). Ecological bias in studies of the short-term effects of air pollution on health. *International Journal of Applied Earth Observation and Geoinformation, 22*, 65–74.

Shaddick, G., Lee, D., Zidek, J. V., & Salway, R. (2008). Estimating exposure response functions using ambient pollution concentrations. *The Annals of Applied Statistics, 2*(4), 1249–1270.

Shaddick, G., & Wakefield, J. (2002). Modelling daily multivariate pollutant data at multiple sites. *Journal of the Royal Statistical Society: Series C (Applied Statistics), 51*, 351–372.

Shaddick, G., Yan, H., Salway, R., Vienneau, D., Kounali, D., & Briggs, D. (2013). Large-scale Bayesian spatial modelling of air pollution for policy support. *Journal of Applied Statistics, 40*(4), 777–794.

Shaddick, G., & Zidek, J. V. (2014). A case study in preferential sampling: Long term monitoring of air pollution in the UK. *Spatial statistics*, 51-65.

Shannon, C. E. (2001). A mathematical theory of communication. *ACM SIGMOBILE Mobile Computing and Communications Review, 5*(1), 3–55.

Sheppard, L., & Damian, D. (2000). Estimating short-term PM effects accounting for surrogate exposure measurements from ambient monitors. *Environmetrics, 11*(6), 675–687.

Shewry, M. C., & Wynn, H. P. (1987). Maximum entropy sampling. *Journal of Applied Statistics, 14*(2), 165–170.

Silverman, B. W. (1986). *Density Estimation*. London: Chapman and Hall.

Silvey, S. D. (1980). *Optimal Design*. Springer.

Sirois, A., & Fricke, W. (1992). Regionally representative daily air concentrations of acid-related substances in canada; 1983–1987. *Atmospheric Environment. Part A. General Topics, 26*(4), 593–607.

Smith, A. F. M., & Roberts, G. O. (1993). Bayesian computation via the Gibbs sampler and other related Markov Chain Monte Carlo methods. *Journal of the Royal Statistical Society, Series B (Statistical Methodology), 55*, 3–23.

Smith, K. (1918). On the standard deviations of adjusted and interpolated values of an observed polynomial function and its constants and the guidance they give towards a proper choice of the distribution of observations. *Biometrika*, 1–85.

Smith, R. L. (1989). Extreme value analysis of environmental time series: an application to trend detection in ground-level ozone. *Statistical Science*, 367–377.

Sobol, I. M. (1994). *A Primer for the Monte Carlo Method*. CRC press.

Spix, C., Heinrich, J., Dockery, D., Schwartz, J., Völksch, G., Schwinkowski, K., ... Wichmann, H. E. (1993). Air pollution and daily mortality in Erfurt, East Germany, 1980-1989. *Environmental Health Perspectives, 101*(6), 518.

Stehman, S. V., & Overton, W. S. (1994). Comparison of variance estimators of the Horvitz–Thompson estimator for randomized variable probability systematic sampling. *Journal of the American Statistical Association, 89*(425), 30–43.

Stein, M. (1987). Large sample properties of simulations using latin hypercube sampling. *Technometrics, 29*(2), 143–151.

Stein, M. (1999). *Interpolation of Spatial Data: Some Theory for Kriging*. Springer Verlag.

Sullivan, L., Dukes, K., & Losina, E. (1999). An introduction to heirarchical linear modelling. *Statistics in Medicine*, *18*, 855-888.

Sun, L., Zidek, J. V., Le, N. D., & Ozkaynak, H. (2000). Interpolating Vancouver's daily ambient PM_{10} field. *Environmetrics*, *11*, 651-663.

Sun, W. (1998). Comparison of a cokriging method with a Bayesian alternative. *Environmetrics*, *9*(4), 445–457.

van der Sluijs, J. P., Craye, M., Funtowicz, S., Kloprogge, P., Ravetz, J., & Risbey, J. (2005). Combining quantitative and qualitative measures of uncertainty in model-based environmental assessment: the nusap system. *Risk Analysis*, *25*(2), 481–492.

VanderWeele, T. J., & Robins, J. M. (2012). Stochastic counterfactuals and stochastic sufficient causes. *Statistica Sinica*, *22*(1), 379.

van Loon, F. P. L., Holmes, S. J., Sirotkin, B. I., Williams, W. W., Cochi, S. L., Hadler, S. C., & Lindegren, M. L. (1995). Mumps surveillance - United States, 1988-1993. *CDC Surveillance Summaries: Morbidity and Mortality Weekly Report*, *44*(3), 1–14.

Vedal, S., Brauer, M., White, R., & Petkau, J. (2003). Air pollution and daily mortality in a city with low levels of pollution. *Environmental Health Perspectives*, *111*(1), 45.

Verhoeff, A. P., Hoek, G., Schwartz, J., & van Wijnen, J. H. (1996). Air pollution and daily mortality in Amsterdam. *Epidemiology*, *7*(3), 225–230.

Wackernagel, H. (2003). *Multivariate geostatistics: an introduction with applications*. Springer Verlag.

Wahba, G. (1990). *Spline Models for Observational Data* (Vol. 59). Siam.

Wakefield, J., & Salway, R. (2001). A statistical framework for ecological and aggregate studies. *Journal of the Royal Statistical Society, Series A (Statistics in Society)*, *164*, 119-137.

Wakefield, J., & Shaddick, G. (2006). Health-exposure modeling and the ecological fallacy. *Biostatistics*, *7*(3), 438–455.

Wakefield, J. C., Best, N. G., & Waller, L. A. (2000). Bayesian approaches to disease mapping. In P. Elliott, J. C. Wakefield, N. G. Best, & D. J. Briggs (Eds.), *Spatial Epidemiology: Methods and Applications*. Oxford: Oxford University Press.

Walker, W. E., Harremoës, P., Rotmans, J., van der Sluijs, J. P., van Asselt, M. B., Janssen, P., & Krayer von Krauss, M. P. (2003). Defining uncertainty: a conceptual basis for uncertainty management in model-based decision support. *Integrated assessment*, *4*(1), 5–17.

Waller, L. A., & Gotway, C. A. (2004). *Applied Spatial Statistics for Public Health Data* (Vol. 368). Wiley-Interscience.

Waternaux, C., Laird, N., & Ware, J. (1989). Methods for analysis of longitudinal data: blood lead concentrations and cognitive development. *Journal of the American Statistical Association*, *84*, 33-41.

Webster, R., & Oliver, M. A. (2007). *Geostatistics for Environmental Scientists*. John Wiley & Sons.

West, M., Harrison, J., & Migon, H. (1985). Dynamic generalized linear models and Bayesian forecasting. *Journal of the American Statistical Association, 80*, 73-90.

Whittle, P. (1954). On stationary processes in the plane. *Biometrika*, 434–449.

Wikle, C. K., Berliner, L. M., & Cressie, N. (1998). Hierarchical Bayesian space-time models. *Environmental and Ecological Statistics, 5*, 117-154.

Wikle, C. K., & Cressie, N. (1999). A dimension reduction approach to space-time Kalman filtering. *Biometrika, 86*, 815-829.

Wood, S. (2006). *Generalized Additive Models: An Introduction with R.* Chapman and Hall/CRC.

Yu, O., Sheppard, L., Lumley, T., Koenig, J. Q., & Shapiro, G. G. (2000). Effects of Ambient Air Pollution on Symptoms of Asthma in Seattle-Area Children Enrolled in the CAMP Study. *Environmental Health Perspectives, 108*, 1209-1214.

Zannetti, P. (1990). *Air Pollution Modeling: Theories, Computational Methods, and Available Software.* Computational Mechanics Southampton.

Zanobetti, A., Wand, M. P., Schwartz, J., & Ryan, L. M. (2000). Generalized additive distributed lag models: quantifying mortality displacement. *Biostatistics, 1*(3), 279-292.

Zeger, S. L., & Liang, K.-Y. (1992). An overview for the analysis of longitudinal data. *Statistics in Medicine, 11*, 1825-1839.

Zeger, S. L., Liang, K.-Y., & Albert, P. S. (1988). Models for longitudinal data: A generalized estimating approach. *Biometrika, 44*(405), 1049-1060.

Zeger, S. L., Thomas, D. C., Dominici, F., Samet, J. M., Schwartz, J., Dockery, D., & Cohen, A. (2000). Exposure measurement error in time–series studies of air pollution: concepts and consequences. *Environmental Health Perspectives, 108*(5), 419.

Zhu, L., Carlin, B. P., & Gelfand, A. E. (2003). Hierarchical regression with mis-aligned spatial data: relating ambient ozone and pediatric asthma ER visits in atlanta. *Environmetrics, 14*(5), 537–557.

Zhu, Z., & Stein, M. L. (2006). Spatial sampling design for prediction with estimated parameters. *Journal of Agricultural, Biological, and Environmental Statistics, 11*(1), 24–44.

Zidek, J. V., Le, N. D., & Liu, Z. (2011). Combining data and simulated data for space–time fields: application to ozone. *Environmental and Ecological Statistics*, 1–20.

Zidek, J. V., Meloche, J., Shaddick, G., Chatfield, C., & White, R. (2003). A computational model for estimating personal exposure to air pollutants with application to London's PM10 in 1997. *2003 Technical Report of the Statistical and Applied Mathematical Sciences Institute.*

Zidek, J. V., Shaddick, G., Meloche, J., Chatfield, C., & White, R. (2007). A framework for predicting personal exposures to environmental hazards. *Environmental and Ecological Statistics, 14*(4), 411–431.

Zidek, J. V., Shaddick, G., & Taylor, C. G. (2014). Reducing estimation bias in adaptively changing monitoring networks with preferential site selection. *The Annals of Applied Statistics, 8*(3), 1640–1670.

Zidek, J. V., Shaddick, G., White, R., Meloche, J., & Chatfield, C. (2005). Using a probabilistic model (pCNEM) to estimate personal exposure to air pollution. *Environmetrics*, *16*(5), 481–493.

Zidek, J. V., Sun, W., & Le, N. D. (2000). Designing and integrating composite networks for monitoring multivariate gaussian pollution fields. *Journal of the Royal Statistical Society: Series C (Applied Statistics)*, *49*(1), 63–79.

Zidek, J. V., & van Eeden, C. (2003). Uncertainty, entropy, variance and the effect of partial information. *Lecture Notes-Monograph Series*, 155–167.

Zidek, J. V., White, R., Sun, W., Burnett, R. T., & Le, N. D. (1998). Imputing unmeasured explanatory variables in environmental epidemiology with application to health impact analysis of air pollution. *Environmental and Ecological Statistics*, *5*(2), 99–105.

Zidek, J. V., Wong, H., Le, N. D., & Burnett, R. (1996). Causality, measurement error and multicollinearity in epidemiology. *Environmetrics*, *7*(4), 441–451.

Zidek, J. V., & Zimmerman, D. L. (2010). Monitoring network design. *Handbook of Spatial Statistics*, 131–148.

Index

Author Index

Printed and bound by CPI Group (UK) Ltd, Croydon, CR0 4YY

01/11/2024

01782623-0010